BIOCHEMISTRY RESEARCH TRENDS

OMEGA-3 FATTY ACIDS

CHEMISTRY, DIETARY SOURCES AND HEALTH EFFECTS

BIOCHEMISTRY RESEARCH TRENDS

Additional books in this series can be found on Nova's website
under the Series tab.

Additional e-books in this series can be found on Nova's website
under the e-book tab.

BIOCHEMISTRY RESEARCH TRENDS

OMEGA-3 FATTY ACIDS

CHEMISTRY, DIETARY SOURCES AND HEALTH EFFECTS

MD.WASIM KHAN, PH.D.
EDITOR

New York

For permission to use material from this book please contact us:
Telephone 631-231-7269; Fax 631-231-8175
Web Site: http://www.novapublishers.com

NOTICE TO THE READER

Library of Congress Cataloging-in-Publication Data

ISBN: 978-1-62948-516-4

Library of Congress Control Number: 2012953810

Published by Nova Science Publishers, Inc. † New York

Contents

Preface

Hippocrates (460-377B.C), the father of medicine recommended, *"Let thy food be thy medicine and thy medicine be thy food"*. Such an idea reflected the importance of dietary supplements for their therapeutic and preventive bioactive components due to their elevated margin of safety and desired range of efficacy.

The last two decades have witnessed a major drift in the interests of the scientific community towards providing better means to containing the health risks of the human race. The century old chemotherapies against various disorders have never been a success, albeit not a total failure. Such therapies have a major drawback of side effects that give rise to unseen disorders that emerge as a new challenge. In this regard, the concept of foodstuffs as natural medicines i.e. nutraceuticals has become very attractive.

Interest in the omega-3 fatty acids as health-promoting nutrients has increased exponentially in the recent decade. Early clues to the possible importance of unsaturated fatty acids especially polyunsaturated fatty acids (PUFA) came from studies showing that young humans and experimental animals experienced impaired growth when all fatty acids were removed from the diet. Small amounts of ω-3 (α-linolenic acid) or ω-6 (linoleic acid) fatty acids prevented that impaired growth and thus were termed as "essential". Highly unsaturated/polyunsaturated fatty acids (HUFA/PUFA) were relatively found to be several-fold more effective.

Further studies have revealed that ω-3 PUFA have numerous health benefits. ω-3 PUFA especially from fish/marine foods were found to be effective in lowering incidence of various pathologies. This stems from the observations of the difference in prevalence of coronary heart disease and other chronic diseases (including psoriasis, bronchial asthma, diabetes mellitus and thyrotoxicosis) in the Greenland Inuit (Eskimo) population relative to Western populations. ω-3 PUFA (e.g. EPA and DHA) from marine fish and mammals were indicated as the main dietary factor responsible for such differences. The consumption of fish has now been linked with the rate of depression among various populations. In countries where people eat the least fish the rate of depression is highest, and vice versa. This correlation holds true across the world.

The most dramatic change in what we eat has happened in the past century, with industrialization and development of the food industry, the intake of saturated fat, trans fat and especially ω-6 enriched refined oils have enormously increased whereas the consumption of ω-3 PUFA has considerably declined. These dietary imbalances in fat intake, in fact, are prime cause of modern sufferings like cancer, hypertension, diabetes, depression, cardiovascular and renal disorders.

Sea fish are considered the best sources of omega-3 fatty acids, but consumption of fish is too low to meet the requirements. Efforts to supplement foods with omega-3 fatty acids have had drawbacks mainly due to the “unappealing taste” associated with the oxidation products of PUFA due to the highly labile nature of raw materials. Thus, the scientific community was presented with the difficult challenge of delivering highly unsaturated fatty acid foods that are appealing to taste buds. The stability of omega-3 fatty acids is dictated not only by the number of double bonds present in the molecules but also by the endogenous and exogenous antioxidants employed as well as the system in which they are incorporated. Microencapsulation of products as a means of delivering these highly labile materials to foods has thus been perfected and introduction of PUFA to bread, cereal, and dairy based products and spreads has slowly been taking place. Furthermore, production of omega-3 capsules as well as concentrates and structured lipids containing omega-3 fatty acids has led to a large number of over the- counter supplements and nutraceuticals.

It is the purpose of this book to present a detailed assessment of the current research about the chemistry, nutrition, and health aspects of omega-3 fatty acids and to address stability issues and the potential for their delivery in functional foods. We are highly indebted to participating authors who provided authoritative views and results of their latest investigations on different aspects of omega-3 fatty acids.

Dr. Md. Wasim Khan,
DST-INSPIRE Faculty,
Room No.250
Cell Biology & Physiology Division,
Indian Institute of Chemical Biology,
4 Raja SC Mullick Road,
Jadavpur, Kolkata, India 700032
+919748529884
wasimkhan786@gmail.com

In: Omega-3 Fatty Acids
Editor: Md. Wasim Khan

ISBN: 978-1-62948-516-4

Chapter 1

The Link between Omega-3 Fatty Acids and Rheumatoid Arthritis: Properties, Mechanisms and Therapeutic Efficacy

Gerard G. Dumancas[1,*], Rangika S. Hikkaduwa Koralege[2], Elmer-Rico E. Mojica[3], Befrika S. Murdianti[4], and Patrisha J. Pham-Bugayong[5]

[1]Arthritis and Clinical Immunology Research Program, MS, Oklahoma Medical Research Foundation, Oklahoma City, OK, US
[2]Department of Chemistry, Physical Science, Oklahoma State University, Stillwater, OK, US
[3]Department of Chemistry and Physical Sciences, Pace University, One Place Plaza, New York, NY, US
[4]Department of Physical Sciences, Arkansas Tech University, Russellville, AR, US
[5]Dave C. Swalm School of Chemical Engineering, Mississippi State University, MS, US

Abstract

Omega-3 fatty acids (FAs) are long chain polyunsaturated fatty acids characterized by a double (C=C) bond starting after the third carbon atom from the end of the carbon chain. These FAs are known to be primarily abundant in fish and plant oil products. Numerous studies indicate that omega-3 FAs have anti-inflammatory properties and, therefore, might be useful in the management of inflammatory and autoimmune diseases such as Rheumatoid Arthritis (RA). Rheumatoid arthritis (RA) is one of the systemic

* Corresponding author: Gerard G. Dumancas, Arthritis and Clinical Immunology Research Program, MS58, Oklahoma Medical Research Foundation, 825 N.E. 13th Street, Oklahoma City, OK 73104. E-mail: gerard-dumancas@omrf.org.

inflammatory disorders that gained widespread importance in every aspect of research due to its controversy of reducing the lifespan of patients from 3 to 12 years. It is a common disorder that affects both men and women at the prime of their lives with extra-articular manifestations that include vasculitis, lung disease, pericarditis, neuropathy, and subcutaneous nodules. This chapter will discuss the pharmacological properties, mechanisms, and therapeutic efficacy of various omega-3 FAs as it relates to RA, as well as the importance of omega-6/omega-3 ratio balance to such disease.

Keywords: Omega-3 fatty acids, rheumatoid arthritis, eicosanoids, inflammation, omega-6/ omega-3 ratio

Introduction

Polyunsaturated fatty acids (PUFAs) are known to play critical roles for the proper functioning of the human body. A group of PUFAs called omega-3 fatty acids (FAs) have become important due to its diverse physiological effects in cellular functions and processes. Among all fatty acids, it is these fatty acids, which possess the most immunomodulatory activities. It has been shown through animal experiments and clinical intervention studies that that omega-3 FAs have anti-inflammatory properties, thus, might be beneficial for the management of inflammatory diseases (Simopoulos, 2002).

Although not yet confirmed as an approved health claim, current research suggests that the anti-inflammatory activity of long chain omega-3 FAs may translate into clinical effects (Wall, Ross, Fitzgerald, and Stanton, 2010). As an example, evidence shows that rheumatoid arthritis (RA) sufferers taking omega-3 fatty acids from fish sources have reduced pain compared to those receiving standard non-steroidal anti-inflammatory drugs (NSAIDs) (Ruggiero et al., 2009).

This chapter will discuss the importance of omega-3 fatty acids as it relates to RA. As a starting point in the discussion and to facilitate readers, we briefly discussed the symptoms, causes, and pathogenesis of the disease.

Rheumatoid Arthritis

Symptoms and Causes

Rheumatoid arthritis (RA) is a common chronic inflammatory and destructive arthropathy that cannot be cured and that has substantial personal, social, and economic costs (Choy and Panayi, 2001). It is considered as the most common inflammatory arthritis and is a major cause of disability (Firestein, 2003). The disease is characterized by synovial inflammation and hyperplasia ("swelling"), autoantibody production (rheumatoid factor and anti-citrullinated protein antibody, cartilage and bone destruction ("deformity"), and systemic features, including cardiovascular, pulmonary, psychological, and skeletal disorders (McInnes and Schett, 2011).

A prevalence of 0.5 and 1.0% for the occurrence of RA among several Europeans (Aho, Kaipiainen-Seppanen, Heliovaara, and Klaukka, 1998; Carmona et al., 2002; Cimmino,

Parisi, Moggiana, Mela, and Accardo, 1998; Kvien et al., 1997; Power, Codd, Ivers, Sant, and Barry, 1999; Riise, Jacobsen, and Gran, 2000; Saraux et al., 1999; Simonsson, Bergman, Jacobsson, Petersson, and Svensson, 1999) and North-American (Gabriel, 2001; Gabriel, Crowson, and O'Fallon, 1999) populations has been reported to be relatively constant.

It has also been noted that American-Indians, specifically have the highest recorded occurrence of RA. A high prevalence of RA has been documented in Pima Indians (5.3%) (Del Puente, Knowler, Pettitt, and Bennett, 1989) and in the Chippewa Indians (6.8%) (Harvey, Lotze, Stevens, Lambert, and Jacobson, 1981). In contrast, populations from China and Japan have been reported to have low occurrences (Silman and Pearson, 2002).

RA is known to have been caused by both genetic and environmental factors. Genetically, the role of HLA DRB1 alleles as a risk factor of RA has been established for 25 years. Besides this, the non-MHC genes such as the CYP19, IFN-γ, and other cytokines are known to be genetic risk factors in RA (Silman and Pearson, 2002). Environmental factors also play roles as susceptibility factors for development of RA. In particular, hormonal and pregnancy factors were investigated in association to RA due to the increased risk of RA in females (Silman and Pearson, 2002).

It has been found that in general, male sex hormones, specifically testosterone, are lower in men who have RA (Silman and Hochberg, 2001) whereas the levels of female sex hormones are not different between RA cases and controls (Heikkila et al., 1998). Investigation of pregnancy as a risk factor RA development, on the other hand, led to contradicting results (Silman and Pearson, 2002). Infection is also another environmental factor known to be associated with RA suggesting that exposure to infectious agents such as Epstein-Barr and parovirus maybe triggers for RA (Silman and Pearson, 2002). Further, research shows that Vitamin D deficiency is common in those with RA and may be causally associated (Wen and Baker, 2011).

Pathogenesis

The RA pathogenesis is poorly understood. Approximately half of patients with RA have specific serologic abnormalities several years before the onset of symptoms. IgM rheumatoid factor (IgM-RF) and anti- anti–cyclic citrullinated peptide (anti-CCP) testing with appropriately high specificity may assist in the early detection of RA in high-risk populations (Nielen et al., 2004).

Although widely diverse mechanisms have been responsible for damage in RA, it is unlikely that immune complexes are the underlying cause. The distinguishing feature of RA in a patient constitute self-aggregating complexes of 7s RF while circulating complexes of 19s RF directed against the hinge region of 7s immunoglobulins are considered less specific. Other autoimmune complexes, such as those antibodies mentioned above are considered more specific for RA but have not been fully characterized. The immune complexes together with the neutrophils are essential for RA pathogenesis. The mechanism involve the release of mediators of inflammation and joint destruction (Weissmann, 2004).

Anti-Inflammatory Properties of Omega-3 Fatty Acids

Inflammation is part of the body's immediate response to infection or injury. As a result of an inflammation occurring in uncontrolled or inappropriate manner, excessive damage to host tissues or disease can occur. Rheumatoid arthritis is a chronic inflammatory disease which triggers hyper-expression of cytokines such as tumor necrosis factor α (TNF-α), interleukin 1β (IL-1β) and interleukin 6 (IL-6).

Omega-3 polyunsaturated fatty acids (PUFA) have shown clear evidence of having anti-inflammatory properties through animal experiments and clinical intervention studies. Long-chain omega-3 PUFAs decrease the production of inflammatory eicosanoids, cytokines, and reactive oxygen species and the expression of adhesion molecules. Long-chain omega-3 PUFAs also give rise to a family of anti-inflammatory mediators known as resolvins (P. C. Calder, 2006).

The key relation between PUFAs and inflammation is that eicosanoids, which are among the mediators and regulators of inflammation and they are involved in modulating the intensity and duration of inflammatory responses.

Recent studies have shown that the increased consumption of long-chain omega-3 PUFAs, such as eicosapentaenoic acid (EPA) and docosahexaenoic acid (DHA) resulted in elevated proportions of those fatty acids in inflammatory cell phospholipids. The inclusion of EPA and DHA into human inflammatory cells is occurring partly at the expense of arachidonic acid (AA) which is usually the major substrate for eicosanoid synthesis. Due to less substrate availability for the synthesis of eicosanoids from AA, a decreased production of prostaglandin E_2 (PGE_2), thromboxane B_2 (TXB_2), leukotriene B_4 (LTB_4), leukotriene E_4 (LTE_4) and 5-hydroxyeicosatetraenoic acid by inflammatory cells have been resulted which in turn reduced the inflammation. EPA can also act as a substrate for cyclooxygenase (COX) and 5-lipoxygenase (5-LOX) resulting eicosanoids that have slightly different structure from those derived from AA, and these are less potent than those formed from AA as inflammatory agents. The reduction in generation of AA derived mediators that associated with the consumption of fish oil had led to the idea that fish oil is anti-inflammatory (P. C. Calder, 2006).

Recent studies have identified a novel group of mediators which are resulted from EPA by COX-2, known as E-series resolvins that showed anti-inflammatory actions (C. N. Serhan, C. B. Clish, J. Brannon, S. P. Colgan, N. Chiang, et al., 2000; C. N. Serhan, C. B. Clish, J. Brannon, S. P. Colgan, K. Gronert, et al., 2000; C. N. Serhan et al., 2002). In addition to EPA, DHA-derived mediators known as D-series resolvins, docosatrienes and neuroprotectins which are resulted by COX-2 activity have also shown anti-inflammatory actions (Hong, Gronert, Devchand, Moussignac, and Serhan, 2003; Marcheselli et al., 2003; Mukherjee, Marcheselli, Serhan, and Bazan, 2004). Apart from these observations, some other studies have shown that the dietary fish oil consumption resulted in decreased leukocyte chemotaxis, decreased production of reactive oxygen species and proinflammatory cytokines, and decreased adhesion molecule expression, which decreased the inflammatory effects (P. C. Calder, 2006).

A dietary supplementation study using 3.1 g of EPA and 14.4 g of DHA has shown a time dependent decrease in chemotaxis of human neutrophils and monocytes toward variety of chemoattractants, including LTB_4, bacterial peptides and human serum. Several studies that included cell culture and animal feeding studies reported decreased production of some adhesion molecules on the surface of monocytes, macrophages or endothelial cells upon exposure to long-chain omega-3 PUFAs. Another supplementation study that provided 3.1 g EPA and 8.4 g DHA has shown decreased production of reactive oxygen species by 30-55%. EPA and DHA are also shown to be capable of inhibiting the production of IL-1β and TNF-α by monocytes and the production of IL-6 and IL-8 by venous endothelial cells (P. C. Calder, 2006).

Fish oil has shown anti-inflammatory effects in several studies which involved patients with rheumatoid arthritis.

Anti-inflammatory effects such as decreased LTB_4 production by monocytes and neutrophils, decreased IL-1 production by monocytes, decreased plasma IL-1β concentrations and serum C-reactive protein concentrations have been reported (Cleland, French, Betts, Murphy, and Elliott, 1988; Espersen, nGrunnet, et al., 1992; Kremer et al., 1985; Kremer et al., 1990; Tulleken, Limburg, Muskiet, and Rijswijk, 1990; H van der Tempel, J E Tulleken, P C Limburg, F A Muskiet, and M H van Rijswijk, 1990). Literature reports on randomized, placebo-controlled, double-blind studies of fish oil in rheumatoid arthritis have shown promising results. Long-chain omega-3 PUFAs used in the dose range of 1.6-7.1 g/d and average of about 3.5 g/d in above studies have shown beneficial effects such as reduced duration of morning stiffness, reduced number of tender or swollen joints, reduced joint pain, reduced time to fatigue, increased grip strength and decreased use of nonsteroidal anti-inflammatory drugs (P. C. Calder, 2006). Furthermore long-chain omega-3 PUFAs may also act as anti-inflammatory agents by competing with arachidonic acid for incorporation in to inflammatory cell membranes and for metabolism by enzymes of eicosanoid synthesis (P. C. Calder, 2006).

Another important anti-inflammatory effect of omega-3 PUFAs is mediated at the level of altered inflammatory gene expression through their effects on transcription factors such as nuclear factor kappa B (NFκB) and peroxisome proliferator-activated receptors (PPARs). NFκB plays a major role in various inflammatory signaling pathways and it controls cytokines such as IL-1, IL-2, IL-6, IL-12, TNF-α and chemokines such as IL-8 and monocyte chemoattractant protein-1. Furthermore NFκB controls adhesion molecules and inducible effector enzymes (Wall et al., 2010). By inhibiting the activity of this pivotal pro-inflammatory transcription factor, NFκB, omega-3 PUFAs help in regulating inflammation. The NFκB pathway includes three major events which are mediated by several proteins and there is evidence that omega-3 PUFAs block each of these three stages independently (Singer et al., 2008). Omega-3 PUFAs and their eicosanoid derivatives act as ligands for PPARs and when activated PPARs bind to the PPAR-response element and repress or induce the transcription of target genes, for example PPARs have shown to inhibit NFκB and therefore it plays a major role in inflammatory processes. EPA and DHA have shown to be down-regulating the lipopolysaccharide –induced activation of NFκB via a PPAR-γ-dependent pathway in human kidney-2-cells (Wall et al., 2010).

Pharmacological Properties of Omega-3 Fatty Acids

Omega-3 FAs includes docosahexanoic acid (DHA), eicosapentanoic acid (EPA), and alpha-linoleic acid (ALA). DHA and EPA are derived from fish or fish oil, while ALA is derived from plants (the main sources are canola and soy bean oils, flaxseeds, and walnut). In the body, ALA can undergo conversion to EPA and DHA, however, this conversion is usually very low. (Burdge, 2004) Omega-3 FA from plant sources may have similar benefits to omega-3 FAs from fish oil, however the benefits are less well-known and only DHA and EPA have the more established benefits.

Regular consumption of fish or fish oil supplement contains omega-3 FAs has been reported to have many beneficial effects and have been used in preventing and managing conditions such as cardiovascular disease (Albert et al., 2002; Hu et al., 2002), diabetes (Brostow et al., 2011; Hendrich, 2010; Storlien et al., 1987), mood change and psychiatric disorder (Magnusson, Axelsson, Karlsson, and Oskarsson, 2000; Peet and Stokes, 2005), cancer chemopreventive agents (Rose and Connolly, 1999), and arthritis (H. van der Tempel, J. E. Tulleken, P. C. Limburg, F. A. Muskiet, and M. H. van Rijswijk, 1990).

The benefits of fish fatty acids on atherosclerosis, which can lead to several serious problems including heart attack, have been postulated since 1940s by Hugh Sinclair (Sinclair, 1956). Several studies on Greenland Eskimos who consume large amount of food which contains high concentration of omega-3 FAs in their diet showed that they have lower rate of coronary heart diseases (CHD) (Bjerregaard and Dyerberg, 1988; Bjerregaard, Mulvad, and Pedersen, 1997; Dyerberg, 1989). Long chain of omega-3 FAs have been shown to decrease triglyceride concentrations, to lower blood pressure, to decrease thrombosis and cardiac arrhythmias, and to increase heart rate variability, where these mechanisms most likely explain the cardiovascular protection by omega-3 FAs (P. C. Calder, 2004).

Omega-3 DHA and EPA have also been studied for their activities in insulin resistance on high-fat feeding in rats, and it is reported that replacement of only 6 percent of omega-6 FAs in diet with omega-3 FAs from fish oil prevented the development of insulin resistance. (Storlien et al., 1987)

Although the benefits of nonmarine omega-3 FAs are less well-known, recent finding by Brostow et al. suggested that omega-3 FAs form nonmarine sources (ALA) play a role in the prevention of type-2 diabetes in Chinese Singaporean (Brostow et al., 2011).

A study by Magnusson et al. suggested that there is a lack of seasonal disorders in Iceland which may be due to the high content of fish in the Icelandic diet (Magnusson et al., 2000). Similar lack of disorders was also observed in Japan which also has a high content of fish in their diet. The mechanism through which EPA and DHA decrease the depressive symptoms is suggested to include the modulation of serotonin turnover, phosphoinositol-mediated signal transduction, and L-type calcium channel regulation (Hibbeln et al., 2000). Peet et al. also reported the importance of the two omega-3 FAs, EPA and DHA, with depression (Peet and Stokes, 2005). EPA, most importantly, is suggested to give benefits in treatment of depression and schizophrenia.

Recent study by MacLennan et al. showed that omega-3 FAs from marine oils may reduce breast cancer risk (MacLennan et al., 2013). In their study, the *fat-1* and MMTV-neu (ndl)-YD5 mouse models were crossed to create a novel double-hybrid mouse used for investigation of the effect of omega-3 FAs exposure on mammary tumor development. The result from this study provides clear evidence that the size and the number of tumors were significantly decreased in group of mice treated with omega-3 FAs.

In relation to RA, different studies show that omega-3 is responsible in important decrease in the synthesis of leukotriene B_4 (LTB_4) (T. H. Lee, Hoover, and Williams, 1985; Sperling et al., 1987), as well as interleukin-1β (IL-1β) and interleukin-1α (IL-1α), which was associated with clinical improvement in patients with rheumatoid arthritis. Van der Tempel et al. reported that patients with RA treated with fish oil show significant improvement for joint swelling index (decrease of six points in swelling index) and the duration of morning stiffness (decrease by a mean of 35 minutes) (H. van der Tempel et al., 1990).

Meta-analysis study by Fortin et al. also shows that the use of fish oil significantly reduced tender joint count and duration of morning stiffness (Fortin et al., 1995). Espersen et al. reported the effects of dietary supplementation with omega-3 FAs on the level of cytokines and complement activation in plasma from the patients with RA (Espersen, Grunnet, et al., 1992). Their findings suggested that there is a significantly reduced level of plasma IL-1β, as well as a decrease in cytokine productions.

Omega-3 Fatty Acids and RA

Research involving omega-3 fatty acids, eicosapentaenoic acid (EPA) and docosahexaenoic acid (DHA) in rheumatoid arthritis began in the mid-eighties, following the demonstration in several autoimmune strains of mice [NZB (NZB x NZW) F_1, MRL/1pr, and B x SB/Mpj) that omega-3 fatty acids reduced the severity of diffuse proliferative glomerulonephritis (DPGN) (Simopoulos, 2002). Animal experiments and clinical intervention studies indicate that omega-3 fatty acids have anti-inflammatory properties and, therefore, might be useful in the management of inflammatory and autoimmune diseases. Omega-3fatty acids may be particularly helpful in managing rheumatoid arthritis (RA) because they reduce production of compounds that cause inflammation (inflammation mediators), while increasing the production of biochemicals that hinder inflammation. Studies have reported anti-inflammatory effects of omega-3 fatty acids in patients with RA, such as decreased leukotriene B_4 (LTB_4) production by neutrophils, decreased monocyte production of platelet-activating factor, decreased PGE_2 production by mononuclear cells, decreased plasma of pro-inflammatory cytokine interleukin-1 beta (IL-1,2,6 and 8β) (Darlington and Stone, 2001) concentrations, decreased serum C-reactive protein concentrations, decreased serum pro-inflammatory cytokine tumor necrosis factor-alpha (TNF-α) and soluble receptor activator of NFκB ligand concentrations, and normalization of the neutrophil chemotactic response (Ariza-Ariza, Mestanza-Peralta, and Cardiel, 1998). The suggested mechanism for the improving effects of omega-3 fatty acids (DHA plus EPA) on RA is a decrease in inflammation due to EPA's ability to competitively inhibit omega-6 prostaglandins and thus increase the production of LTB5, a non- inflammatory eicosanoid and decreasing the production of pro-inflammatory eicosanoids (eg. PGE2, thromboxane and LTB4). The result is decreased vasodilation, neutrophil degranulation, vascular permeability and hyperalgesia. This combination effectively decreases swelling, redness and loss of function which are the hallmark signs and symptoms of RA. The authors conclude there is strong and consistent scientific evidence that omega-3 fatty acid supplementation is effective at reducing inflammation and improving symptoms of RA in randomized controlled trials and that dietary alteration provides an added opportunity to give patients some control over their health and management of their disease. Recently, new substances derived from the omega-3 fatty acids that resolve and protect against inflammation (Ariel, Li, and Wang, 2005; Chiang and Serhan, 2006) aptly called *resolvins* and *protectins*, help provide arthritis relief without the side effects of conventional arthritis drugs and does not seem to suppress the immune system. These are a novel class of bioactive lipid mediators, which are enzymatically biosynthesized in vivo from EPA and DHA. First, from eicosapentaenoic acid called the E-series resolvins (RvEs) and second, docosahexaenoic acid, termed the D-series resolvins (RvDs) and protectin D1 (PD1). These lipid mediators exert anti-inflammatory and pro-resolving properties and are log-orders more potent than their precursors (L. V. Norling and Perrett, 2013). Initial animal studies showed that compared with vegetable oil, mice fed with fish oil delayed the onset (mean 34 vs 25 days), reduced the incidence (69% versus 93%) and severity (mean peak severity score 6.7 vs 9.8) of type II collagen-induced arthritis (Leslie et al., 1985). There are at least 13 double-blind, randomized, placebo- controlled clinical trials that show benefit from fish oil supplements in patients with RA (M.J. James and Cleland, 1997). A common feature of the studies has been a reduction in symptoms and in the number of tender joints. Also, there was a reduction in the dose of analgesic anti-inflammatory drugs. In a subsequent meta-analysis, morning stiffness decreased, as well as the number of tender joints (Fortin et al., 1995). The Japanese population consumes a larger quantity of omega-3 fatty acids in their diet compared to North Americans. A case-controlled study showed that subjects who ate 2 or more fish (omega-3 fatty acid source) meals per week had a 43% reduced risk of developing RA as compared to those who consumed less than one serving of fish per week (Stamp, James, and Cleland, 2005). Cleland compared outcomes among patients with RA who did not consume fish oil supplements and those who did. They found that fish oil users were more likely to reduce use of NSAIDs and were more likely to be in remission (Cleland, Caughey, James, and Proudman, 2006). There have been a number of clinical trials of fish oil in patients with RA. Most of these trials report clinical improvements (e.g. improved patient assessed pain, decreased morning stiffness, fewer painful or tender joints, and decreased use of NSAIDs), and when the trials have been pooled in meta analyses statistically, significant clinical benefit has emerged (Y. H. Lee, Bae, and Song, 2012). Evidence for clinical efficacy of omega-3 PUFAs in RA is robust. A recent study, involved 37 individuals with RA. All the volunteers completed a food frequency questionnaire and kept a daily diary of their food intake for three days. Researchers analyzed levels of adipokines and oxidative stress markers in serum and saliva and found that consuming omega-3 fish oil and monounsaturated fatty acid had a beneficial effect on RA and reduced inflammation (Hayashi et al., 2012). In another study, results of a meta- and mega- analysis involving 10 double-blind, placebo-controlled trials that included 183 patients with rheumatoid arthritis of 10-11 years duration and 187 patients who were treated with placebo, showed that patients who took omega-3 fatty acids (more than 2.7- >6 grams daily for more than 3 months) needed to take less non-steroidal anti-inflammatory drugs to treat their symptoms (M.J. James and Cleland, 1997). In addition, the number of tender and swollen joints, level of morning stiffness, and physical function also showed a trend toward improvement among patients who took omega-3 fatty acids when compared with controls. No statistically significant changes were observed for the other measured indicators of disease severity. Little difference in the magnitude of effect between 2.9 and >6 grams/d of omega-3 fatty acids. A systematic review of 23 trials conducted in England concluded that the two main omega-3 fatty acids, EPA and DHA reduced the level of arachidonic acid in cells involved in immune response and thus reduced the production of inflammatory substances (Miles and Calder, 2012). Use of the essential fatty acids also had a beneficial impact on the use of non-steroidal anti-inflammatory drugs. The anti-inflammatory and immunosuppressant drugs used in current treatment of rheumatoid arthritis have long-term adverse secondary effects. In a double-blind, placebo-controlled study conducted at the University of Newcastle in Australia, 50 people with RA were randomly assigned to take either 40 mg/kg body weight of fish oil (containing 60 percent omega-3 fatty acids) or placebo daily for 15 weeks.

The fish oil group had an average age of 54 and average disease duration of 12 years, while the placebo group (50/50 corn/olive oil) was 60 years old on average and had RA for an average of 15 years. Because omega-6 fatty acids, found in most vegetable oils, compete with omega-3 fatty acids, the researchers included only subjects who generally ate less than 10 g/day of omega-6 fatty acids. Participants were monitored for several measures of disease severity before and during the study. There was no difference between the two groups at 4 and 8 weeks. At the 15 week evaluation, however, the average measures of the fish oil group had decreased for swollen joint count, duration of early-morning stiffness, pain, health assessment, and patient and physician assessment of arthritis severity. The level of improvement was meager, because only five fish oil patients showed an improvement of 20 percent or more compared with three control patients who experienced the same rate of improvement. The research however, did not mention whether the subjects were men or women. Moreover, the difficulty of conducting blind studies with fish oil because of its taste, as well as the initial difference in age and disease history between the two groups was not addressed (Volker, Fitzgerald, Major, and Garg, 2000). In a double-blind, placebo-controlled study involving 66 patients with active RA, the researchers concluded that some RA patients using fish oil supplementation may be able to discontinue NSAIDs without experiencing a flare-up of RA. Moreover, the benefits achieved from supplementing with 9 grams/ day of fish oil were no greater than those observed with studies using only 3 to 6 grams/day (Kremer et al., 1995). A clinical trial that investigated cod liver oil supplementation versus placebo in RA (2.2 g EPA per 10 g cod liver oil) reported that cod liver oil supplements containing omega-3 fatty acids demonstrated NSAID-sparing effects in RA patients (Galarraga et al., 2008).

Omega-6/Omega-3 Fatty Acid Ratio and RA

Dose-response studies demonstrated that DHA is more effective than EPA, and that diets with combinations of these two omega-3 fatty acids are synergistic. Not all animal studies of dietary supplementation with omega-3 fatty acids have resulted in beneficial effects. As an example, a study on type II collagen–induced arthritis worsened in animals fed with fish oil compared with those fed with beef tallow (Prickett, Trentham, and Robinson, 1984). Interestingly, recent studies of purified EPA and DHA indicate that the mixture of these 2 major omega-3 fatty acid constituents of fish oil may be more effective than either fatty acid by itself (Robinson et al., 1989).

All of the published randomized controlled trials involving omega-3 supplementation with DHA/EPA from fish oil have shown a moderate benefit when combined with standard RA pharmacotherapy (Stamp et al., 2005). Furthermore, there have been reports that symptoms worsened after fish oil supplementation was ceased. The supplement dose in these trials varied from 0.4 to 6.0 g of DHA/ EPA (combined) per day, with the mean dose being approximately 3g (3000 mg)/day. Overall, these studies have established that a dose of 2.6 g/day or more of omega-3 fatty acid supplementation (DHA+EPA) generally will reduce the symptoms of RA after about 12 weeks; consuming a larger dose will reduce this latency period. Omega-6 fatty acids (found in cooking oil and margarines), on the other hand, have been found to exacerbate RA symptoms.

Several sources of information suggest that human beings evolved on a diet with a ratio of omega-6 to omega-3 essential fatty acids (EFA) of approximately 1 whereas in Western diets the ratio is 15:1-16.7:1. Western diets are deficient in omega-3 fatty acids, and have excessive amounts of omega-6 fatty acids compared with the diet on which human beings evolved and their genetic patterns were established. A low omega-6 to omega-3 ratio of 2:1-3:1 was associated with less severe inflammation in patients with rheumatoid arthritis. In order to improve the ratio, it will be necessary to decrease the intake of omega-6 fatty acids by using oils rich in omega-3 fatty acids and increase the intake of fish two to three times per week or take supplements (1).

Mechanisms of Anti-Inflammatory Properties of Omega-3 Fatty Acids in RA

The beneficial effects of omega-3 fatty acids in arthritis have been known as early as 1783 when Percival reported that cod liver oil should be given in "obstinate chromic rheumatisms sciaticas or long standing and in those cases of premature decrepitude" (Percival, 1783). Epidemiological observations showed that people whose main diets are fish that are rich in omega-3 fatty acids were found to have low incidence of autoimmune and inflammatory disorders such as psoriasis, asthma, heart disease and rheumatoid arthritis (Horrobin, 1987; Kromann and Green, 1980; Simopoulos, 2002). Evidences from clinical studies have at least clarified the role of the omega-3 fatty acids on rheumatoid arthritis as being an agent that reduce inflammation and cartilage destruction (Hurst, Zainal, Caterson, Hughes, and Harwood, 2010).

There has been much interest with regards to the mechanisms on how the omega-3 fatty acids exert their beneficial effects. The possible cellular and molecular mechanisms for this effect have been explored and summarized in Figure 1.

Eicasanoids that comprises of prostaglandins, leukotrienes, thromboxanes and other oxidized derivatives are among the mediators regulators of inflammation and are generated from 20 carbon fatty acids. It served as the linked between inflammation and the fatty acids particularly polyunsaturated fatty acid (PUFA) (P.C. Calder, 2006). The inflammatory cells contain a high proportion of the omega-6 fatty acid, arachidonic acid (AA 20:4omega-6) and low proportions of other 20-carbons PUFAs like omega-3 fatty acid, eicosapentaenoic acid (EPA) (P.C. Calder, 2006).

The 20-carbon fatty acids served as biosynthetic precursors for omega-6 and omega-3 eicasanoids where in prostaglandins (PGs) and leukotrienes (LTs) are formed from the action of cyclooxygenase (COX-2) and 5-lipooxygenase (5-LOX) (P.C. Calder, 2006; M.J. James and Cleland, 1997). AA produced 2-series PGs and 4-series LTs while EPA produced 3-series PGs and 5-series LTs.

However, there is a distinct difference in terms of the activities of the eicasanoids of the omega-6 and omega-3 groups (M.J. James and Cleland, 1997). The omega-6 eicasanoids such as PGE_2 and LTB_4 have potent pro-inflammatory properties and largely responsible for the signs and symptoms of joint tenderness and swelling in inflammatory arthritis (M.J. James, Proudman, and Cleland, 2003). PGE_2 has been found to induce COX-2 hence upregulating its own production (Bagga, Wang, Farias-Eisner, Glaspy, and Reddy, 2003).

However, recent studies showed that it can also act as an anti-inflammatory agent since it inhibits 5-LOX hence decreasing the 4-series LTs and induces 15-LOX which promotes the formation of lipoxins (Levy, Clish, Schmidt, Gronert, and Serhan, 2001; Vachier et al., 2002).

The omega-3 eicosanoids on the other hand, which are entirely homologous with the omega-6 eicasanoids except for the additional double bond in the omega-3 position, have less inflammatory activity which is the case for LTB_5 (Goldman, Pickett, and Goetzl, 1983; T.H. Lee et al., 1988) and have same activity but are poorly synthesized from EPA, which is the case for PGE_3 (Hawkes, James, and Cleland, 1991, 1992). Dietary omega-3 fatty acids can increase cellular omega-3 content and decrease omega-6 eicosanoid synthesis (M.J. James and Cleland, 1997). This results in EPA acting as substrate for both COX-2 and 5-LOX and producing the less potent eicasanoids. EPA is readily metabolized to LTB_5 and the ratio of the LTB_5/LTB_4 directly proportional to the ratio of EPA/AA (M.J. James, Cleland, Gibson, and Hawkes, 1991). Because less AA is available for eicasanoid synthesis, there is a decrease production of the PGE_2 and LTB_4. The reduction of these metabolites from AA has led to the idea that omega-3 fatty acids possess anti-inflammatory properties.

Another anti-inflammatory mechanism is the ability of omega-3 fatty acids to suppress production of peptide mediators involved in inflammation. Cytokines such tumor necrosis factor-α (TNF-α) and interleukin-1β (IL-1β) have both pro-inflammatory properties as well as long term deleterious effects on cartilage and bone within and adjacent joints (Arend, 2002; M.J. James et al., 2003). Both have been implicated in the tissue destruction seen in RA. These cytokines stimulate PGE_2 synthesis, collagenase production, aggrecanase activity and production of matrix metalloproteinases (MMPs), enzymes that can degrade all components of the extracellular matrix (Arend and Dayer, 1990; Curtis et al., 2000; Dinarello, 1989). This action has been the target for therapies in RA. In general, dietary omega-3 fatty acids suppress the production of these cytokines. Dietary fish oil supplementation decreased mononuclear cell TNF-α and IL-1β in different studies including those that have RA. *In vitro* studies have shown that the omega-3 fatty acids (EPA, DHA, and ALA) were able to decrease the collagenase and aggrecanase activities and to decrease the expression of aggracanase -1 and -2 by bovine chondrocytes as well as MMP-3 and -13 by human articular explants (Curtis et al., 2000; Curtis et al., 2002). The same studies showed that omega-6 fatty acids have no effect on these systems (Curtis et al., 2000; Curtis et al., 2002).

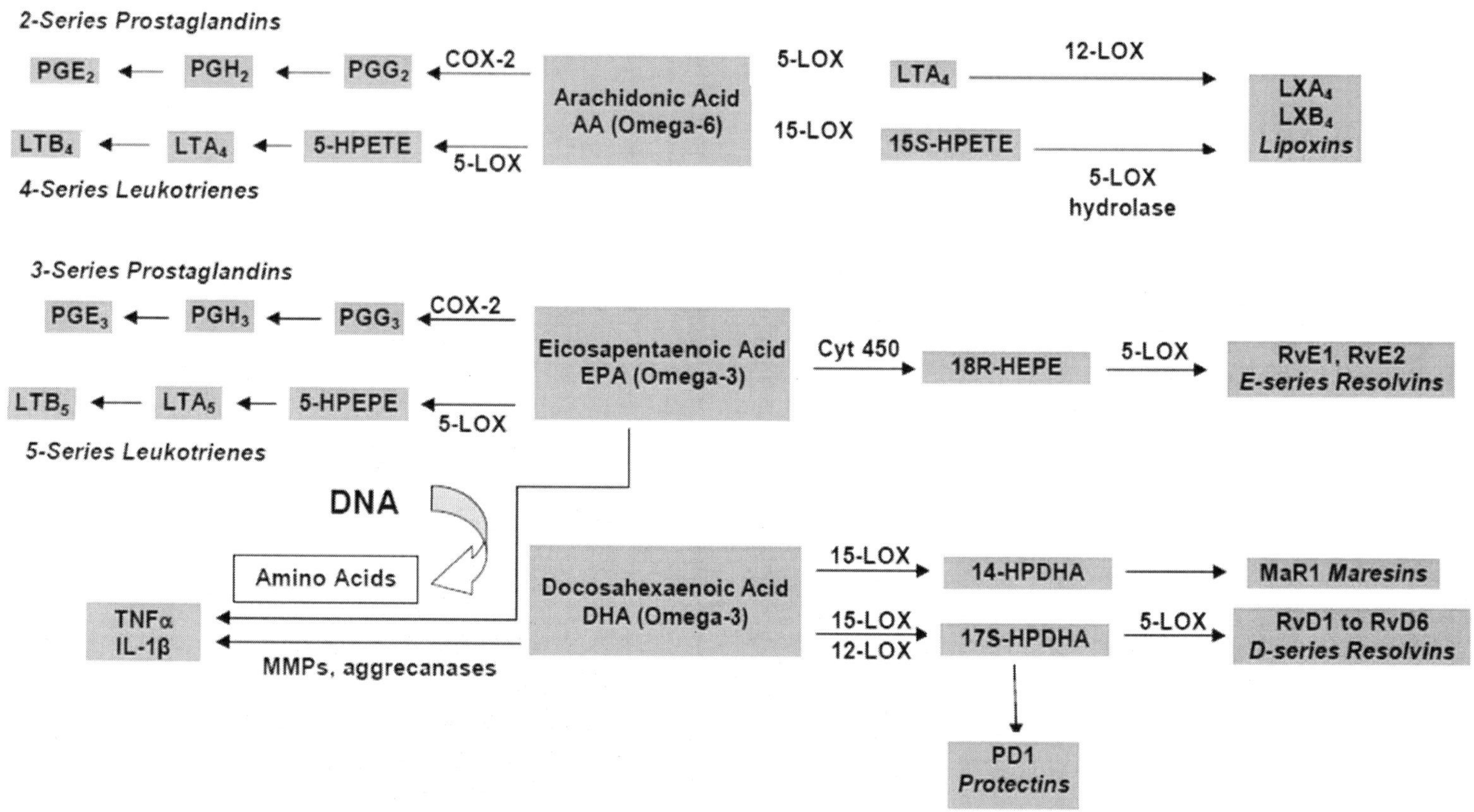

Figure 1. Summary of the possible mechanisms on the anti anti-inflammatory properties of omega-3 fatty acids.

Recent discoveries by Serhan et al. (C.N. Serhan, Chiang, and van Dyke, 2008; C.N. Serhan et al., 2000; C.N. Serhan et al., 2002) indicate another mechanism where omega-3 fatty acids can act as anti-inflammatory agents (Norling LV, 2013). Systematic metabolomics studies of lipid mediators in the course of acute inflammatory responses revealed that omega-3 fatty acids-derived mediators are produced within inflammatory exudates (Seki, Tani, and Arita, 2009; C.N. Serhan et al., 2000; C.N. Serhan et al., 2002). These mediators include E-series resolvins from EPA, D-series resolvins from DHA and other related compounds (protectins and maresins) which were produce enzymatically through the COX and LOX pathways in the presence and absence of aspirins (Figure 1) (P.C. Calder, 2012; Norling LV, 2013). The biological effects of resolvins, protectins and maresins were tested extensively in cell culture and animal models of inflammation (P.C. Calder, 2012). These mediators were found to be anti-inflammatory and promote the resolution of inflammation both *in vivo* and *in vitro* (Seki et al., 2009). They have been proven to be log-orders more potent than their lipid precursors (L.V. Norling and Serhan, 2010; Norling LV, 2013). For instance, resolving D1 inhibited IL-1β production while protectin D1 inhibited TNF-α and IL-1β production (C.N. Serhan et al., 2008; C.N. Serhan et al., 2000; C.N. Serhan et al., 2002; Serhan CN and N., 2000).

Conclusion

Omega-3 FAs have become controversially important due to its diverse physiological effects in cellular functions and processes. One of the properties of omega-3 FAs that gain intense interests is its anti-inflammatory properties, thus, might be beneficial for the management of inflammatory diseases. The key relation between PUFAs and inflammation is the eicosanoids, which are among the mediators and regulators of inflammation. Eicosanoids are involved in modulating the intensity and duration of inflammatory responses. Long-chain omega-3 FAs may also act as anti-inflammatory agents by competing with arachidonic acid for incorporation in to inflammatory cell membranes and for metabolism by enzymes of eicosanoid synthesis.

Omega-3 FAs are responsible in important decrease in the synthesis of leukotriene B_4 (LTB_4), as well as interleukin-1β (IL-1β) and interleukin-1α (IL-1α), which was associated with clinical improvement in patients with rheumatoid arthritis. The suggested mechanism for the improving effects of EPA and DHA omega-3 FAs on RA is a decrease in inflammation due to EPA's ability to competitively inhibit omega-6 prostaglandins and thus increase the production of LTB5, a non-inflammatory eicosanoid, and decreasing the production of pro-inflammatory eicosanoids. A low omega-6 to omega-3 ratio of 2:1-3:1 was associated with less severe inflammation in patients with rheumatoid arthritis.

Another anti-inflammatory mechanism of omega-3 FAs is the ability to suppress production of peptide mediators involved in inflammation. Omega-3 FAs can also act as anti-inflammatory agents, where omega-3 FAs-derived mediators are produced within inflammatory exudates. These mediators include E-series resolvins from EPA, D-series resolvins from DHA and other related compounds (protectins and maresins), which are found to be anti-inflammatory and promote the resolution of inflammation both *in vivo* and *in vitro*.

References

Aho, K., Kaipiainen-Seppanen, O., Heliovaara, M., and Klaukka, T. (1998). Epidemiology of rheumatoid arthritis in Finland. *Semin. Arthritis Rheum.,* 27(5), 325-334.

Albert, C. M., Campos, H., Stampfer, M. J., Ridker, P. M., Manson, J. E., Willett, W. C., and Ma, J. (2002). Blood levels of long-chain n-3 fatty acids and the risk of sudden death. *N Engl. J. Med.,* 346(15), 1113-1118. doi: 10.1056/NEJMoa012918 346/15/1113 [pii]

Arend, W. P. (2002). The mode of action of cytokine inhibitors. *J. Rheumatol.,* 29, 16-21.

Arend, W. P. and Dayer, J. M. (1990). Cytokines and cytokine inhibitors or antagonists in rheumatoid arthritis. *Arthritis Rheum.,* 33, 305-315.

Ariel, A., Li, P. L. and Wang, W. (2005). The docosatriene protectin D1 is produced by TH2 skewing and promotes human T cell apoptosis via lipid raft clustering. *J. Biol. Chem.,* 280(52), 43079-43086.

Ariza-Ariza, R., Mestanza-Peralta, M. and Cardiel, M. H. (1998). Omega-3 fatty acids in rheumatoid arthritis: an overview. *Semin. Arthritis Rheum.,* 27(6), 366-370.

Bagga, D., Wang, L., Farias-Eisner, R., Glaspy, J. A., and Reddy, S. T. (2003). Differential effects of prostaglandin derived from v-6 and v-3 polyunsaturatedfatty acids on COX-2 expression and IL-6 secretion. *Proc. Natl. Acad. Sci. US,* 100, 1751-1756.

Bjerregaard, P. and Dyerberg, J. (1988). Mortality from ischaemic heart disease and cerebrovascular disease in Greenland. *Int. J. Epidemiol.,* 17(3), 514-519.

Bjerregaard, P., Mulvad, G. and Pedersen, H. S. (1997). Cardiovascular risk factors in Inuit of Greenland. *Int. J. Epidemiol.,* 26(6), 1182-1190.

Brostow, D. P., Odegaard, A. O., Koh, W. P., Duval, S., Gross, M. D., Yuan, J. M., and Pereira, M. A. (2011). Omega-3 fatty acids and incident type 2 diabetes: the Singapore Chinese Health Study. *Am. J. Clin. Nutr.,* 94(2), 520-526. doi: 10.3945/ajcn.110.009357 ajcn.110.009357 [pii]

Burdge, G. (2004). Alpha-linolenic acid metabolism in men and women: nutritional and biological implications. *Curr. Opin. Clin. Nutr. Metab. Care,* 7(2), 137-144. doi: 000751 97-200403000-00006 [pii]

Calder, P. C. (2004). n-3 Fatty acids and cardiovascular disease: evidence explained and mechanisms explored. *Clin. Sci. (Lond.),* 107(1), 1-11. doi: 10.1042/CS20040119CS2004 0119 [pii]

Calder, P. C. (2006). n-3 polyunsaturated fatty acids, inflammation, and inflammatory diseases. *American Journal of Clinical Nutrition,* 83(6), 1505s-1519s.

Calder, P. C. (2006). n-3 Polyunsaturated fatty acids, inflammation and inflammatory diseases. *Am. J. Clin. Nutr.,* 83, 1505S-1519S.

Calder, P. C. (2012). Omega-3 polyunsaturated fatty acids and inflammatory processes: nutrition or pharmacology? *Brit. J. Clin. Pharmacol.,* 75, 645-662.

Carmona, L., Villaverde, V., Hernandez-Garcia, C., Ballina, J., Gabriel, R., and Laffon, A. (2002). The prevalence of rheumatoid arthritis in the general population of Spain. *Rheumatology (Oxford),* 41(1), 88-95.

Chiang, N. and Serhan, C. N. (2006). Cell-cell interaction in the transcellular biosynthesis of novel omega-3-derived lipid mediators. *Methods Mol. Biol.,* 341, 227-250. doi: 1-59745-113-4:227 [pii] 10.1385/1-59745-113-4:227

Choy, E. H. and Panayi, G. S. (2001). Cytokine pathways and joint inflammation in rheumatoid arthritis. *N Engl. J. Med.,* 344(12), 907-916. doi: 10.1056/NEJM200103223 441207

Cimmino, M. A., Parisi, M., Moggiana, G., Mela, G. S., and Accardo, S. (1998). Prevalence of rheumatoid arthritis in Italy: the Chiavari Study. *Ann. Rheum. Dis.,* 57(5), 315-318.

Cleland, L. G., Caughey, G. E., James, M. J., and Proudman, S. M. (2006). Reduction of cardiovascular risk factors with longterm fish oil treatment in early rheumatoid arthritis. *J. Rheumatol.,* 33(10), 1973-1979. doi: 06/13/081 [pii]

Cleland, L. G., French, J. K., Betts, W. H., Murphy, G. A., and Elliott, M. J. (1988). Clinical and biochemical effects of dietary fish oil supplements in rheumatoid arthritis. *J. Rheumatol.,* 15(10), 1471-1475.

Curtis, C. L., Hughes, C. E., Flannery, C. R., Little, C. B., Harwood, J. L., and Caterson, B. (2000). n-3 Fatty acids specifically modulate catabolic factors involved in articular cartilage degradation. *J. Biol. Chem.,* 275, 721-724.

Curtis, C. L., Rees, S. G., Little, C. B., Wilson, C., Dent, C. E., Otterness, I. G.,... Caterson, B. (2002). Pathologic indicators of degradation and inflammation in human osteoarthritic cartilage are abrogated by exposure to n-3 fatty acids. *Arthritis Rheum.,* 46, 1544-1553.

Darlington, L. G. and Stone, T. W. (2001). Antioxidants and fatty acids in the amelioration of rheumatoid arthritis and related disorders. *Br. J. Nutr.,* 85(3), 251-269. doi: S0007114501 00037X [pii]

Del Puente, A., Knowler, W. C., Pettitt, D. J., and Bennett, P. H. (1989). High incidence and prevalence of rheumatoid arthritis in Pima Indians. *Am. J. Epidemiol.,* 129(6), 1170-1178.

Dinarello, C. A. (1989). Interleukin-1 and its biologically related cytokines. *Adv. Immunol.,* 44, 153-205.

Dyerberg, J. (1989). Coronary heart disease in Greenland Inuit: a paradox. Implications for western diet patterns. *Arctic Med. Res.,* 48(2), 47-54.

Espersen, G. T., Grunnet, N., Lervang, H. H., Nielsen, G. L., Thomsen, B. S., Faarvang, K. L.,... Ernst, E. (1992). Decreased interleukin-1 beta levels in plasma from rheumatoid arthritis patients after dietary supplementation with n-3 polyunsaturated fatty acids. *Clin. Rheumatol.,* 11(3), 393-395.

Espersen, G. T., nGrunnet, N., Lervang, H. H., Nielsen, G. L., Thomsen, B. S., Faarvang, K. L.,... Ernst, E. (1992). Decreased Interleukin-1 beta levels in plasma from rheumatoid arthritis patients after dietary supplementation with n-3 polyunsaturated fatty acids. *Clinical Rheumatology,* 11(3), 393-395. doi: 10.1007/bf02207200

Firestein, G. S. (2003). Evolving concepts of rheumatoid arthritis. *Nature,* 423(6937), 356-361. doi: 10.1038/nature01661

Fortin, P. R., Lew, R. A., Liang, M. H., Wright, E. A., Beckett, L. A., Chalmers, T. C., and Sperling, R. I. (1995). Validation of a meta-analysis: the effects of fish oil in rheumatoid arthritis. *J. Clin. Epidemiol.,* 48(11), 1379-1390. doi: 0895-4356(95)00028-3 [pii]

Gabriel, S. E. (2001). The epidemiology of rheumatoid arthritis. *Rheum. Dis. Clin. North Am.,* 27(2), 269-281.

Gabriel, S. E., Crowson, C. S. and O'Fallon, W. M. (1999). The epidemiology of rheumatoid arthritis in Rochester, Minnesota, 1955-1985. *Arthritis Rheum.,* 42(3), 415-420. doi: 10. 1002/1529-0131(199904)42:3<415::AID-ANR4>3.0.CO;2-Z

Galarraga, B., Ho, M., Youssef, H. M., Hill, A., McMahon, H., Hall, C.,... Belch, J. J. (2008). Cod liver oil (n-3 fatty acids) as an non-steroidal anti-inflammatory drug sparing agent in rheumatoid arthritis. *Rheumatology (Oxford),* 47(5), 665-669. doi: 10.1093/ rheumatology/ken024 ken024 [pii]

Goldman, D. W., Pickett, W. C. and Goetzl, E. J. (1983). Human neutrophil chemotactic and degranulating activities of leukotriene B5 (LTBs) derived from eicosapentaenoic acid. *Biochem. Biophys. Res. Commun.,* 117, 282-288.

Harvey, J., Lotze, M., Stevens, M. B., Lambert, G., and Jacobson, D. (1981). Rheumatoid arthritis in a Chippewa Band. I. Pilot screening study of disease prevalence. *Arthritis Rheum.,* 24(5), 717-721.

Hawkes, J. S., James, M. J. and Cleland, L. G. (1991). Separation and quantification of PGE3 following derivitization with panacyl bromide by high pressure liquid chromatography with fluorometric detection. *Prostaglandins,* 42, 355-368.

Hawkes, J. S., James, M. J. and Cleland, L. G. (1992). Biological activity of prostaglandin E 3 with regard to oedema formation in mice. *Agents Actions,* 35, 85-87.

Hayashi, H., Saito-Mito, N., Kaburagi, T., Yoshino, H., Higaki, M., Nishimoto, K., and Sato, K. (2012). Nutritional status in relation to adipokines and oxidative stress is associated with disease activity in patients with rheumatoid arthritis. *Nutrition,* 28(11-12), 1109-1114.

Heikkila, R., Aho, K., Heliovaara, M., Knekt, P., Reunanen, A., Aromaa, A.,... Palosuo, T. (1998). Serum androgen-anabolic hormones and the risk of rheumatoid arthritis. *Ann. Rheum. Dis.,* 57(5), 281-285.

Hendrich, S. (2010). (n-3) Fatty Acids: Clinical Trials in People with Type 2 Diabetes. *Adv. Nutr.,* 1(1), 3-7. doi: 10.3945/an.110.1003 1003 [pii]

Hibbeln, Joseph R., Umhau, John C., George, David T., Shoaf, Susan E., Linnoila, Markku, and Jr, Norman Salem. (2000). Plasma total cholesterol concentrations do not predict cerebrospinal fluid neurotransmitter metabolites: implications for the biophysical role of highly unsaturated fatty acids. *Am. J. Clin. Nutr.* 2000, 71 (suppl.), 331S-338S.

Hong, Song, Gronert, Karsten, Devchand, Pallavi R., Moussignac, Rose-Laure, and Serhan, Charles N. (2003). Novel Docosatrienes and 17S-Resolvins Generated from Docosahexaenoic Acid in Murine Brain, Human Blood, and Glial Cells: Autacoids in Anti-Inflammation. *Journal of Biological Chemistry, 278*(17), 14677-14687. doi: 10. 1074/jbc.M300218200

Horrobin, D. F. (1987). Low prevalences of coronary heart disease, psoriasis, asthma and rheumatoid arthritis in Eskimos. *Med. Hypotheses,* 22, 421-428.

Hu, F. B., Bronner, L., Willett, W. C., Stampfer, M. J., Rexrode, K. M., Albert, C. M.,... Manson, J. E. (2002). Fish and omega-3 fatty acid intake and risk of coronary heart disease in women. *JAMA,* 287(14), 1815-1821. doi:joc11695 [pii]

Hurst, S., Zainal, Z., Caterson, B., Hughes, C. E., and Harwood, J. L. (2010). Dietary fatty acids and arthritis. *Prostag. Leukotr. Ess.* 82(4-6), 315-318.

James, M. J., Cleland, L. G., Gibson, R. A., and Hawkes, J. S. (1991). Interaction between fish and vegetable oils in relation to rat leucocyte leukotriene production. *J. Nutr.,* 121, 631-637.

James, M. J. and Cleland, L. G. (1997). Dietary n-3 fatty acids and therapy for rheumatod arthritis. *Semin. Arthritis Rheu.,* 27(2), 85-97.

James, M. J., Proudman, S. M. and Cleland, L. G. (2003). Dietary n-3 fats as adjunctive therapy in a prototypic inflammatory disease: issues and obstacles for use in rheumatoid arthritis. *Prostag. Leukotr. Ess.* 68, 399-405.

Kremer, J. M., Bigauoette, J., Michalek, A. V., Timchalk, M. A., Lininger, L., Rynes, R. I.,... Bartholomew, L. E. (1985). Effects of manipulation of dietary fatty acids on clinical manifestations of rheumatoid arthritis. *Lancet,* 1(8422), 184-187.

Kremer, J. M., Lawrence, D. A., Jubiz, W., DiGiacomo, R., Rynes, R., Bartholomew, L. E., and Sherman, M. (1990). Dietary fish oil and olive oil supplementation in patients with rheumatoid arthritis. Clinical and immunologic effects. *Arthritis Rheum.,* 33(6), 810-820.

Kremer, J. M., Lawrence, D. A., Petrillo, G. F., Litts, L. L., Mullaly, P. M., Rynes, R. I., et al. (1995). Effects of high-dose fish oil on rheumatoid arthritis after stopping nonsteroidal antiinflammatory drugs. Clinical and immune correlates. *Arthritis Rheum.,* 38(8), 1107-1114.

Kromann, N. and Green, A. (1980). Epidemiological studies in the Upernavik district, Greenland. Incidence of some chronic diseases 1950-1974. *Acta Med. Scand.* 208, 401-406.

Kvien, T. K., Glennas, A., Knudsrod, O. G., Smedstad, L. M., Mowinckel, P., and Forre, O. (1997). The prevalence and severity of rheumatoid arthritis in Oslo. Results from a county register and a population survey. *Scand. J. Rheumatol.,* 26(6), 412-418.

Lee, T. H., Hoover, R. L. and Williams, J. D. (1985). Effect of dietary enrichment with eicosapentaenoic and docosahexaenoic acids on in vitro neutrophil and monocyte leukotriene generation and neutrophil function. *N Engl. J. Med.,* 312, 1217-1223.

Lee, T. H., Sethi, T., Crea, A. E., Peters, W., Arm, J. P., Horton, C. E.,... Spur, B. W. (1988). Characterization of leukotriene B3: Comparison of its biological activities with leukotriene B4 and leukotriene B5 in complement receptor enhancement lysozyme release and chemotaxis of human neutrophils. *Clin. Sci.* 74, 467-475.

Lee, Y. H., Bae, S. C. and Song, G. G. (2012). Omega-3 polyunsaturated fatty acids and the treatment of rheumatoid arthritis: a meta-analysis. *Arch. Med. Res.,* 43(5), 356-362. doi: 10.1016/j.arcmed.2012.06.011 S0188-4409(12)00166-X [pii]

Leslie, C. A., Gonnerman, W. A., Ullman, M. D., Hayes, K. C., Franzblau, C., and Cathcart, E. S. (1985). Dietary fish oil modulates macrophage fatty acids and decreases arthritis susceptibility in mice. *J. Exp. Med.,* 162(4), 1336-1349.

Levy, B. D., Clish, C. B., Schmidt, B., Gronert, K., and Serhan, C. N. (2001). Lipid mediator class switching during acute inflammation: signals in resolution. *Nat. Immunol.,* 2, 612-619.

MacLennan, Mira B., Clarke, Shannon E., Perez, Kate, Wood, Geoffrey A., Muller, William J., Kang, Jing X., and Ma, David W. L. (2013). Mammary tumor development is directly inhibited by lifelong n-3 polyunsaturated fatty acids. *The Journal of Nutritional Biochemistry,* 24(1), 388-395. doi%20http/::dx.doi.org:10.1016:j.jnutbio.2012.08.002

Magnusson, A., Axelsson, J., Karlsson, M. M., and Oskarsson, H. (2000). Lack of seasonal mood change in the Icelandic population: results of a cross-sectional study. *Am. J. Psychiatry,* 157(2), 234-238.

Marcheselli, Victor L., Hong, Song, Lukiw, Walter J., Tian, Xiao Hua, Gronert, Karsten, Musto, Alberto,... Bazan, Nicolas G. (2003). Novel Docosanoids Inhibit Brain Ischemia-Reperfusion-mediated Leukocyte Infiltration and Pro-inflammatory Gene Expression. *Journal of Biological Chemistry,* 278(44), 43807-43817. doi: 10.1074/jbc.M305841200

McInnes, I. B. and Schett, G. (2011). The pathogenesis of rheumatoid arthritis. *N Engl. J. Med.,* 365(23), 2205-2219. doi: 10.1056/NEJMra1004965

Miles, E. A. and Calder, P. C. (2012). Influence of marine n-3 polyunsaturated fatty acids on immune function and a systematic review of their effects on clinical outcomes in rheumatoid arthritis. *Brit. J. of Nutrition,* 107: Suppl. 2, S171-184.

Mukherjee, Pranab K., Marcheselli, Victor L., Serhan, Charles N., and Bazan, Nicolas G. (2004). Neuroprotectin D1: A docosahexaenoic acid-derived docosatriene protects human retinal pigment epithelial cells from oxidative stress. *Proceedings of the National Academy of Sciences of the US,* 101(22), 8491-8496. doi: 10.1073/pnas.0402531101

Nielen, M. M., van Schaardenburg, D., Reesink, H. W., van de Stadt, R. J., van der Horst-Bruinsma, I. E., de Koning, M. H.,... Dijkmans, B. A. (2004). Specific autoantibodies precede the symptoms of rheumatoid arthritis: a study of serial measurements in blood donors. *Arthritis Rheum.,* 50(2), 380-386. doi: 10.1002/art.20018

Norling, L. V. and Perrett, M. (2013). The role of omega-3 derived resolvins in arthritis. *Curr. Opin. Pharmacol.*

Norling, L. V. and Serhan, C. N. (2010). Profiling in resolving inflammatory exudates identifies novel anti-inflammatory and pro-resolving mediators and signals for termination. *J. Intern. Med.,* 268, 15-24.

Norling, L. V., Perretti, M. (2013). The role of omega-3 derived resolvins in arthritis. *Curr. Opin. Pharmacol.*

Peet, M. and Stokes, C. (2005). Omega-3 fatty acids in the treatment of psychiatric disorders. *Drugs,* 65(8), 1051-1059. doi: 6582 [pii]

Percival, T. (1783). Observations on the medical uses of the Oleum Jecons Aselli, or cod liver oil, in the chronic rheumatism and other painful disorders. *London Med. J.,* 3, 393-401.

Power, D., Codd, M., Ivers, L., Sant, S., and Barry, M. (1999). Prevalence of rheumatoid arthritis in Dublin, Ireland: a population based survey. *Ir. J. Med. Sci.,* 168(3), 197-200.

Prickett, J. D., Trentham, D. E. and Robinson, D. R. (1984). Dietary fish oil augments the induction of arthritis in rats immunized with type II collagen. *J. Immunol.,* 132(2), 725-729.

Riise, T., Jacobsen, B. K. and Gran, J. T. (2000). Incidence and prevalence of rheumatoid arthritis in the county of Troms, northern Norway. *J. Rheumatol.,* 27(6), 1386-1389.

Robinson, D. R., Tateno, S., Knoell, C., Olesiak, W., Xu, L., Hirai, A.,... Colvin, R. B. (1989). Dietary marine lipids suppress murine autoimmune disease. *J. Intern. Med. Suppl.,* 731, 211-216.

Rose, David P. and Connolly, Jeanne M. (1999). Omega-3 fatty acids as cancer chemopreventive agents. *Pharmacology and Therapeutics,* 83(3), 217-244. doi: http://dx.doi.org/10.1016/S0163-7258(99)00026-1

Ruggiero, C., Lattanzio, F., Lauretani, F., Gasperini, B., Andres-Lacueva, C., and Cherubini, A. (2009). Omega-3 polyunsaturated fatty acids and immune-mediated diseases: inflammatory bowel disease and rheumatoid arthritis. *Curr. Pharm. Des.,* 15(36), 4135-4148.

Saraux, A., Guedes, C., Allain, J., Devauchelle, V., Valls, I., Lamour, A.,... Le Goff, P. (1999). Prevalence of rheumatoid arthritis and spondyloarthropathy in Brittany, France. Societe de Rhumatologie de l'Ouest. *J. Rheumatol.,* 26(12), 2622-2627.

Seki, H., Tani, Y. and Arita, M. (2009). Omega-3 PUFA derived anti-inflammatory mediator resolvin E1. *Prostaglandins Other Lipid Mediat.,* 89, 126-130.

Serhan, C. N., Clish, C. B., Brannon, J., Colgan, S. P., Chiang, N., and Gronert, K. (2000). Novel functional sets of lipid-derived mediators with antiinflammatory actions generated from omega-3 fatty acids via cyclooxygenase 2-nonsteroidal antiinflammatory drugs and transcellular processing. *J. Exp. Med.,* 192(8), 1197-1204.

Serhan, C. N., Clish, C. B., Brannon, J., Colgan, S. P., Gronert, K., and Chiang, N. (2000). Anti-microinflammatory lipid signals generated from dietary N-3 fatty acids via cyclooxygenase-2 and transcellular processing: a novel mechanism for NSAID and N-3 PUFA therapeutic actions. *J. Physiol. Pharmacol.,* 51(4 Pt 1), 643-654.

Serhan, C. N., Hong, S., Gronert, K., Colgan, S. P., Devchand, P. R., Mirick, G., and Moussignac, R. L. (2002). Resolvins: a family of bioactive products of omega-3 fatty acid transformation circuits initiated by aspirin treatment that counter proinflammation signals. *J. Exp. Med.,* 196(8), 1025-1037.

Serhan, C. N., Chiang, N. and van Dyke, T. E. (2008). Resolving inflammation: dual antiinflammatory and pro-resolution lipid mediators. *Nat. Rev. Immunol.*, 8, 349-361.

Serhan, C. N., Clish, C. B., Brannon, J., Colgan, S. P., Chiang, N., and Gronert, K. (2000). Novel functional sets of lipid-derived mediators with antiinflammatory actions generated from omega-3 fatty acids via cyclooxygenase 2-nonsteroidal antiinflammatory drugs and transcellular processing. *J. Exp. Med.,* 192(8), 1197-1204.

Serhan, C. N., Hong, S., Gronert, K., Colgan, S. P., Devchand, P. R., Mirick, G., and Moussignac, R. L. (2002). Resolvins: a family of bioactive products of omega-3 fatty acid transformation circuits initiated by aspirin treatment that counter proinflammation signals. *J. Exp. Med.,* 196(8), 1025-1237.

Serhan, C. N., Clish, C. B., Brannon, J., Colgan, S. P., Gronert, K., and N. Chiang. (2000). Anti-inflammatory lipid signals generated from dietary n-3 fatty acids via cyclooxygenase-2 and transcellular processing: a novel mechanism for NSAID and n-3 PUFA therapeutic actions. *J. Physiol. Pharmacol.* 4, 643-654.

Silman, A. J. and Hochberg, M. C. (2001). Rheumatoid arthritis. In: A. J. Silman (Ed.), *Epidemiology of the rheumatic diseases* (pp. 31-71). Oxford: Oxford University Press.

Silman, A. J. and Pearson, J. E. (2002). Epidemiology and genetics of rheumatoid arthritis. *Arthritis Res.,* 4 Suppl. 3, S265-272.

Simonsson, M., Bergman, S., Jacobsson, L. T., Petersson, I. F., and Svensson, B. (1999). The prevalence of rheumatoid arthritis in Sweden. *Scand. J. Rheumatol.,* 28(6), 340-343.

Simopoulos, A. P. (2002). Omega-3 fatty acids in inflammation and autoimmune diseases. *J. Am. Coll. Nutr.,* 21(6), 495-505.

Sinclair, H. M. (1956). Deficiency of essential fatty acids and atherosclerosis, etcetera. *Lancet., i,* 381-383.

Singer, Pierre, Shapiro, Haim, Theilla, Miryam, Anbar, Ronit, Singer, Joelle, and Cohen, Jonathan. (2008). Anti-inflammatory properties of omega-3 fatty acids in critical illness: novel mechanisms and an integrative perspective. *Intensive Care Medicine,* 34(9), 1580-1592. doi: 10.1007/s00134-008-1142-4

Sperling, R. I., Weinblatt, M., Robin, J. L., Ravalese, J., Hoover, R. L., House, F.,... Austen, K. F. (1987). Effects of Dietary Supplementation with Marine Fish Oil on Leukocyte Lipid Mediator Generation and Function in Rheumatoid-Arthritis. *Arthritis and Rheumatism,* 30(9), 988-997. doi: DOI 10.1002/art.1780300905

Stamp, L. K., James, M. J. and Cleland, L. G. (2005). Diet and rheumatoid arthritis: a review of the literature. *Semin. Arthritis Rheum.,* 35(2), 77-94. doi: S0049-0172(05)00087-9 [pii] 10.1016/j.semarthrit.2005.05.001

Storlien, L. H., Kraegen, E. W., Chisholm, D. J., Ford, G. L., Bruce, D. G., and Pascoe, W. S. (1987). Fish oil prevents insulin resistance induced by high-fat feeding in rats. *Science,* 237(4817), 885-888.

Tulleken, Jacob E., Limburg, Pieter C., Muskiet, Frits A. J., and Rijswijk, Martin H. Van. (1990). Vitamin e status during dietary fish oil supplementation in rheumatoid arthritis. *Arthritis and Rheumatism,* 33(9), 1416-1419. doi: 10.1002/art.1780330914

Vachier, I., Chanez, P., Bonnans, C., Godard, P., Bousquet, J., and Chavis, C. (2002). Endogenous anti-inflammatory mediators from arachidonate in human neutrophils. *Biochem. Biophys. Res. Commun.,* 290, 219-224.

Van der Tempel, H., Tulleken, J. E., Limburg, P. C., Muskiet, F. A., and van Rijswijk, M. H. (1990). Effects of fish oil supplementation in rheumatoid arthritis. *Annals of the Rheumatic Diseases,* 49(2), 76-80. doi: 10.1136/ard.49.2.76

Van der Tempel, H., Tulleken, J. E., Limburg, P. C., Muskiet, F. A., and van Rijswijk, M. H. (1990). Effects of fish oil supplementation in rheumatoid arthritis. *Ann. Rheum. Dis.,* 49 (2), 76-80.

Volker, D., Fitzgerald, P., Major, G., and Garg, M. (2000). Efficacy of fish oil concentrate in the treatment of rheumatoid arthritis. *J. Rheumatol.,* 27(10), 2343-2346.

Wall, R., Ross, R. P., Fitzgerald, G. F., and Stanton, C. (2010). Fatty acids from fish: the anti-inflammatory potential of long-chain omega-3 fatty acids. *Nutr. Rev.,* 68(5), 280-289. doi: 10.1111/j.1753-4887.2010.00287.x

Weissmann, G. (2004). Pathogenesis of rheumatoid arthritis. *J. Clin. Rheumatol.,* 10 (3 Suppl.), S26-31. doi: 10.1097/01.rhu.0000130687.75646.44

Wen, H. and Baker, J. F. (2011). Vitamin D, immunoregulation, and rheumatoid arthritis. *J. Clin. Rheumatol.,* 17(2), 102-107. doi: 10.1097/RHU.0b013e31820edd18

In: Omega-3 Fatty Acids
Editor: Md. Wasim Khan

ISBN: 978-1-62948-516-4

Chapter 2

Omega-3 Fatty Acids Target NF-κB to Modulate Breast and Prostate Cancer Progression

D. A. Cavazos, S. A. Apte, C. H. Chen and L. A. deGraffenried
Department of Nutritional Sciences,
University of Texas at Austin, Austin, Texas, US

Abstract

Background: Global epidemiological trends suggest that consumption of dietary sources enriched in omega-3 fatty acids correlate with improved clinical benefit in the prevention and treatment of certain cancers. In this chapter, we will focus primarily on the chemopreventive and chemotherapeutic properties of cold-water fish-derived omega-3 fatty acids. Numerous studies have found that omega-3 fatty acids suppress the formation and growth of a variety of solid tumors, including carcinomas of the breast and prostate. Here we will review the current literature and present our *in vitro* mechanistic findings which support the role of omega-3 fatty acids in decreasing risk of breast and prostate cancer development through modulation of nuclear factor kappa-light-chain-enhancer of activated B cells (NF-κB).

Key Findings: NF-κB is a transcription factor involved in regulating expression of pro-inflammatory and pro-survival genes in response to various forms of cellular stress, and is commonly dysregulated in solid tumors. Previous studies by our group and others provide evidence that docosahexaenoic acid (DHA), an omega-3 fatty acid, suppresses NF-κB transcriptional activity and precludes translocation into the nucleus in two human prostate tumor-derived cell lines, LNCaP and PacMetUT1. Additionally, inhibition of NF-κB was associated with several parameters of physiological response, including decreased proliferation and survival. In addition to the beneficial effects of DHA in the prostate, we observe complimentary and potent chemotherapeutic effects in breast cancer tumor cell lines. MCF-7 (ER+) and MDA-MB-231 (ER-) cells treated with omega-3 fatty acids also display nuclear exclusion of NF-κB and reduced proliferation and survival.

Together these findings suggest that dietary omega-3 fatty acids elicit antiproliferative and pro-apoptotic effects in prostate and breast tumor cell lines.

Conclusions: Preclinical and epidemiologic studies indicate that omega-3 fatty acids improve clinical outcome in prostate and breast cancers. We provide *in vitro* evidence that omega-3 fatty acids function as a chemotherapeutic agent in breast and prostate carcinomas, in part through inhibition of NF-κB.

Keywords: Omega-3 Fatty Acids, NF-κB, Prostate Cancer, Breast Cancer

Introduction

According to the American Cancer Society, malignant carcinomas of the breast and prostate are the second leading causes of cancer-related death for women and men in the US. The National Cancer Institute projects that in 2013, an estimated 232,000 women will be diagnosed with breast cancer (BCa), resulting in over 39,000 deaths. Also in 2013, almost 238,000 men will be diagnosed with prostate cancer (PCa), resulting in over 29,000 deaths. Despite recent advances in drug development and accelerated approval programs for clinical screening, the incidence and mortality rates of breast and prostate cancer have only marginally dropped relative to lung and colon cancer. Even while a large number of new chemical entities (NCEs) have shown promise in preclinical and Phase I trials, the side effects associated with many of these pharmaceuticals present considerable obstacles to bringing candidate NCEs to market.

Additionally, prolonged use of these preventative or adjuvant therapy drugs is often associated with multiple organ toxicities. For this reason, development of nutraceutical compounds derived from natural food sources is a promising field of growing research. Nutraceutical supplements that decrease cancer risk, augment cytotoxic therapies or function as adjuvant drug therapies, yet display minimal side effects, could radically improve outcome.

Despite promising results in preclinical trials, many natural compounds have failed to demonstrate clinical benefit in the prevention or treatment of BCa or PCa. Possible explanations for these shortcomings may arise due to the design of randomized controlled trials in which large heterogeneous populations of ethnically and socioeconomically diverse individuals are enrolled onto study. Confounding factors inherent to such studies include the possibility that polymorphisms contained within certain genes may underlie drug sensitivity or resistance.

Additionally, epigenetic changes may influence response to nutraceutical compounds. A growing body of literature supports the notion that environmental exposures and lifestyle, including: diet, exercise, alcohol consumption, or smoking may alter the epigenetic profile of individuals to mask the beneficial effects of nutraceutical compounds. Moreover, natural compounds generally lack the drug potency of pharmaceutical drugs, therefore these trials may require long-term supplementation to impact disease development or progression. Furthermore, the difficulty of detecting low to moderate clinical response over long periods of time may prove to be prohibitive.

One case in point is the recent Centrum Silver Prevention Study in which 14,641 male physicians 50 years and over were randomized onto a daily multivitamin, or placebo and tracked for the incidence of total or site-specific cancers. After over a decade-long trial, no significant association between multivitamin use and prostate cancer were detected. Only after 15 years was a small benefit observed in outcome (Gaziano et al.). Similar multivitamin studies have been conducted in women with similar results (Ishitani et al.). Moreover, in some cases the beneficial effects of nutraceutical supplements may not be detectable until years after the conclusion of the trial, if at all. Despite these limitations, some naturally occurring compounds have demonstrated significant clinical benefits and are regularly prescribed to improve health.

2. Omega-3 Fatty Acids and Cancers of Breast and Prostate

On the basis that the global occurrence rates of BCa and PCa can vary by a factor of 5-fold (Kamangar, Dores and Anderson; Quinn and Babb), and are influenced by changes in diet and lifestyle that accompany migration to different regions of the world, several studies compared international dietary patterns with risk of BCa and PCa (Quinn and Babb; Sharma et al.; Kolonel, Altshuler and Henderson).

From these early epidemiological studies, it was determined that certain cultures at lower risk of developing BCa or PCa were more likely to consume dietary items higher in omega-3 fatty acids (n-3 FAs) (Willett; Kim et al.). In a recent report investigating the association of three diet quality indices - the Healthy Eating Index-2005 (HEA-2005), Alternative Healthy Eating Index-2010 (AHEI-2010), and alternative Mediterranean diet score (aMED) with PCa risk, the authors found no significant association between any of the indices and total PCa, advanced PCa or fatal PCa. Although overall diet displayed no impact on PCa in this study, the authors observed that in individual component analyses, the fish component of aMED and omega-3 FAs component of AHEI-2010 were in fact inversely correlated with fatal prostate cancer (Bosire et al.). This is supported by several large epidemiological studies that have observed a strong inverse correlation between consumption of n-3 FAs and risk of BCa (Kim et al.) and PCa (Leitzmann et al.).

These studies provide compelling evidence suggesting that the incidence and mortality of BCa and PCa are modifiable according to diet and lifestyle, and are not solely predicated upon genetic factors alone.

Given the epidemiological data suggesting that consumption of food items rich in n-3 FAs may decrease risk or improve clinical outcome in BCa and PCa patients, greater emphasis has been made to elucidate the molecular targets by which n-3 FAs impact tumorigenesis. A large number of preclinical and *in vitro* investigations now provide compelling evidence that n-3 FAs are indeed very potent chemopreventive and chemotherapeutic agents (Rose and Connolly; Hardman et al.; Berquin, Edwards and Chen; Berquin et al.). Further, some studies indicate that combination therapies using n-3 FAs enhance antitumor therapies to extend survival compared with therapy alone (Hardman; Hardman et al.). Our group has contributed to this body of knowledge through investigation of the *in vitro* effects of the n-3 FAs on breast and prostate cancer cell survival and growth.

3. Omega-3 Fatty Acids

n-3 FAs are one form of polyunsaturated fatty acids (PUFAs). PUFAs contain two or more double bonds within the molecule's hydrocarbon tail. All fatty acids have a carboxyl group at one end, and a methyl group at the other end. The terminal carbon at the methyl-end of the molecule is referred to as the *omega* carbon. n-3 FAs contain a double bond between the 3rd and 4th carbons when counting from the *omega* end of the carbon chain. Certain fatty acids are required for normal physiology, but cannot be synthesized *de novo* in humans due to the lack of specific desaturase enzymes. These fatty acids are therefore considered 'essential' fatty acids. α-Linolenic acid (ALA) is the shortest-chained member of the essential n-3 FAs family. Other members of the n-3 FAs family, such as docosahexaenoic acid (DHA) and Eicosapentaenoic acid (EPA), can be biochemically synthesized from ALA in small, often insufficient quantities through the addition of hydrocarbon groups to the fatty acid chain and through addition or rearrangement of the double bonds. Therefore, DHA and EPA are considered 'conditionally essential' fatty acids.

Dietary sources rich in ALA include certain nuts and seeds, particularly flaxseed and walnuts (Bourre; Simopoulos), while the DHA and EPA are encountered at high levels in the fat of certain coldwater fish such as salmon, tuna and sardines (Bourre; Nettleton). The majority of the DHA and EPA found in these fish are procured from their diet of microscopic marine invertebrates, including plankton, krill and microalgae. Therefore, the concentration of n-3 FAs in fish oil may vary widely depending on the diet of farmed and wild-caught fish (Cahu, Salen and de Lorgeril).

n-3 FAs are digested and absorbed similarly to other dietary lipids. Along with cholesterol, triacylglycerides, diacylglycerides and monoacylglycerides and free fatty acids, n-3 FAs coalesce with bile salts and digestive enzymes to form micelles in the proximal sections of the small intestines. Pancreatic lipase and colipase facilitate the release of many of the lipid constituents from micelles for absorption into the enterocytes lining the distal portion of the duodenum and jejunum. Absorption of fatty acids from the diet is very efficient. The small intestines can absorb large quantities of triacylglycerides (nearly 660 grams per day at 95% efficiency), with the majority of triacylglyceride digestion and absorption occurring within the jejunum. Upon release from the micelles, triacylglycerides, diacylglycerides and monoacylglycerides and free fatty acids can diffuse passively across the cells of the mucosal membrane or be passively transported by a family of fatty acid transport proteins (FATP-1,-4). Once absorbed, fatty acids are processed by the smooth endoplasmic reticulum and packaged into lipid transport vesicles called chylomicrons. From the intestines, the majority of chylomicrons are released directly into the intestinal lymphatics for sustained distribution to peripheral cells throughout the body. Alternatively, a fraction of chylomicrons carrying the dietary n-3 FAs are released directly into the superior mesenteric vein to join with the hepatic portal vein to transport nutrients from the small intestines to the liver. In the liver, n-3 FAs encounter one of several fates: n-3 FAs may be stored and utilized for structural functions within the hepatocytes, used as an anabolic fuel source in a process called β-oxidation, or oxidized into one of many metabolites. Metabolites of n-3 FAs are physiologically active molecules which serve as ligands for receptors, or as substrate for the synthesis of eicosanoids. Eicosanoids are a broad group of PUFA-derived signaling molecules which regulate systemic inflammation (Chapkin et al.; Pischon et al.; von Schacky, Fischer and

Weber), blood platelet aggregation (Lands et al.; von Schacky, Fischer and Weber), cellular immunity (von Schacky, Fischer and Weber; Kinsella and Lokesh), proliferation (Begin et al.) and differentiation (Schmitz and Ecker). n-6 FAs typically are converted into eicosanoids that promote inflammation and vascular constriction. Some enzymes, such as the cyclooxygenase family, that use n-6 FAs as substrate to produce proinflammatory eicosanoids also convert n-3 FAs into other series of eicosanoids (Kobayashi et al.). While commonly reported that n-3 FAs directly yield eicosanoids that reduce inflammation, this is a misnomer in that these products do not possess anti-inflammatory properties.

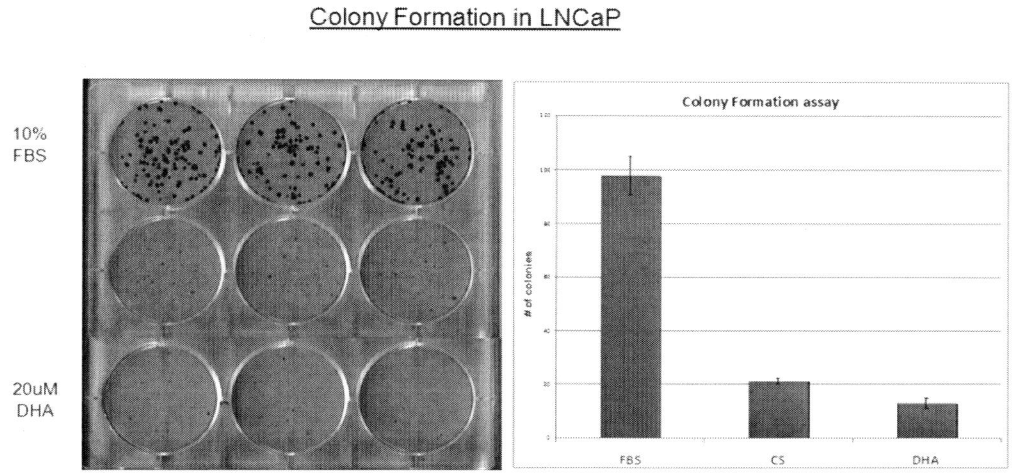

Figure 1. Colony formation assay (CFC).LNCaP cells treated with n-6 (AA) or n-3 (DHA) fatty acids in charcoal-stripped (CS) medium.

By simple competition, when n-3 FAs are abundant, fewer n-6 FAs are converted into proinflammatory eicosanoids. However, when n-3 FAs are consumed in conjunction with non-steroidal anti-inflammatory drugs (NSAIDS), such as aspirin, novel lipids such as resolvins which do contain potent anti-inflammatory properties are synthesized (Arita, Clish and Serhan; Serhan et al.).

Given the inherent dependence of BCa and PCa cells on the series 2 eicosanoids produced from n-6 FAs, and because epidemiological and clinical trials suggest that n-3 FAs may improve outcome, we sought to determine the effect of DHA, on the ability of prostate cancer cells to survive and form colonies *in vitro*. Clonogenic (CFC) assays are designed to measure two biological parameters of response: 1) replicative capacity – or the ability of individual cells to proliferate. 2) survival – or the ability to withstand the experimental conditions and avoid necrosis, apoptosis and autophagy. As such, we performed a CFC assay using LNCaP prostate cancer cells seeded at a concentration of 500 cells per well. Clearly visible in Figure 1, are the number of colonies grown under normal tissue culture conditions (10% fetal bovine serum - FBS). Charcoal stripped (CS) serum was used to remove the lipid fraction from fetal bovine serum to directly assess the effect of n-3 FAs on cancer cell survival. CS medium with 20 uM DHA, an n-3 FA, results in fewer and smaller colonies than CS medium alone – indicating the antiproliferative or cytotoxic properties of DHA.

Next we determined the effect of DHA on cell viability. Using a colorimetric MTT assay, in which a soluble tetrazolium salt is reduced to a purple precipitate by NAD(P)H-dependent oxioreductase enzymes, we were able to measure the number of metabolically respiring cells after treatment with DHA. Figure 2 compares the viability of LNCaP cells grown under standard conditions to those grown in CS and CS + DHA. Although removal of all lipids from the medium dramatically suppressed viability, the addition of DHA to the CS medium further suppressed viability.

These data support the results observed in the CFC assay and confirm that DHA is cytotoxic to human prostate cancer cells.

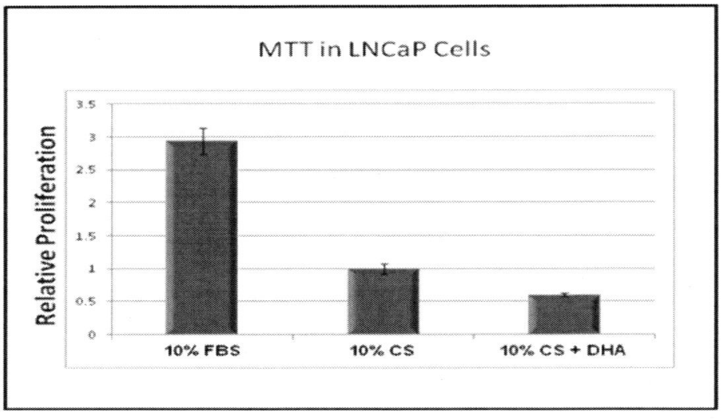

Figure 2. MTT assay in LNCaP cells. LNCaPs grown in charcoal-stripped medium (CS) to remove lipids and hormones from the serum were treated with DHA. DHA suppressed the growth of LNCaP cells by 50%.

4. Role of Omega-3 Fatty Acids as Antioxidants

There is evidence that long-chained fatty acids possess indirect antioxidant properties by way of activating cellular antioxidant pathways, such as SOD and catalase (Poprzecki et al.). These physiologically active lipids may also be directly oxidized in cells, however this does not confer antioxidant activity. All cellular molecules, including protein, DNA and lipids are potential targets of reactive oxygen and nitrogen species, but this does not qualify them as antioxidants as they are active participants in the REDOX reactions that generate additional free radicals.

As such, we sought to determine the extent to which DHA directly or indirectly functions as an antioxidant in prostate cancer cells. Figure 3 demonstrates that 30 uM DHA, a concentration that is physiologically achievable in the blood serum of individuals consuming fish oil supplements, is a potent antioxidant. LNCaP cells grown in standard 10% fetal bovine serum-supplemented medium display low basal levels of γ-H2AX foci when observed by immunofluorescence. γ-H2AX, a histone variant that becomes phosphorylated and participates in repair of DNA double-strand breaks (DSBs), is a robust indicator of overall oxidative stress and accompanying DSBs. In Figure 3, we observed that treatment with a low physiological pulse of 62.5 uM hydrogen peroxide (H_2O_2) for 40 minutes caused massive induction of γ-H2AX foci, while pretreatment overnight with DHA almost entirely suppressed the H_2O_2-induced DSBs. This data provides evidence that n-3 FA supplementation may protect against oxidative stress.

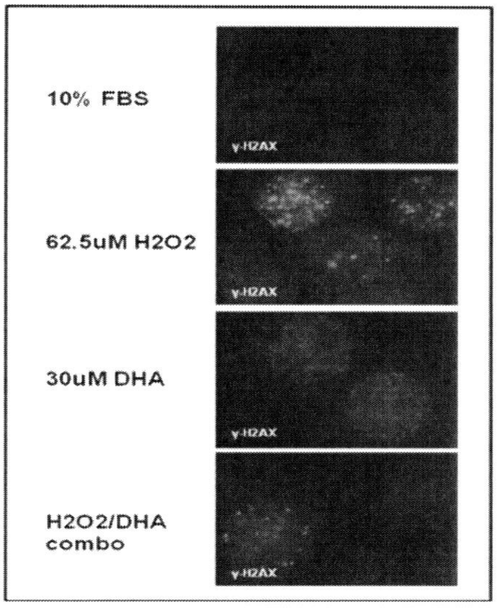

Figure 3. DHA functions as an antioxidant in LNCaP cells. Immunofluorescent staining for γ-H2AX (green). Formation of γ-H2AX foci after treatment with H_2O_2 is attenuated by DHA.

5. Nuclear Factor Kappa-Light-Chain-Enhancer of Activated B-Cells (NF-κB) and Cancer

NF-κB is a heteromeric protein complex that is normally sequestered in the cytoplasm of unstimulated cells by a family of Inhibitors of κB (IκB). Under conditions of cellular stress, IκB is phosphorylated by IkappaB kinase (IKK) and marked for proteasomal degradation, thereby allowing rapid nuclear localization of NF-κB (Li et al.; Beg et al.). NF-κB is a nuclear transcription factor capable of modulating the expression of genes that enable cells to cope with various cellular stresses (Dolcet et al.). In fact, NF-κB activity has been shown to be upregulated in tumors and tumor cell lines from BCa and PCa patients (Santini, Perrone et al.; Santini, Schiavon et al.; Ismail et al.; Lessard et al.). Further, the ionizing radiation therapy (IR) and chemotherapies used to treat these cancers are known to upregulate NF-κB (Camp et al.; Jung and Dritschilo; Chopra et al.), thereby helping cancer cells evade apoptosis (Maxwell et al.). Figure 4 demonstrates the effectiveness of DHA in suppressing NF-κB transcriptional activity using an NF-κB luciferase reporter plasmid (Stratagene) and pNull-renilla reporter plasmid (Promega) transiently transfected into LNCaP cells, then grown in charcoal-stripped medium (CS) and treated with DHA or a potent NF-κB inhibitor, parthenolide. That a single overnight treatment with 30uM DHA effectively suppressed NF-κB transcriptional activity at an intermediate level to parthenolide is very encouraging as this demonstrates that DHA is a fast, potent and well tolerated inhibitor of NF-κB. Based on these *in vitro* results, further investigation in the preclinical and clinical settings are warranted to elucidate the effects of n-3 FAs on NF-κB transcriptional activity in tumor cells. The results of these studies may provide the basis for the development of new clinical approaches, whereby large doses of DHA are administered 24 hours prior to IR or chemotherapy to inhibit NF-κB and enhance outcome. To this end, a study by Shaikh et al. demonstrates that DHA, a component of fish oil, enhances the efficacy of docetaxel in the prostate through modulation of NF-κB (Shaikh et al.). More recently, our lab reported that DHA selectively sensitizes prostate cancer cells, but not non-malignant cells, to cell death through modulation of NF-κB. We propose that nuclear exclusion of NF-κB by DHA can be used to exploit the endogenous oxidative stress and growth factor signaling inherent to prostate tumor cells in the preventative setting (Cavazos et al.). The rational for this approach is that cancer cells, which are constantly exposed to elevated levels of oxidative stress and cytokine signaling, are much more dependent on NF-κB activity compared to non-malignant cells.

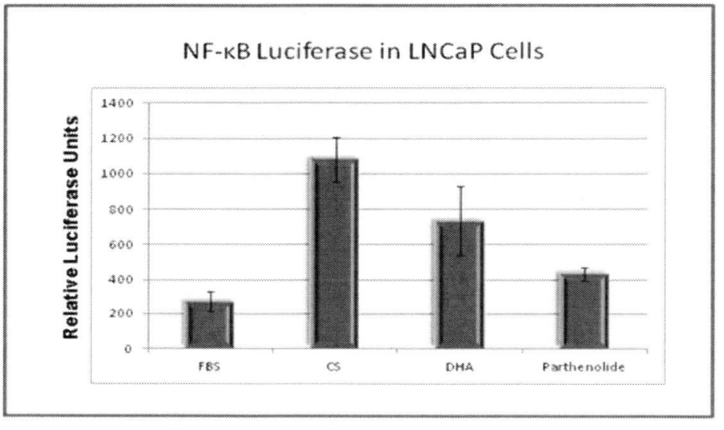

Figure 4. NF-κB Luciferase assay. LNCaP cells grown in CS medium and treated with DHA. A single 24 hour dose of 30 uM DHA suppresses NF-κB transcriptional activity by 35%. Parthenolide, a potent and highly toxic proteasome inhibitor suppressed NF-κB activity by 65%.

In the prolonged absence of NF-κB, the heightened accumulation of endogenous oxidative stress would selectively kill malignant cells – which have lost their mechanism to cope. In contrast, non-malignant cells, which typically experience low basal levels of oxidative stress, would maintain the ability to cope irrespective of NF-κB activity. Figure 4 demonstrates DHA suppresses NF-κB transcriptional activity in LNCaP cells. Here, $2^{\wedge 5}$ cells were cotransfected with NF-κB reporter plasmid and pNull-renilla standardizing plasmid. After overnight treatment, cell lysates were prepared and firefly luminescence quantified and standardized to renilla and reported as relative luciferase units. As clearly indicated in figure 4, DHA is sufficient to reduce NF-κB transcriptional activity in LNCaP cells. While parthenolide, a proteasome-inhibitor that potently inhibits NF-κB, was found to be more effective, this pharmaceutical drug is not well tolerated. We are encouraged by the strong effect of DHA in suppressing NF-κB transcriptional activity at low physiological doses with minimal toxicity. Supporting evidence of the potent ability of DHA to inhibit NF-κB is demonstrated in MCF-7 cells, an ER+ human breast cancer cell line. Figure 5 shows immunofluorescence staining patterns of p65, a heteromeric component of NF-κB, after growth in standard conditions, CS or CS + omega-3-acid ethyl esters. Omega-3-acid ethyl esters are an esterified form of n-3 FAs and are commonly prescribed as Lovaza®. We determined the effects of n-3-acid ethyl esters on subcellular location of NF-κB. In the MCF-7 cells grown with FBS, a nuclear, perinuclear and cytoplasmic pattern of staining was clearly visible. As expected, TNF-α, an NF-κB agonist, resulted in a strong nuclear staining pattern of NF-κB. Addition of 10 uM n-3-acid ethyl esters dramatically suppressed overall p65 staining and resulted in a clear pattern of exclusion from the nucleus. These results confirm that various forms of n-3 FAs are capable of eliciting their NF-κB-inhibitory effects in both breast and prostate cancer cells. However, more studies are warranted. Of particular interest is the potential use of omega-3-acid ethyl esters in combination with cancer therapies that target the various subtypes of breast cancer, including: ER+, PR+, HER2 overexpressers and triple negative disease. Further investigations using this approach show promise in the treatment of breast cancer. Ghosh-Choudhury et al. had demonstrated similar mechanisms of action when breast cancer cells were exposed *in vitro* to omega-3 fatty acids. Previous studies have also reported suppression of breast cancer growth when cells were subjected to similar treatments with omega-3 fatty acids (Schley et al.).

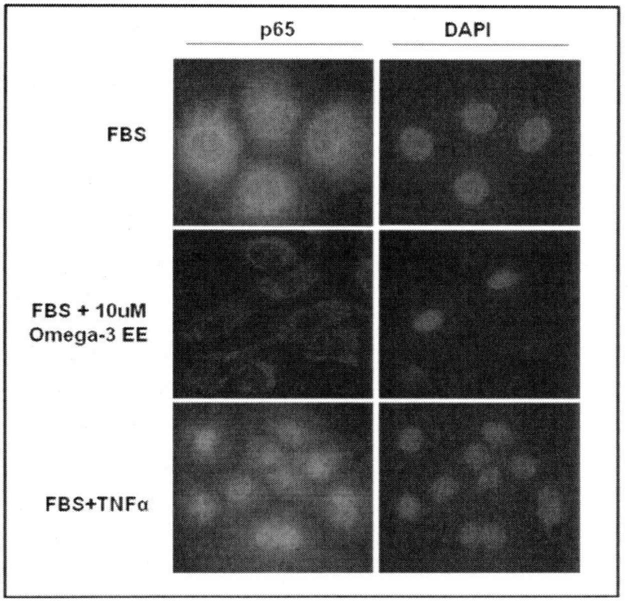

Figure 5. Omega-3-acid Ethyl Esters suppress the expression and nuclear localization of the p65 subunit of NF-κB in MCF-7 breast cancer cells. Immunofluorescent staining of p65 (red) and DAPI-counterstained nuclei (blue) indicate subcellular location of p65.

6. Ratios of Omega-6 to Omega-3 Fatty Acids and Cancer

Linoleic acid (LA) is the shortest-chained n-6 essential fatty acid, from which conditionally essential family members may be derived. The n-6 FAs are frequently encountered at high levels in dairy products, red meat, grains and vegetable oils, all common stables of the Western diet. While ecological, case-control and cohort studies all support a strong positive relationship between the consumption of animal fat and prostate cancer incidence, little direct epidemiological evidence has been reported showing a correlation between n-6 FAs and prostate cancer incidence. While several large human studies have sought to establish a correlation between n-6 FAs in blood serum, adipose tissue or prostate tissue of men with prostate cancer, none have reported a positive association (Willett). The absence, however, of human studies showing any definitive associations between increased tissue levels of n-6 FAs and BCa or PCa is in stark contrast to the vast array of *in situ* and animal studies which overwhelmingly demonstrate the tumorigenic potential of n-6 FAs in animal models of BCa and PCa (Rose et al.; Simonsen et al.; Menendez et al.; Ghosh and Myers). In addition, a large volume of literature supports the role of n-6 FAs in promoting localized inflammation directly within the cells of the breast and prostate tissue (Ghosh and Myers; Hughes-Fulford, Li et al.; Hughes-Fulford, Tjandrawinata et al.; Bagga et al.; Aronson et al.). Although excess consumption of omega-6 fatty acids may contribute to chronic inflammation and ultimately lead to progression of breast and prostate cancer, it is important to keep in mind that these are an essential family of PUFAs that are required for proper physiological functions, including: stimulating immune cell function (Tricon et al.; Calder; Kew et al.), blood clotting (Marcus) and vascular constriction (Roman).

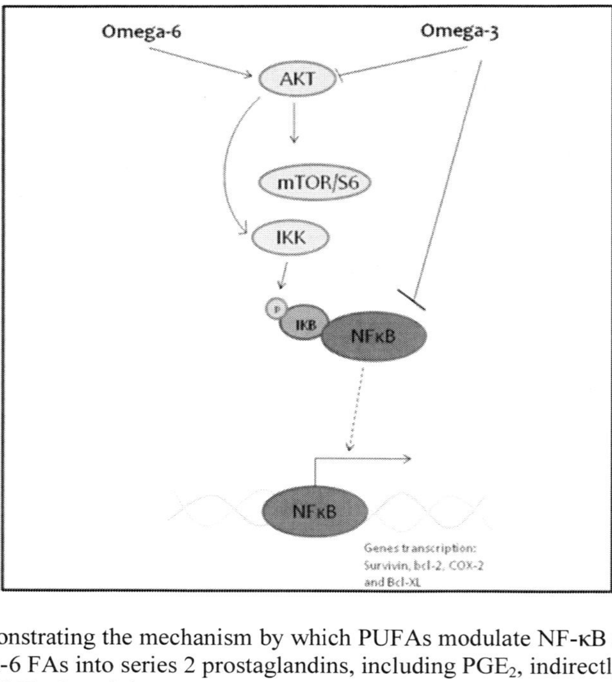

Figure 6. Model demonstrating the mechanism by which PUFAs modulate NF-κB activity in BCa and PCa. Conversion of n-6 FAs into series 2 prostaglandins, including PGE_2, indirectly stimulates Akt, resulting in enhanced NF-κB activity. In contrast, n-3 FAs are converted into series 3 prostaglandins which indirectly inhibit Akt and NF-κB transcriptional activity.

Despite a high intake of n-6 FAs in the Western diet, n-3 FAs are also consumed in a wide variety of food sources; therefore both of these fatty acids are ingested in relative ratios of n-6 to n-3 FAs. Interestingly, certain cultures that are at lower risk of developing BCa or PCa tend to consume dietary items higher in n-3 FAs (Pischon et al.). The American Heart Association (AHA) now recommends limiting the consumption of foods that contain high levels of n-6 FAs and increasing the intake of n-3 FAs to approximately 1g/d DHA and EPA from fish, as determined by the GISSI-Prevention Study. Until recently, the relative intake of n-6 to n-3 ratios consumed by various cultures around the world was largely influenced by nationality, ethnicity and geographical factors. Figure 6 represents our current model in which n-3 FAs indirectly inhibit the activity of NF-κB and upstream activating factors. In contrast, n-6 FAs indirectly activate Akt activity to induce NF-κB.

Conclusion

In aggregate, the data presented herein, together with a wealth of epidemiological studies and randomized controlled trials, support the use of n-3 FAs for the prevention and treatment of BCa and PCa disease. Our *in vitro* studies provide evidence indicating that NF-κB is a molecular target of n-3 FAs in both BCa and PCa tumor cell lines. This work establishes the basis for future studies to examine the efficacy of combination drug therapies using n-3 FAs to suppress IR or chemotherapy-induced NF-κB activation. The rational for this approach is that targeted suppression of NF-κB by n-3 FAs during tumor therapy might improve cytotoxicity and potentially improve outcome. This, in addition to the high bioavailability and low toxicity of n-3 FAs makes them highly suitable for clinical trials. In the event that n-3 FAs indeed prove beneficial in human trials, then perhaps novel, more potent inhibitors of NF-κB can be developed to further improve anticancer therapies.

In the preventative setting, a diet high in n-3 FAs may function similarly to suppress NF-κB in pre-malignant cells. Here we speculate that supplementation with fish oil from an early age might prevent BCa or PCa due to the sustained inhibition of NF-κB. Abnormal cells undergoing malignant transformation produce large amounts of ROS, resulting in genotoxic and cytotoxic stress. Numerous groups and ours, have demonstrated that pre-malignant and malignant BCa and PCa cells rely heavily on NF-κB activity for survival from the elevated levels of endogenous oxidative stress that accompanies tumorigenesis (Dolcet et al.; Lessard et al.; Santini, Schiavon et al.). This dependence on NF-κB may potentially be exploited with fish oil supplements to suppress NF-κB activity, thereby selectively sensitizing pre-malignant or malignant BCa or PCa cells to their endogenous ROS, resulting in lower cancer incidence. To more empirically determine the efficacy of using fish oil supplements for prevention of BCa or PCa, we must engineer translational studies specifically designed to evaluate the status of NF-κB directly within the premalignant lesions from participants consuming food items high in n-3 FAs. The outcome of these trials may prove to have far reaching applications as deregulated NF-κB activity is common to many autoimmune disorders and other metabolic and inflammation-related diseases.

Because fish oil supplementation may show clinical benefit only in certain ethnic groups, randomized trials, whereby enrollment is limited to distinct ethnic backgrounds may improve our ability to detect positive associations between n-3 FAs and health benefit. While some conflicting studies have shown no association with n-3 FA intake and BCa or PCa incidence, such as the EURAMIC study and Health Professionals' Follow Up Study, again these studies were not designed to assess efficacy within genetically distinct groups.

In summary, human studies designed to evaluate the effects of n-3 FAs on cite-specific cancers, including that of the breast and prostate, have shown conflicting results. Despite this, some encouraging studies have reported a positive association between the increased consumption of n-3 FAs and clinical benefit. Considering the overall health benefits of a diet that includes more fish, including improved cardiovascular health, in addition to the potential to reduce BCa or PCa incidence, we should not completely discount these studies. Instead, a refocused effort to design trials that more conclusively determine the role of n-3 FAs in treating or preventing cancer are warranted.

References

Arita, M., C. B. Clish, and C. N. Serhan. "The Contributions of Aspirin and Microbial Oxygenase to the Biosynthesis of Anti-Inflammatory Resolvins: Novel Oxygenase Products from Omega-3 Polyunsaturated Fatty Acids." *Biochem. Biophys. Res. Commun.,* 338.1 (2005): 149-57.

Aronson, W. J., et al. "Modulation of Omega-3/Omega-6 Polyunsaturated Ratios with Dietary Fish Oils in Men with Prostate Cancer." *Urology,* 58.2 (2001): 283-8.

Bagga, D., et al. "Long-Chain N-3-to-N-6 Polyunsaturated Fatty Acid Ratios in Breast Adipose Tissue from Women with and without Breast Cancer." *Nutr. Cancer,* 42.2 (2002): 180-5.

Beg, A. A., et al. "Tumor Necrosis Factor and Interleukin-1 Lead to Phosphorylation and Loss of I Kappa B Alpha: A Mechanism for Nf-Kappa B Activation." *Mol. Cell Biol.,* 13.6 (1993): 3301-10.

Begin, M. E., et al. "Differential Killing of Human Carcinoma Cells Supplemented with N-3 and N-6 Polyunsaturated Fatty Acids." *J. Natl. Cancer Inst.,* 77.5 (1986): 1053-62.

Berquin, I. M., I. J. Edwards, and Y. Q. Chen. "Multi-Targeted Therapy of Cancer by Omega-3 Fatty Acids." *Cancer Lett.,* 269.2 (2008): 363-77.

Berquin, I. M., et al. "Modulation of Prostate Cancer Genetic Risk by Omega-3 and Omega-6 Fatty Acids." *J. Clin. Invest.,* 117.7 (2007): 1866-75.

Bosire, C., et al. "Index-Based Dietary Patterns and the Risk of Prostate Cancer in the Nih-Aarp Diet and Health Study." *Am. J. Epidemiol.,* 177.6: 504-13.

Bourre, J. M. "Where to Find Omega-3 Fatty Acids and How Feeding Animals with Diet Enriched in Omega-3 Fatty Acids to Increase Nutritional Value of Derived Products for Human: What Is Actually Useful ?" *J. Nutr. Health Aging,* 9.4 (2005): 232-42.

Cahu, C., P. Salen, and M. de Lorgeril. "Farmed and Wild Fish in the Prevention of Cardiovascular Diseases: Assessing Possible Differences in Lipid Nutritional Values." *Nutr. Metab. Cardiovasc. Dis.,* 14.1 (2004): 34-41.

Calder, P. C. "N-3 Polyunsaturated Fatty Acids and Immune Cell Function." *Adv. Enzyme Regul., 37* (1997): 197-237.

Camp, E. R., et al. "Inducible Nuclear Factor-Kappab Activation Contributes to Chemotherapy Resistance in Gastric Cancer." *J. Am. Coll Surg.,* 199.2 (2004): 249-58.

Cavazos, D. A., et al. "Docosahexaenoic Acid Selectively Induces Human Prostate Cancer Cell Sensitivity to Oxidative Stress through Modulation of Nf-Kappab." *Prostate,* 71.13: 1420-8.

Chapkin, R. S., et al. "Dietary Docosahexaenoic and Eicosapentaenoic Acid: Emerging Mediators of Inflammation." *Prostaglandins Leukot Essent Fatty Acids,* 81.2-3 (2009): 187-91.

Chopra, D. P., et al. "Tnf-Alpha-Mediated Apoptosis in Normal Human Prostate Epithelial Cells and Tumor Cell Lines." *Cancer Lett.,* 203.2 (2004): 145-54.

Dolcet, X., et al. "Nf-Kb in Development and Progression of Human Cancer." *Virchows Arch.,* 446.5 (2005): 475-82.

Gaziano, J. M., et al. "Multivitamins in the Prevention of Cancer in Men: The Physicians' Health Study Ii Randomized Controlled Trial." *Jama,* 308.18: 1871-80.

Ghosh, J., and C. E. Myers. "Arachidonic Acid Stimulates Prostate Cancer Cell Growth: Critical Role of 5-Lipoxygenase." *Biochem. Biophys. Res. Commun.,* 235.2 (1997): 418-23.

Hardman, W. E. "Omega-3 Fatty Acids to Augment Cancer Therapy." *J. Nutr.,* 132.11 Suppl (2002): 3508S-12S.

Hardman, W. E., et al. "Dietary Omega-3 Fatty Acids and Ionizing Irradiation on Human Breast Cancer Xenograft Growth and Angiogenesis." *Cancer Cell Int.,* 5.1 (2005): 12.

Hughes-Fulford, M., et al. "Arachidonic Acid Activates Phosphatidylinositol 3-Kinase Signaling and Induces Gene Expression in Prostate Cancer." *Cancer Res.,* 66.3 (2006): 1427-33.

———. "Arachidonic Acid, an Omega-6 Fatty Acid, Induces Cytoplasmic Phospholipase A2 in Prostate Carcinoma Cells." *Carcinogenesis,* 26.9 (2005): 1520-6.

Ishitani, K., et al. "A Prospective Study of Multivitamin Supplement Use and Risk of Breast Cancer." *Am. J. Epidemiol.,* 167.10 (2008): 1197-206.

Ismail, H. A., et al. "Expression of Nf-Kappab in Prostate Cancer Lymph Node Metastases." *Prostate,* 58.3 (2004): 308-13.

Jung, M., and A. Dritschilo. "Nf-Kappa B Signaling Pathway as a Target for Human Tumor Radiosensitization." *Semin. Radiat. Oncol.,* 11.4 (2001): 346-51.

Kamangar, F., G. M. Dores, and W. F. Anderson. "Patterns of Cancer Incidence, Mortality, and Prevalence across Five Continents: Defining Priorities to Reduce Cancer Disparities in Different Geographic Regions of the World." *J. Clin. Oncol.,* 24.14 (2006): 2137-50.

Kew, S., et al. "Relation between the Fatty Acid Composition of Peripheral Blood Mononuclear Cells and Measures of Immune Cell Function in Healthy, Free-Living Subjects Aged 25-72 Y." *Am. J. Clin. Nutr.,* 77.5 (2003): 1278-86.

Kim, J., et al. "Fatty Fish and Fish Omega-3 Fatty Acid Intakes Decrease the Breast Cancer Risk: A Case-Control Study." *BMC Cancer,* 9 (2009): 216.

Kinsella, J. E., and B. Lokesh. "Dietary Lipids, Eicosanoids, and the Immune System." *Crit. Care Med.,* 18.2 Suppl (1990): S94-113.

Kobayashi, N., et al. "Effect of Altering Dietary Omega-6/Omega-3 Fatty Acid Ratios on Prostate Cancer Membrane Composition, Cyclooxygenase-2, and Prostaglandin E2." *Clin. Cancer Res.,* 12.15 (2006): 4662-70.

Kolonel, L. N., D. Altshuler, and B. E. Henderson. "The Multiethnic Cohort Study: Exploring Genes, Lifestyle and Cancer Risk." *Nat. Rev. Cancer,* 4.7 (2004): 519-27.

Lands, W. E., et al. "Maintenance of Lower Proportions of (N - 6) Eicosanoid Precursors in Phospholipids of Human Plasma in Response to Added Dietary (N - 3) Fatty Acids." *Biochim. Biophys. Acta.,* 1180.2 (1992): 147-62.

Leitzmann, M. F., et al. "Dietary Intake of N-3 and N-6 Fatty Acids and the Risk of Prostate Cancer." *Am. J. Clin. Nutr.,* 80.1 (2004): 204-16.

Lessard, L., et al. "Nuclear Localization of Nuclear Factor-Kappab P65 in Primary Prostate Tumors Is Highly Predictive of Pelvic Lymph Node Metastases." *Clin. Cancer Res.,* 12.19 (2006): 5741-5.

Li, Z. W., et al. "The Ikkbeta Subunit of Ikappab Kinase (Ikk) Is Essential for Nuclear Factor Kappab Activation and Prevention of Apoptosis." *J. Exp. Med.,* 189.11 (1999): 1839-45.

Marcus, A. J. "The Role of Lipids in Platelet Function: With Particular Reference to the Arachidonic Acid Pathway." *J. Lipid. Res.,* 19.7 (1978): 793-826.

Maxwell, P. J., et al. "Hif-1 and Nf-Kappab-Mediated Upregulation of Cxcr1 and Cxcr2 Expression Promotes Cell Survival in Hypoxic Prostate Cancer Cells." *Oncogene,* 26.52 (2007): 7333-45.

Menendez, J. A., et al. "Novel Signaling Molecules Implicated in Tumor-Associated Fatty Acid Synthase-Dependent Breast Cancer Cell Proliferation and Survival: Role of Exogenous Dietary Fatty Acids, P53-P21waf1/Cip1, Erk1/2 Mapk, P27kip1, Brca1, and Nf-Kappab." *Int. J. Oncol.,* 24.3 (2004): 591-608.

Nettleton, J. A. "Omega-3 Fatty Acids: Comparison of Plant and Seafood Sources in Human Nutrition." *J. Am. Diet Assoc.,* 91.3 (1991): 331-7.

Pischon, T., et al. "Habitual Dietary Intake of N-3 and N-6 Fatty Acids in Relation to Inflammatory Markers among Us Men and Women." *Circulation,* 108.2 (2003): 155-60.

Poprzecki, S., et al. "Modification of Blood Antioxidant Status and Lipid Profile in Response to High-Intensity Endurance Exercise after Low Doses of Omega-3 Polyunsaturated Fatty Acids Supplementation in Healthy Volunteers." *Int. J. Food Sci. Nutr.,* 60 Suppl 2 (2009): 67-79.

Quinn, M., and P. Babb. "Patterns and Trends in Prostate Cancer Incidence, Survival, Prevalence and Mortality. Part Ii: Individual Countries." *BJU Int.,* 90.2 (2002): 174-84.

Roman, R. J. "P-450 Metabolites of Arachidonic Acid in the Control of Cardiovascular Function." *Physiol Rev.,* 82.1 (2002): 131-85.

Rose, D. P., and J. M. Connolly. "Effects of Dietary Omega-3 Fatty Acids on Human Breast Cancer Growth and Metastases in Nude Mice." *J. Natl. Cancer Inst.,* 85.21 (1993): 1743-7.

Rose, D. P., et al. "Influence of Diets Containing Eicosapentaenoic or Docosahexaenoic Acid on Growth and Metastasis of Breast Cancer Cells in Nude Mice." *J. Natl Cancer Inst.,* 87.8 (1995): 587-92.

Santini, D., et al. "Expression Pattern of Receptor Activator of Nfkappab (Rank) in a Series of Primary Solid Tumors and Related Bone Metastases." J. *Cell Physiol.,* 226.3: 780-4.

———. "Receptor Activator of Nf-Kb (Rank) Expression in Primary Tumors Associates with Bone Metastasis Occurrence in Breast Cancer Patients." *PLoS One,* 6.4: e19234.

Schley, P. D., et al. "Mechanisms of Omega-3 Fatty Acid-Induced Growth Inhibition in Mda-Mb-231 Human Breast Cancer Cells." *Breast Cancer Res. Treat.,* 92.2 (2005): 187-95.

Schmitz, G., and J. Ecker. "The Opposing Effects of N-3 and N-6 Fatty Acids." *Prog. Lipid. Res.,* 47.2 (2008): 147-55.

Serhan, C. N., et al. "Resolvins, Docosatrienes, and Neuroprotectins, Novel Omega-3-Derived Mediators, and Their Endogenous Aspirin-Triggered Epimers." *Lipids,* 39.11 (2004): 1125-32.

Shaikh, I. A., et al. "Docosahexaenoic Acid Enhances the Efficacy of Docetaxel in Prostate Cancer Cells by Modulation of Apoptosis: The Role of Genes Associated with the Nf-Kappab Pathway." *Prostate,* 68.15 (2008): 1635-46.

Sharma, S., et al. "Diet Impacts Mortality from Cancer: Results from the Multiethnic Cohort Study." *Cancer Causes Control.,* 24.4: 685-93.

Simonsen, N., et al. "Adipose Tissue Omega-3 and Omega-6 Fatty Acid Content and Breast Cancer in the Euramic Study. European Community Multicenter Study on Antioxidants, Myocardial Infarction, and Breast Cancer." *Am. J. Epidemiol.,* 147.4 (1998): 342-52.

Simopoulos, A. P. "Omega-3 Fatty Acids and Antioxidants in Edible Wild Plants." *Biol. Res.,* 37.2 (2004): 263-77.

Tricon, S., et al. "Effects of Cis-9,Trans-11 and Trans-10,Cis-12 Conjugated Linoleic Acid on Immune Cell Function in Healthy Humans." *Am. J. Clin. Nutr.,* 80.6 (2004): 1626-33.

von Schacky, C., S. Fischer, and P. C. Weber. "Long-Term Effects of Dietary Marine Omega-3 Fatty Acids Upon Plasma and Cellular Lipids, Platelet Function, and Eicosanoid Formation in Humans." *J. Clin. Invest.,* 76.4 (1985): 1626-31.

Willett, W. C. "Specific Fatty Acids and Risks of Breast and Prostate Cancer: Dietary Intake." *Am. J. Clin. Nutr.,* 66.6 Suppl (1997): 1557S-63S.

In: Omega-3 Fatty Acids
Editor: Md. Wasim Khan

ISBN: 978-1-62948-516-4

Chapter 3

Omega-3 Fatty Acids and Cardiovascular Diseases

Federico Guerra*[*], *Luca Piangerelli, Andrea Romandini, Simone Maffei and Alessandro Capucci
Cardiology and Arrhythmology Clinic, Center for the Study of Obesity, Marche Polytechnic University, Italy

Abstract

Since the expeditions to Greenland by Bang and Dyerberg starting in the late 1960s, a diet rich of omega-3 fatty acids was suggested as the main responsible for the very low incidence of myocardial infarction in the Inuit population when compared to Danish controls. A few decades after, omega-3 fatty acids have been reported as associated with an antiatherogenic blood lipid pattern, reduced platelet reactivity, and fewer cardiovascular events. Proposed mechanisms for the protective role of omega 3 fats against cardiovascular diseases include blood pressure lowering; altered lipid profile, especially reduced serum triglyceride concentration; reduced thrombotic tendency; anti-inflammatory effects; anti-arrhythmic effects including heart rate reduction; improved vascular endothelial function; increased plaque stability; increased paraoxonase levels and improved insulin sensitivity. Unfortunately, while previous randomized clinical trials presented some enthusiastic results regarding a strong positive role of omega-3 fatty acids in preventing death, myocardial infarction, stroke and ventricular arrhythmias, recent ones failed to confirm such benefits on broader populations. Moreover, different dosages of either eicosapentaenoic acid (EPA), docosahexaenoic acid (DHA) or both, administered either through dietary counseling or tablet supplement, were used while collecting evidences. Systematic reviews and meta-analyses added further confusion to the existing controversy as they often reported conflicting findings.

Recent guidelines issued by major cardiology societies broadly recommend their use, either as supplements or through dietary counseling, but for different settings. While is

[*] Cardiology and Arrhythmology Clinic, Center for the Study of Obesity, Marche Polytechnic University; Email: f.guerra@univpm.it.

widely accepted the use of omega-3 fatty acids for secondary prevention in patients with a recent myocardial infarction, some European countries have approved the omega-3 supplements for lowering cardiovascular risk whereas the US Food and Drug Administration has approved their administration only as triglyceride-lowering agents in patients with overt hypertriglyceridemia. In the present chapter the authors will summarize currently available evidence in order to clarify the role of omega-3 fatty acids in the cardiovascular setting and define the best spot for this new therapeutics.

Keywords: n-3 PUFA, cardiovascular risk, lipid metabolism, atherosclerosis, endothelial dysfunction, blood pressure, coagulation, omega-3, ischemic heart disease, prevention of cardiovascular disease

Introduction

The first evidence regarding n-3 polyunsaturated fatty acids (PUFA) rose from 1960s observational reports regarding a very low incidence of cardiovascular disease (CVD) among the Inuit population.

In order to confirm those reports, Bang and Dyerberg undertook several expedition to Greenland starting in the late 1960s. They confirmed the very low incidence of myocardial infarction in the Inuit population when compared to Danish controls, and suggested the Inuit diet, which was rich of omega-3 fatty acids, as the main responsible (Bang, Dyerberg, & Nielsen, 1971; Dyerberg, Bang, & Aagaard, 1980). A few decades after, n-3 PUFAs have been reported as associated with an antiatherogenic blood lipid pattern, reduced platelet reactivity, and fewer cardiovascular events (Leaf, & Weber, 1988). Nowadays, proposed mechanisms for the protective role of omega- 3 fats against cardiovascular diseases are many and include blood pressure lowering, improved lipid profile, reduced thrombotic tendency, anti-inflammatory effects, anti-arrhythmic effects, improved vascular endothelial function, increased plaque stability, increased paraoxonase levels and improved insulin sensitivity (De Caterina, 2011).

Unfortunately, clinical evidences are not as strong as those derived from in vitro or biological models. While previous randomized clinical trials presented some enthusiastic results regarding a strong positive role of omega-3 fatty acids in preventing death, myocardial infarction, stroke and ventricular arrhythmias (Gruppo Italiano per lo Studio della Sopravvivenza nell'Infarto Miocardico, 1999), recent ones failed to confirm such benefits on broader populations (Kromhout, Giltay, & Geleijnse, 2010). Moreover, different dosages of either eicosapentaenoic acid (EPA), docosahexaenoic acid (DHA) or both, administered either through dietary counseling or tablet supplement, were used while collecting evidences. Systematic reviews and meta-analyses added further confusion to the existing controversy as they often reported conflicting results (Rizos, Ntzani, Eftychia Bika, Kostapanos, Moses S. Elisaf, FASA, & FRSH, 2012). Recent guidelines issued by major cardiology societies now broadly recommend their use, although different advices are given for different settings. Aim of the authors is to summarize in the present chapter currently available evidence on the role of n-3 PUFAs in cardiovascular medicine, in order to define the best spot for this new therapeutics in current everyday practice.

n-3 PUFAs and Cardiovascular Prevention

Since the first observational studies within the general population showed an inverse relationship between fish consumption and cardiovascular diseases (Daviglus, Stamler, Orencia, Dyer, Liu, Greenland, Walsh, Morris, & Shekelle, 1997), CV mortality and sudden death (Albert, Campos, Stampfer, Ridker, Manson, Willett, & Ma, 2002), the potential benefits of n-3 PUFAs on cardiovascular system has been widely researched. Growing evidence suggests that n-3 PUFAs exert pleiotropic cardiometabolic actions because of their positive effect on vasculature and endothelial function, lipid and glucose metabolism, platelets function, myocardial arrythmogenics and hemodynamics.

A combination of these mechanisms, discussed below, explains how n-3 PUFA may affect several components of the cardiovascular continuum, preventing cardiovascular events.

Lipid Metabolism

Hypertriglyceridemia is widely associated with increased cardiovascular risk in numerous prospective and randomized trials (Murad, Hazem, Coto-Yglesias, Dzyubak, Gupta, Bancos, Lane, Erwin, Berglund, Elraiyah, & Montori, 2012; Jacobson, Miller, & Schaefer, 2007). Nowadays, lowering plasma triglycerides (TG) is perhaps the most recognized effect of n-3 PUFAs. Controlled trials reported that supplementation of more than 4 grams per day of EPA and more than 2 grams per day of DHA reduces serum TGs concentrations up to 20% and 33% respectively (Woodman, Mori, Burke, Puddey, Watts, & Beilin, 2002; Nestel, Shige, Pomeroy, Cehun, Abbey, & Raederstorff, 2002; Mori, Burke, Puddey, Watts, O'Neal, Best, & Beilin, 2000; Grimsgaard, Bonaa, Hansen, & Nordøy, 1997; Yamamoto, Yoshimura, Noma, Suzuki, Kai, Tajimi, Sugihara, & Kikuchi, 1995; Agren, Hänninen, Julkunen, Fogelholm, Vidgren, Schwab, Pynnönen, & Uusitupa, 1996; Conquer, & Holub, 1996; Mori, & Woodman, 2006; Mozaffarian, & Wu, 2012). Sensitivity analyses showed that dose of n-3 PUFAs and baselines concentrations of TGs had a cumulative impact on magnitude of changes in TGs blood levels (EFSA Panel on Dietetic Products, Nutrition and Allergies, 2012). The TGs lowering effect of omega-3 fatty acids is dose dependent but show an individual response correlated to TG baseline levels. It must nonetheless be noted that typical dietary doses produce only modest effects on TGs serum levels (EFSA Panel on Dietetic Products, Nutrition and Allergies, 2012).

Effects on high-density lipoprotein (HDL) and low-density lipoprotein (LDL) were less consistent and generally small (Woodman, Mori, Burke, Puddey, Watts, & Beilin, 2002; Nestel, Shige, Pomeroy, Cehun, Abbey, & Raederstorff, 2002; Mori, Burke, Puddey, Watts, O'Neal, Best, & Beilin, 2000; Grimsgaard, Bonaa, Hansen, & Nordøy, 1997; Yamamoto, Yoshimura, Noma, Suzuki, Kai, Tajimi, Sugihara, & Kikuchi, 1995; Agren, Hänninen, Julkunen, Fogelholm, Vidgren, Schwab, Pynnönen, & Uusitupa, 1996; Conquer, & Holub, 1996; Mori, & Woodman, 2006; Mozaffarian, & Wu, 2012). A recent meta-analysis of randomized controlled trials including twenty-two studies (DHA versus EPA, DHA alone and EPA alone) reported that DHA-containing supplements increase LDL-cholesterol by about 3% but with a significant decrease in TGs (-22.4%), a significant increase in HDL-cholesterol (+7.3%) and no changes in non-HDL-cholesterol (Jacobson, Glickstein, Rowe, & Soni,

2012). EPA supplementation do not have a significant effect on LDL or HDL (Jacobson, Glickstein, Rowe, & Soni, 2012). Additionally, in those trials evaluating lipid subfractions, an anti-atherogenic shift distribution of HDL and LDL subclasses have been reported. Even when HDL-cholesterol concentrations were not significantly altered, n-3 PUFAs can promote an increase in HDL size with DHA raising HDL_2–cholesterol serum level and both DHA and EPA lowering HDL_3 – cholesterol serum level (Mori, Burke, Puddey, Watts, O'Neal, Best, & Beilin, 2000; Kelley, Siegel, Vemuri, & Mackey, 2007; Neff, Culiner, Cunningham-Rundles, Seidman, Meehan, Maturi, Wittkowski, Levine, & Breslow, 2011). Moreover, the LDL subclasses distributions revealed an increase in cholesterol carried by larger and less atherogenic LDL_1 and LDL_2 particles, resulting in a significant decline in cholesterol carried by smaller, dense particles (Mori, Burke, Puddey, Watts, O'Neal, Best, & Beilin, 2000; Kelley, Siegel, Vemuri, & Mackey, 2007; Cazzola, Russo-Volpe, Miles, Rees, Banerjee, Roynette, Wells, Goua, Wahle, Calder, & Cestaro, 2007). For this reason, it has been suggested that the LDL-cholesterol increase seen in most studies could be primarily driven by an increase in particle size rather than number (Mori, Burke, Puddey, Watts, O'Neal, Best, & Beilin, 2000; Neff, Culiner, Cunningham-Rundles, Seidman, Meehan, Maturi, Wittkowski, Levine, & Breslow, 2011; Maki, Van Elswyk, McCarthy, Hess, Veith, Bell, Subbaiah, & Davidson, 2005).

Evidence suggests that the mechanisms responsible for the changes in the lipid profile associated with n-3 PUFAs may be related to a reduced hepatic very-low-density lipoprotein (VLDL) and TG synthesis, along with an accelerated TG clearance from circulating chylomicrons and VLDL particles. The reduced TG formation seems to be related to a downregulation of de novo lipogenesis and to a simultaneous increase in mithocondrial/peroxisomal beta-oxidation. Furthermore, n-3 PUFAs may directly impair activity of key TG biosynthesis enzymes (Bays, Tighe, Sadovsky, & Davidson, 2008).

Atherosclerosis and Plaque Stability

Inflammation and endothelial dysfunction play a central role in the development and progression of CHD (Libby, 2013). N-3 PUFAs have been shown to have a range of anti-inflammatory actions altering inflammatory pathway throughout several mechanisms. Increased consumption of omega-3 fatty acids is associated with a dose-dependent incorporation into cell phospholipids at the expense of arachidonic acid. The resulting effect is a changed pattern of eicosanoids production with less inflammatory potential (Rees, Miles, Banerjee, Wells, Roynette, Wahle, & Calder, 2006; Calder, 2009; Serhan, Clish, Brannon, Colgan, Gronert, & Chiang, 2000; Serhan, Clish, Brannon, Colgan, Chiang, & Gronert, 2000). In addition, recent studies reported novel pathways by which n-3 PUFAs are also converted into local mediators, called resolvins, who seem to have potent local anti-inflammatory and immunomodulatory actions on neutrophil granulocytes, macrophages, dendritic cells and T cells (Serhan, 2010). Together, these anti-inflammatory actions might be beneficial in several clinical setting, such as atherosclerosis and acute coronary syndrome.

Omega-3 fatty acids directly affect endothelial activation. Several studies on endothelial function reported evidence suggesting such hypothesis. Inhibition of cytokines production (IL-1, TNF-α) (Endres, Ghorbani, Kelley, Georgilis, Lonnemann, van der Meer, Cannon,

Rogers, Klempner, & Weber, 1989), decreased expression of adhesion molecules (De Caterina, & Massaro, 2005; De Caterina, Cybulsky, Clinton, Gimbrone, & Libby, 1994; Mori, Woodman, Burke, Puddey, Croft, & Beilin, 2003); increased endothelial nitric oxide (NO) release (Harris, Rambjør, Windsor, & Diederich, 1997; López, Möller, Denicola, Casós, Rubbo, Ruiz-Sanz, & Mitjavila, 2008) and reduction of smooth muscle cell growth and proliferation (Fox, & DiCorleto, 1988; Kaminski, Jendraschak, Kiefl, & von Schacky, 1993) were all described as possible mechanisms.

A recent systematic review, including an heterogeneous set of trials testing n-3 PUFAs effect on endothelial function, concluded that markers of endothelial function might be improved by n-3 PUFAs supplementation in subjects with CVD or cardiovascular risk factors while no additional benefit on healthy subjects have been demonstrated (Egert, & Stehle, 2011). In this setting, studies evaluating omega-3 fatty acids effect on biomarkers of endothelial dysfunction (E-selectin, VCAM-1, ICAM-1) have brought mixed results, precluding strong conclusions on beneficial effect of n-3 PUFA on endothelial dysfunction (Mozaffarian, & Wu JH, 2012).

In clinical setting, modern observational studies in high intake seafood population suggested that n-3 PUFA could prevent development or progression of atherosclerosis and its related cardiovascular disease (Albert, Hennekens, O'Donnell, Ajani, Carey, Willett, Ruskin, & Manson, 1998). Indeed, epidemiological and autopsy studies confirmed a significant lesser extent of atherosclerotic lesions in population exposed to high dietary intake of n-3 fatty acids (Sekikawa, Ueshima, Kadowaki, El-Saed, Okamura, Takamiya, Kashiwagi, Edmundowicz, Murata, Sutton-Tyrrell, Maegawa, Evans, Kita, & Kuller, 2007; Sekikawa, Curb, Ueshima, El-Saed, Kadowaki, Abbott, Evans, Rodriguez, Okamura, Sutton-Tyrrell, Nakamura, Masaki, Edmundowicz, Kashiwagi, Willcox, Takamiya, Mitsunami, Seto, Murata, White, Kuller, & ERA JUMP (Electron-Beam Tomography, Risk Factor Assessment Among Japanese and U.S. Men in the Post-World War II Birth Cohort) Study Group, 2008;Newman, Middaugh, Propst, & Rogers, 1993). Such findings were also consistent in young cohort subjects, suggesting that n-3 PUFAs exert their anti-atherogenic effects also during the early stages of the developing lesions (Newman, Middaugh, Propst, & Rogers, 1993).

Vascular and Cardiac Hemodynamics

Several randomized trials have demonstrated an improved flow-mediated arterial dilatation after n-3 PUFA supplementation. A recent meta-analysis of 16 randomized clinical trials evaluating endothelial function by flow-mediated dilatation technique reports that omega-3 PUFAs significantly improve endothelial-mediated dilatation in subjects with CVD or CVD risk factors, without a reported dose-dependent effect (Wang, Liang, Wang, Lu, Huang, Cao, Li, & Gu, 2012). Besides increasing NO production, omega-3 PUFAs exhibit a wide range of biological actions by which may influence vasomotor tone and renal sodium excretion, thus lowering blood pressure. In short-term trials PUFAs reduced angiotensin-converting enzyme activity and angiotensin-II formation, mitigated vasoconstrictive response, enhanced the production of vasodilators eicosanoids, activated the parasympathetic nervous system, and improved arterial compliance (Nestel, Shige, Pomeroy, Cehun, Abbey, & Raederstorff, 2002; O'Keefe, Abuissa, Sastre, Steinhaus, & Harris, 2006; Kenny, Warltier,

Pleuss, Hoffmann, Goodfriend, & Egan, 1992; Chin, Gust, Nestel, & Dart, 1993; Stirban, Nandrean, Götting, Tamler, Pop, Negrean, Gawlowski, Stratmann, & Tschoepe, 2010; Dangardt, Osika, Chen, Nilsson, Gan, Gronowitz, Strandvik, & Friberg, 2010; Mori, Watts, Burke, Hilme, Puddey, &Beilin, 2000). The extent of the antihypertensive effect seems to be dependent on the baseline blood pressure levels. A previous meta-regression reported a modest reduction in systolic blood pressure of 2.1 mmHg and in diastolic blood pressure of 1.6 mmHg with a median fish oil intake of 3.7 g/day (Geleijnse, Giltay, Grobbee, Donders, & Kok, 2002).

Moreover n-3 PUFAs have a documented effect on lowering heart rate. This finding is probably mediated by different mechanisms. Omega-3 fatty acids directly affect electro-physiological pathways and probably affect hear rate even indirectly, by improving left ventricular diastolic filling and enhancing parasympathetic tone (Mori, Bao, Burke, Puddey, & Beilin, 1999). Overall, evidence suggests that DHA may be more effective than EPA in improving cardiac hemodynamics. In several trials DHA lowers both blood pressure and heart rate while EPA does not significantly alter blood pressure and has inconstant effect on heart rate (Mozaffarian, Lemaitre, King, Song, Spiegelman, Sacks, Rimm, & Siscovick, 2011; Theobald, Goodall, Sattar, Talbot, Chowienczyk, & Sanders, 2007).

Coagulation

Some evidence suggests that n-3 PUFAs may have anti-thrombotic effect. Several uncontrolled studies showed a reduced ex-vivo collagen-induced platelet aggregation (Mori, & Woodman, 2006) which also seems to be dose-related (Mann, O'Connell, Baldwin, Singh, & Meyer, 2010). Most of the current evidence of anti-thrombotic effect of n-3 PUFAs comes from population undergoing percutaneous coronary intervention (PCI). In those patients, n-3 PUFAs have been associated with a decreased thrombin formation and a favorable alteration of clot properties, at least in stable coronary artery disease (Gajos, Zalewski, Rostoff, Nessler, Piwowarska, & Undas, 2011). Biological effects of n-3 PUFAs supplementation in those patients include an improvement of the platelet response to clopidogrel and a reduction of aspirin resistance (Gajos, Rostoff, Undas, & Piwowarska, 2010; Lev, Solodky, Harel, Mager, Brosh, Assali, Roller, Battler, Kleiman, & Kornowski, 2010), possibly reducing recurrent events in patients resistant to clopidogrel and positively affecting gastrointestinal tolerability of aspirin. Positive effects of n-3 PUFAs on coagulation could also be seen in some inherited prothrombotic conditions such as sickle cell disease (Okpala,, Ibegbulam, Duru, Ocheni, Emodi, Ikefuna, Umar, Asinobi, Madu, Okoye, Nwagha, Oquonu, Uamai, Aqwu, Nonyelu, Anike, Agu, Anigbo, Chukwura, Ugwu, & Herrada, 2011).

Ischemic Heart Disease

As said before, the first spark of interest for n-3 PUFAs effects on cardiovascular system came from the first reports of a possible link between omega-3 fatty acids consumptions and a reduced incidence of cardiovascular events, mainly driven by a very low incidence of coronary artery disease and its related clinical manifestations, such as angina, myocardial infarction (MI), and sudden cardiac death (Bang, Dyerberg, & Nielsen, 1971; Dyerberg, Bang, & Aagaard, 1980; Leaf, & Weber, 1988). Since then, a series of epidemiological studies and randomized controlled clinical trials followed, all with the aim to evaluate the protective efficacy of n-3 PUFAs on ischemic heart disease.

The Nurses' Health Study (Hu, Cho, Rexrode, Albert, & Manson, 2002) was the first study to confirm the positive effects of high consumption of fish oil and n-3 PUFAs in women. In this study 84.688 women were recruited, aged between 34 and 59 years and with no history of cardiovascular events or cancer. The consumption of fish oil and omega-3 fatty acid supplements was inversely associated with the incidence of acute coronary events after a median follow up of 16 years.

The DART study (Diet And Reinfarction Trial) was the first randomized clinical trial on n-3 PUFAs and ischemic heart disease, enrolling more than 2000 patients with a previous acute MI (Burt, Fehily, Gilbert, Rogers, Holliday, Sweetnam, et al., 1989). The enrolled subjects were divided into 3 groups: the first group experienced only a reduction of fatty foods in the assigned diet, the second group had a diet rich which was rich in fish and group 3 had a diet rich in whole foods. The DART trial showed a favorable effect in the n-3 PUFAs treatment arm, with a significant reduction in total mortality and fatal ischemic events after 2 years of follow-up. Patients randomized in the second group had an incidence of death for any cause 29% lower than patients from the first and in the third group.

The efficacy of n-3 PUFAs as cardioprotective agents in secondary prevention was then somewhat confirmed by the GISSI-Prevenzione (Marchioli, Barzi, Bomba, Chieffo, Di Gregorio, Di Mascio, et al.; GISSI-Prevenzione Investigators., 2002) which enrolled 11324 patients with recent acute MI. Patients enrolled were then randomized into omega-3 fatty acids supplements (1 gram per day), vitamin E supplements (300 mg per day), both or none in a double-way blinded fashion. The 4-way factorial analysis showed a significant reduction in the primary endpoint (total mortality, nonfatal MI, non-fatal stroke) only in the group treated with n-3 PUFAs and, although to a lesser extent, in the group receiving combined treatment. In particular, a significant reduction in total mortality (20%), cardiovascular mortality (30%), and sudden death (45%) was detected in the n-3 PUFAs group, supporting the hypotesis of a direct protective effect of n-3 PUFAs on the myocardium, even independently of the anti-atherogenic and anti-thrombotic effects. In light of these results a great enthusiasm arose, and in many countries n-3 PUFAs started to be widely prescribed. In Italy these drugs rose from 754th place in 2001 to eighth in 2004 for total annual national health system cost.

The first doubts arose with the DART-2 trial (Ness, Ashfield-Watt, Whiting, Smith, Hughes, & Burr, 2004; Burr, Ashfield-Watt, Dunstan, Fehily, Breay, Ashton, Zotos, Haboubi& Elwood, 2003). In the DART-2, 3114 men with angina were randomly allocated to four diet groups and followed for three to nine years. Members of the first group ($n = 764$) were instructed to eat at least two portions of fish each week or up to 3 g of fish oils as a partial or total substitutes. Other groups were advised to eat more fruits and vegetables or to eat a specific diets. All-cause mortality was not reduced in the group advised to eat more fish. Risk of cardiac death was higher in men given fish advice (adjusted HR = 1.26; 95% CI, 1.00 1.58; $p = 0.047$) and even higher for risk of sudden cardiac death (adjusted HR = 1.54; 95% CI, 1.06 2.23; $p = 0.025$). The authors offered several possible explanations for these unexpected findings, such as an adverse effect of fish oils in patients with angina, interaction with certain drugs, or changes in the patients' health behavior (Burr, Ashfield-Watt, Dunstan, Fehily, Breay, Ashton, Zotos, Haboubi,& Elwood, 2003). The authors also suggested that fish oils could be protective after MI, but increased the risk of cardiac death in men with angina. The adverse effects occurred only in men who took fish oil capsules and not in men who ate fish. The excess in mortality in the fish group was shown in men who took supplements but were not taking ß-blockers. Digoxin was the only drug that showed a significant interaction with advice to eat more fish. The authors proposed that fish oils could have pro-arrhythmic effects in patients with chronic disease (DART-2) but anti-arrhythmic effects in patients with acute disease (DART, GISSI-P) (Burr, Ashfield-Watt, Dunstan, Fehily, Breay, Ashton, Zotos, Haboubi, & Elwood, 2003). The same authors also hypothesized that certain drugs, such as ß-blockers and calcium channel inhibitors, could exert a protective effect against the pro-arrhythmic effects of fish oils in stable ischemic patients, while digoxin could exacerbate the arrhythmic effects in recent acute MI (Burr, Dunstan, & George, 2005).

In order to shed some light on this apparent paradox, a systematic review was performed on 48 randomized clinical trials (included the ones just mentioned) and 41 cohort studies to examine the efficacy of omega-3 on total mortality and on the incidence of combined cardiovascular events and cancer (Hooper, Thompson, Harrison, Summerbell, Ness, Moore, Worthington, Durrington, Higgins, Capps, Riemersma, Ebrahim, & Smith, 2006). Unfortunately, no evidence was found regarding a clear benefit of n-3 PUFAs on total mortality, cardiovascular events, cancer and stroke. The authors concluded that it is probably not appropriate to recommend a high intake of omega-3 fatty acids for people who have angina but have not had a MI. Similar results has been found by a more recent meta-analysis on a similar pool of studies, in which n-3 PUFAs supplementation was not associated with a lower risk of all-cause of mortality, cardiac death, sudden death, myocardial infarction, or stroke based on relative and absolute measures of association (Rizos, Ntzani, Bika, Kostapanos, & Elisaf, 2012).

More recently, two other trials were published in order to tackle this issue. OMEGA was a randomized, placebo-controlled, double-blind, multicenter trial testing the effects of long-term omega-3 fatty acids supplementation (1 gram per day for 1 year) on the rate of sudden cardiac death in survivors of acute MI, if given in addition to current guideline-adjusted treatment (Rauch, Schiele, Schneider, Diller, Victor, Gohlke, Gottwik, Steinbeck, Del Castillo, Sack, Worth, Katus, Spitzer, Sabin, Senges, & OMEGA Study Group, 2010). Secondary end points were total mortality and nonfatal clinical events. Nearly four thousands patients were randomized in 104 German centers 3 to 14 days after acute MI. Acute coronary angiography was performed in 93.8% and acute percutaneous interventions in 77.8% of all patients. During a follow-up of 365 days, the event rates did not differ between n-3 PUFAs and control groups: sudden cardiac death prevalence was 1.5% in both groups and secondary endpoint rates were also similar between the two groups. The conclusion of this trial was that guideline-adjusted treatment of acute myocardial infarction results in a low rate of sudden cardiac death and other clinical events within 1 year of follow-up, which could not be further reduced by the administration of omega-3 fatty acids (Rauch, Schiele, Schneider, Diller, Victor, Gohlke, Gottwik, Steinbeck, Del Castillo, Sack, Worth, Katus, Spitzer, Sabin, Senges, & OMEGA Study Group, 2010). Another multicenter, double-blind, placebo-controlled trial, randomly assigned one of four trial margarines: a margarine supplemented with a combination of EPA and DHA (with a targeted additional daily intake of 400 mg of EPA and DHA), a margarine supplemented with alpha-linolenic acid (with a targeted additional daily intake of 2 g of alpha-linolenic acid), a margarine supplemented with EPA–DHA and alpha-linolenic acid, or a placebo margarine. The study enrolled 4837 patients, 60 through 80 years of age (78% men), who had had a myocardial infarction and were receiving state-of-the-art antihypertensive, antithrombotic, and lipid modifying therapy (Kromhout, Giltay, & Geleijnse, 2010). The primary end point was the rate of major cardiovascular events, which comprised fatal and nonfatal cardiovascular events as well as cardiac interventions. During the 40 months follow-up period, a major cardiovascular event occurred in 671 patients (13.9%). Neither n3-PUFAs nor alpha-linolenic acid reduced this primary end point (HR with n-3 PUFAs, 1.01; HR with alpha-linolenic acid, 0.91; all p non-significant). In the pre-specified subgroup of women, alpha-linolenic acid, as compared with placebo and EPA–DHA alone, was associated with a reduction in the rate of major cardiovascular events that approached the statistical significance (HR 0.73; 95% CI, 0.51 to 1.03; P=0.07). The rate of adverse events did not differ significantly among the study groups. Even in this case, the authors concluded that low-dose supplementation with EPA–DHA or alpha-linoleic acid did not significantly reduce the rate of major cardiovascular events among patients who had had a myocardial infarction and who were receiving state-of-the-art therapy (Kromhout, Giltay, & Geleijnse, 2010).

The ORIGIN trial more recently sought to investigate the usefulness of n-3 PUFAs in preventing cardiovascular death in patients with dysglycemia (Bosch, Gerstein, Dagenais, Díaz, Dyal, Jung, Maggiono, Probstfield, Ramachandran, Riddle, Rydén, &Yusuf, 2012). In this double-blind study with a 2-by-2 factorial design, 12.536 patients who were at high risk for cardiovascular events and had impaired fasting glucose, impaired glucose tolerance, or diabetes were randomly assigned to receive a 1-gram capsule containing at least 900 mg of n-3 PUFAs or placebo daily and to receive either glargine insulin or standard care. The primary outcome was death from cardiovascular causes. During a median follow up of 6.2 years, n–3 fatty acids intake was not associated with a reduced incidence of the primary outcome. Patients in the active treatment group had similar rates of major vascular events, death from any cause, or death from arrhythmia when compared to patients taking placebo. Omega-3 fatty acid supplementations were associated with a reduction of TGs serum levels (-14.5 mg per deciliter), without a significant effect on other lipids. Adverse effects were similar in the two groups.

Heart Failure

Growing evidence suggests n-3 PUFAs consumption may also improve cardiac filling and myocardial efficiency. Animal and humans studies showed improving in both in energy-dependent and passive component of the diastolic phase. Such evidence suggests some early rapid metabolic effects over than a long-term favorable ventricular compliance due to lower systemic resistance (Mozaffarian, Lemaitre, King, Song, Spiegelman, Sacks, Rimm, &Siscovick, 2012). Some evidence from small trials reported positive effect on hemodynamic parameters in heart failure patients. A recent controlled trial involving 43 patients with severe non-ischemic heart failure (HF) patients reported a dose-dependent increase in left ventricular ejection fraction after n-3 PUFAs treatment for 3 months with the best results obtained with dosage of 4 g/day (Moertl, Hammer, Steiner, Hutuleac, Vonbank, & Berger, 2011). Moreover, a recent substudy of the GISSI-HF trial on 608 patients with severe heart failure reported a small but statistically significant increase of left ventricular ejection fraction

(LVEF) of 8.1% at 1 year, 11.1% at 2 years, and 11.5% at 3 years with just 1 gram per day n-3 PUFAs supplement (Ghio, Scelsi, Latini, Masson, Eleuteri, Palvarini, Vriz, Pasotti, Gorini, Marchioli, Maggioni, &Tavazzi, 2010). Therefore, it should be hypothesized that n-3 PUFAs could have a direct protective effect on myocardial systolic function (Bei, Frigiola, &Masuelli 2011). A recent randomized trial showed that one year of supplementation with 2 gram per day of omega-3 fatty acids was associated with a significant improvement of LVEF, exercise maximum duration, clinical functional class, maximum oxygen consumption and hospitalization rates for heart failure (Nodari, Triggiani, Campia, Manerba, Milesi, Cesana, Gheorghiade, & Dei Cas, 2011).

Also large epidemiological studies provide evidence for the effectiveness of n-3 PUFAs in preventing the development of heart failure. During a follow-up of 12 years the Cardiovascular Health Study reported that consumption of baked/broiled fish or supplemental n-3 PUFAs was associated with a lower incidence of new heart failure diagnosis among a cohort of over four thousands elderly subjects. This risk was inversely proportional to the quantity of fish/supplement intake with a 32% relative risk reduction in patients consuming the equivalent of $\geq$ 1 gram per day of omega-3 fatty acids (Mozaffarian, Bryson, Lemaitre, &Siscovick, 2005). In contrast, patients consuming much lower quantities of fish (the equivalent of 28-313 mg/day of omega-3 PUFAs) in the Rotterdam study showed no significant risk reduction (Dijkstra, Brouwer, van Rooij, Hofman, Witteman, &Geleijnse, 2011).

Cardiac Arrhythmias

Although antiarrhythmic therapy, implanted device therapy and ablation techniques are now all common and promptly available strategies in management of cardiac arrhythmia, lots are the problems in term of low efficacy rate, surgical complications, quality of life impairment and increased pro-arrhythmic and hemorrhagic risk. Consequently, useful alternatives are being sought. Between those, n-3 PUFAs have been studied in clinical trials involving treatment of atrial fibrillation (AF), ventricular tachycardias and sudden cardiac death in a wide range of populations (Guerra, Shkoza, Scappini, Roberti, & Capucci, 2013)

Post-Operative Atrial Fibrillation

Post-operative AF is a frequent complication of cardiac surgery, occurring in 30% to 50% of all patients (Andrews, Reimold, Berlin, &Antman, 1991; Mathew, Parks, Savino, Mathew, Parks, Savino, Friedman, Koch, Mangano, & Browner, 1996) and more often within 2 to 4 days after the procedure (Aranki, Shaw, Adams, Rizzo, Couper, VanderVliet, Collins, Cohn, & Burstin, 1996). It is associated with an increased risk of long-term mortality (Villareal, Hariharan, Liu, Kar, Lee, Elayda, Lopez, Rasekh, Wilson, & Massumi, 2004), longer hospital stay and additional healthcare-related costs (Aranki, Shaw, Adams, Rizzo, Couper, VanderVliet, Collins, Cohn, & Burstin, 1996). A randomized, single-center, double-blind study randomized 108 consecutive patients undergoing coronary artery bypass graft (CABG) to placebo or n-3 PUFAs (2 g capsules per day of 85% fish oil concentrate) (Saravanan,

Bridgewater, West, O'Neill, Calder, & Davidson, 2010). Patients with previous or current history of any atrial arrhythmia were excluded, as were patients under treatment with any anti-arrhythmic drugs or fish oil supplements. Heart rhythm was monitored for 5 days after cardiac surgery and an AF of at least 30 seconds was considered as clinically significant. AF developed in 43% of the placebo group and 56% in the n-3 PUFAs group with no significant different between the two groups. None of the other secondary endpoint were also different. Similar findings was shown by Heidarsdottir and colleagues in a randomized, double-blind, single-centre trial which enrolled 170 patients admitted for CABG or valve repair surgery (Heidarsdottir, Arnar, Skuladottir,Torfason, Edvardsson, Gottskalksson, Palsson, & Indridason, 2010). Patients were started 5-7 days prior to surgery and post-operatively to hospital discharge on two capsules containing 1240 mg of EPA and 1000 mg of DHA in the active group or olive oil in the placebo group. Onset of post-operative AF (single episode >5 minutes) was evaluated with continuous electrographic monitoring. Although an higher plasmatic level of both EPA and DHA was found in the treated group there was no evidence of reduced incidence of postoperative AF between active and control groups (54.2% in the intervention group versus 54.1% in the placebo group). In a substudy conducted on patients undergoing only CABG, it was shown that PUFAs implementation could be useful in preventing AF only in patients with low baseline plasma levels of EPA and DHA (Skuladottir, Heidarsdottir, Arnar, Torfason, Edvardsson, Gottskalksson, Palsson, & Indridason, 2011).

A similar study, but with a bigger pre-interventional treatment period (3 weeks) was performed by Farquharson and colleagues. In this single center, randomized, double-blind study 200 patients undergoing CABG or valve repair were randomized either to high dose n-3 PUFAs supplements (4.6 grams per day) or to linoleic acid (Farquharson, Metcalf, Sanders, Stuklis, Edwards, Gibson, Cleland, Sullivan, James, & Young, 2011). The primary endpoint was recurrence of AF (defined as any episode with a duration of more than 10 minutes or requiring medical intervention) and was evaluated with continuous monitoring during the first 6 post-operative days. Even in this case, no difference was found in the incidence of post-operative AF in both groups, although a significant decrease in length of stay in the intensive care unit was noted in the omega-3 fatty acids group.

The results of the previous studies are conflicting with those presented by Calò et al. (Calò, Bianconi, Colivicchi, Lamberti, Loricchio, de Ruvo, Meo, Pandozi, Staibano, & Santini, 2005) who randomized 160 patients to a placebo or PUFAs 2 g/day for at least 5 days before CABG and until the day of discharge. The primary endpoint (incidence of post-operative AF) was assessed in 27 patients of the control group and 12 in the intervention group, with a statistically significant difference in term of incidence. This study also demonstrated a significant reduction in hospitalization in the PUFAs patients. Similar results was achieved by Heidt and colleagues, who enrolled 102 patients undergoing CABG. The patients treated with n-3 PUFAs (100 mg fish oil per kilogram of body weight per day via perfusion pump starting on admission to hospital and ending at discharge) developed significantly less AF recurrences than control patients (Heidt, Vician, Stracke, , Stadlbauer, Grebe, Boening, Vogt, & Erdogan, 2009).

A recent meta- analysis (Liu, Korantzopoulos, Shehata, Li, Wang, & Kaul, 2007) conducted on trials for prevention of post-operative AF also failed to show significant effects of n-3 PUFAs supplementation on prevention of AF recurrence (OR 0.81; 95% CI 0.57 to 1.15; $p=0.24$) and claimed for larger trials with higher doses of PUFAs, claim that was recently accepted. The OPERA trial (Omega-3 Fatty Acids for Prevention of Post-Operative Atrial Fibrillation) trial is a double-blind, randomized, clinical trial of n-3 PUFAs versus placebo in patients scheduled for cardiac surgery (Mozaffarian, Marchioli, Macchia, Silletta, Ferrazzi, Gardner, Latini, Libby, Lombardi, O'Gara, Page, Tavazzi, & Tognoni, 2012). Main exclusion criteria were the absence of sinus rhythm at enrollment or regular fish oil supplements consumption. One-thousand and five-hundreds patients were enrolled and randomized to receive either fish oil (1-gram capsules; 840 mg of EPA plus DHA as ethyl-esters) or placebo with preoperative loading of 10 g over 3 to 5 days followed by 2 grams per day until hospital discharge or for a maximum of 10 days after surgery. Primary endpoint was occurrence of postoperative AF of at least 30 seconds in duration and documented by rhythm strip or 12-lead ECG. Secondary endpoints were incidence of sustained AF (>1 hour), electrical cardioversion for postoperative AF, time to first postoperative AF and number of postoperative AF episodes per patient. The primary endpoint occurred in 233 patients (30.7%) in the placebo group and 227 (30.0%) in the n-3-PUFAs group (p=0.74]. This large, randomized trial failed to show any evidence that perioperative n-3-PUFAs supplementation could reduce postoperative AF in patients undergoing cardiac surgery. In conclusions, on the basis of the recent data, there are no sufficient evidences to encourage the use of n-3 PUFAs therapy for prevention of post-operative AF. However we must consider differences in the populations studied like the type of surgery, the definition of arrhythmia and the method of arrhythmia diagnosis that could explain the negative results of some studies (Guerra, Shkoza, Scappini, Roberti, &Capucci, 2013). Another important issue is the use of other therapies, such as beta-blockers or statins, which were differently prescribed in the aforementioned studies and could explain, at least in part, the different results.

Paroxysmal and Persistent Atrial Fibrillation

The first evidences about a possible role of PUFAs in the prevention of recurrent AF came from epidemiological studies. In a prospective cohort of 4815 elderly patients followed for 12 years 980 cases of AF were diagnosed. Multivariate analysis showed an inverse relation between fish consumption and the incidence of AF, and a diet including fish for more than 5 times per week was associated with a 31% lower risk of developing AF (Mozaffarian, Psaty, Rimm, Lemaitre, Burke, Lyles, Lefkowitz, & Siscovick, 2004).

Some positive findings were described at first even for n-3 PUFAs as an add-on therapy for rhythm control. An observational study involving 1500 patients admitted for catheter ablation showed that in 129 patients treated with omega-3 fatty acids only 35 (27.1%) had early recurrence versus 57 (44%) in the control group. Less procedural failures were also reported (Patel, Shaheen, Venkatraman, Armaganijan, Sanchez, Horton, DiBiase, Mohanty, Canby, Bailey, Burkhardt, Gallinghouse, Zagrodzky, Kozeluhova, & Natale, 2010). Another trial assessed the effect of n-3 PUFA implementation on AF recurrence over 1 year of observation in 208 patients hospitalized with a diagnosis of myocardial infarction. Rehospitalization for AF was reduced by 19% with a further reduction in all-cause mortality (Macchia, Monte, Pellegrini, Romero, Ferrante, Doval, D'Ettorre, Maggioni, & Tognoni 2004). Similar results were shown by a more recent randomized, controlled, double-blind, single center trial (Nodari, Triggiani, Campia, Manerba, Milesi, Cesana, Gheorghiade, &

DeiCas, 2011). Of 254 screened patients, 199 were randomized to 2 g capsules containing 85% of EPA and DHA ethyl–ester or olive oil as placebo starting 4 weeks before electrical cardioversion. Inclusion criteria were persistent AF with at least one relapse after a previous cardioversion, an atrial size > 60 mm and treatment with both amiodarone and an inhibitor of the renin-angiotensin-aldosterone system.

In the interventional group fewer patients had a recurrence of AF after 1-year follow-up compared with placebo (37% versus 56%). In addition, therapy with omega-3 fatty acids may have facilitated electrical cardioversion attempts with a reduction of patients with procedural failure (2% versus 6%) and an increase of patients in whom sinus rhythm was successfully restored with a single shock (78% versus 33%). Unfortunately, a more recent meta-analysis conducted on 5 studies in 1179, including the data from previous studies, failed to show any effect of PUFAs on secondary prevention of AF (OR 0.83; 95% CI 0.48-1.45; p=0.51) (Mozaffarian, & Wu, 2012). However a significant heterogeneity between the studies involved was found, mainly related to the different populations, n-3 PUFAs doses and follow-up ranges.

These results were confimed by Bianconi and colleagues in a multi-center, randomized, controlled, double blind trial involving 204 patients with persistent AF undergoing external cardioversion for sinus rhythm restoration (Bianconi, Calò, Mennuni, Santini, Morosetti, Azzolini, Barbato, Biscione, Romano, & Santini, 2011). Patients were randomized to receive either 3 grams per day of n-3 PUFAs until cardioversion and 2 grams per day after or matched placebo. Recurrence incidence was evaluated by trans-telephonic monitoring and clinical visits. Sinus rhythm was restored by successful cardioversion in 91.4% patients in the active group and 92% on placebo. No difference in AF recurrence was found between the two groups (58.9% versus 51.1%) with a mean time to recurrence of 84 days in the n-3 PUFAs group and 106 days in the placebo group. The recent FORWARD trial (Randomized Trial to Assess Efficacy of PUFA for the Maintenance of Sinus Rhythm in Persistent Atrial Fibrillation) enrolled 586 outpatients with a confirmed diagnosis of paroxysmal AF requiring cardioversion or with at least 2 episodes of AF in the last 6 month (Macchia, Grancelli, Varini, Nul, Laffaye, Mariani, Ferrante, Badra, Figal, Ramos, Tognoni, Doval, 2013). Patients were randomly assigned to receive n-3 PUFAs (1g/day) or placebo for 12 months. 56 patients (18.9%) in the placebo group and 69 patients (24.0%) in the n-3 PUFAs group had a recurrent symptomatic AF episode after one month with no significant difference between the two groups. There was no difference between placebo and interventional group for any other secondary endpoint, including all-cause mortality.

In conclusion current data do not suggest the use of n-3 PUFAs for AF recurrence prevention. The data reported offer conflicting evidences about the efficacy of n-3 therapy in the setting of secondary prevention of AF (Guerra, Shkoza, Scappini, Roberti, &Capucci, 2013).

There are many differences in the study populations and design of the trials and the populations that may best benefit from this therapy have not been identified. Some clinical trials evaluated patients with refractory AF (i.e. recurrent AF after previous cardioversion) or more advanced underlying heart disease while others focused on ischemic heart disease. Another important issue is the dosage of PUFAs that may produce the antiarrhythmic effects mentioned above, and the duration of omega-3 fatty acid supplementation.

Ventricular Arrhythmias and Sudden Cardiac Death

The first evidence about the effectiveness of n-3 fatty acids in the prevention of ventricular arrhythmias came from epidemiological observations. In a study involving 260 patients with acute myocardial infarction, an appropriate intake of n-3 PUFAs was inversely correlated with ectopic ventricular beats (Smith, Blumenthal, Babyak, Georgiades, Sherwood, Sketch, & Watkins). Consistent to these data was the observation that a low omega-3 fatty acid intake is associated with an increased risk of ventricular fibrillation during the acute phase of MI (Aarsetoey, Aarsetoey, Lindner, Staines, Harris, & Nilsen, 2011) and the risk of sudden cardiac death (SCD). In two small studies conducted in an electrophysiology laboratory, ventricular tachycardia was less inducible in patients after a 6 week intake EPA and DHA, either oral or intravenous (Metcalf, Sanders, James, Cleland, & Young, 2008; Madsen, Christensen, Thøgersen, Schmidt, & Toft, 2010).

Some studies were conducted on carriers of an implanted cardioverter-defibrillator (ICD) in order to evaluate the incidence of ventricular arrhythmias or device-mediated interventions. Three different trials evaluated the effect of n-3 PUFAs on risk of ventricular arrhythmias, but none of them showed strong evidence in preventing electrical therapies (Raitt, Connor, Morris, Kron, Halperin, Chugh, McClelland, Cook, MacMurdy, Swenson, Connor , Gerhard, Kraemer, Oseran, Marchant, Calhoun, Shnider, & McAnulty 2005; Leaf & Weber, 2005; Brouwer, Raitt, Dullemeijer, Kraemer, Zock, Morris, Katan, Connor, Camm, Schouten, & McAnulty, 2006). However a more recent meta-analysis including all 1148 patients from the aforementioned trials suggested a possible beneficial effect in patients with coronary heart disease (CHD) (Brouwer et al., 2009). Similar results were achieved by Christensen and colleagues, who showed an inverse relationship between levels of fatty acids and incidence of ventricular arrhythmias in 98 patients with ischemic heart disease (Christensen et al., 2005).

Unfortunately all these studies were too small and substantially underpowered to properly evaluate the effects of n-3 fatty acids in ICD carriers. One reason for the results of these trials might have been the different mechanisms of arrhythmia in different subsets of patients. The lack of energy to maintain the membrane potential due to the impaired oxygen supply during acute ischemic event can produce a depolarization gradient across the myocardium. N-3 fatty acids are incorporated in cellular membranes with an indirect effect on sodium and calcium ion channels, thus shifting the resting membrane potential to a more hyperpolarizated state and prolonging the refractory period with a lower probability to develop ventricular arrhythmias. Therefore we can expect a greater effectiveness in ischemic patients than in non-ischemic patients where other mechanisms of arrhythmias, such as reentry, are more frequent. Besides, the results of these trials also support the need to differentiate appropriate versus inappropriate ICD interventions and the therapy mediated by the device (anti-tachycardia pacing or shock).

Regarding SCD, several trials evaluated the effectiveness of PUFAs in cardiovascular events but very few used SCD it as a primary endpoint preferring a combined endpoint, maybe due to the low frequency of SCD even in the high-risk ischemic population. The first available evidence of a possible role of PUFAs in preventing SCD is dated 1989, and has been mentioned before (Burt et al., 1989). The DART trial enrolled 2033 patients post-MI, and showed a statistically significant 29% reduction in fatal MI in those with an increased fish consumption (two to three times per week). These results appeared early in the trial and were hypothesized to be due to a reduction of sudden death (Burt et al., 1989). Some years later Lorgeril et al. showed a decreased mortality and sudden cardiac death in post-MI patients with supplementation of another fatty acid, the alpha-linolenic acid (De Lorgeril et al., 1994).

More recently the GISSI-Prevenzione, a randomized, controlled trial randomized 11323 patients with recent acute MI to n-3 PUFAs (1 gram per day) and/or vitamin E supplements (Kromhout et al., 2010). This double-active, four groups trial demonstrated that n-3 PUFAs supplementation significantly reduced the primary endpoint (composed by death, non-fatal myocardial and non-fatal stroke) by 19%. This reduction was mainly due to fewer fatal cardiovascular events and a reduced incidence of SCD. The protective effect of the omega-3 fatty acids was evident after only 3 months of active treatment. The reduction in SCD was already significant after 4 months, confirming a possible benefit of n-3 PUFAs as anti-arrhythmic agents in patients with recent MI. Moreover, the patients enrolled in this trial were currently treated with standard of care therapy, including several other protective drugs.

Reports on populations at less risk for SCD are however quite different. The JELIS (Japan EPA Lipid Intervention Study) evaluated 18.000 hypercholesterolemic Japanese patients randomized to receive n-3 PUFAs treatment (1.8g/day) along with statin or statin alone (Yokoyama et al., 2007). After a mean follow-up of 4.6 years the authors observed a 19% reduction in non-fatal CHD events in the intervention group but no significant benefit on fatal events, SCD included. Nevertheless, the rate of SCD in the JELIS population was extremely low during the 5 years of follow-up, defining a basic lack of statistical power to detect any possible effect on SCD incidence.

The only trial using SCD as primary endpoint was the OMEGA, a randomized double-blind placebo controlled trial which enrolled 3851 patients with MI in the previous 3 to 14 days (Rauch et al., 2010). This trial compared a 1 gram per day regimen of EPA and DHA as an ethyl-ester to an olive oil placebo. As previously mentioned, the intention-to-treat analysis demonstrated an occurrence of SCD in 1.5% of both groups with no difference in secondary endpoints. The incidence of SCD was lower than the expected one, making the trial underpowered to detect an anti-arrhythmic effect of n-3 PUFAs. This obviously prevents the formulation of any firm conclusion. Even the SU.FOL.OM3 trial (Galan et al., 2010) which randomized 2501 patients with history of CV disease to either 600 mg EPA + DHA, vitamins, both or placebo-only, failed to show any difference between groups in preventing major cardiovascular events. Again the study had low statistical power due to a lower-than-expected rate of events in the whole population.

The Alpha-Omega trial (Kromhout et al., 2010) was a multi-center double-blind placebo controlled trial conducted on 4837 patients with a previous MI. As previously described, patients were randomly assigned to receive one of four types of margarine. The primary endpoint was a reduction in fatal and non-fatal cardiac events and cardiac interventions. Study size estimations was initially based on an estimated 4% mortality per year but had to be adjusted for the lower mortality observed during the study. None of the interventions reduced the incidence of the primary endpoint. N−3 PUFA therapy did not reduce the secondary endpoints like ventricular arrhythmia-related events (SCD, cardiac arrest, and ICD implantation). Reasons for the neutral results may be researched in the concomitant pharmacological treatment, the low-dose of n-3 PUFAs given and the design of the study, which did not compare n-3 PUFAs to a pure placebo group causing difficulties in interpreting the anti-arrhythmic effects of these drugs.

To summarize, the trials presented offered no definitive conclusions about the effectiveness of a PUFA therapy in the setting of prevention ventricular arrhythmias or SCD. When interpreting these data we must consider that the participants were recruited independently of their baseline levels of EPA and DHA and the bioavailability problems regarding EPA and DHA preparations as ethyl-ester used in most studies.

Conclusion

From the evidence collected through basic science studies, n-3 PUFAs offer innumerable advantages in term of lipid-lowering, anti-arrhythmic and anti-thrombothic potential. The plethora of possible applications in the cardiovascular field ranges from cardiovascular prevention to acute myocardial infarction, from atrial fibrillation to sudden cardiac death. Observational studies more often than not confirmed these findings, raising great expectations for a widespread application of n-3 PUFAs as a first-line therapeutic options. Unfortunately, all these promising evidences were often disproven by randomized clinical trial and meta-analyses. So why have clinical trials been so inconclusive and contrasting about a real efficacy of omega-3 fatty acids in everyday clinical practice? A few confounders must be underlined.

First, irrespective of the clinical field and population studied, nearly 100% of all trials and observational studies on n-3 PUFAs comprised few patients and were utterly underpowered when trying to detect a small difference in mortality and morbidity endpoints in patients, who were often under best-of-care therapy in term of anti-ischemic, cardioprotective and anti-arrhythmic drugs. Follow-up times were often within the order of days to week, further reducing event detection and inflating type 2 error. Second, timing of actual administration of omega-3 supplementation varies from study to study, as does the actual dosage, method of administration, and percentage of EPA and DHA in each tested compound. At present date, we lack a single phase II, dose-ranging trial, and all phase III trials available used a lot of different dosages and EPA/DHA ratios, without any solid reason to prefer a formulation over another. Third, in some specific fields, such as AF prevention, important clinical endpoints such as AF recurrence had been often defined only as clinically diagnosed AF, not taking into account all the asymptomatic episodes of AF which did not require hospitalization but still provide a high risk for thromboembolic events and cardiovascular mortality.

In conclusion, albeit promising, nowadays there are no sufficient data to encourage n-3 PUFAs therapy in a lot of different settings, with the notable exceptions of cardiovascular prevention as an effective lipid-lowering drug (class of evidence IIa, level of evidence B according to the European Society of Cardiology guidelines) (Reiner et al., 2011) and in chronic HF in order to prevent cardiovascular mortality on top of standard therapy (class of evidence IIb, level of evidence B according to the European Society of Cardiology guidelines) (McMurray et al., 2012). Phase III trials with good statistical power, hard pre-specified primary end points and real-life population would allow us to make up our mind regarding the other interesting issues.

References

Aarsetoey, H., Aarsetoey, R., Lindner, T., Staines, H., Harris, H.S., & Nilsen DW. (2011). Low levels of the omega-3 index are associated with sudden cardiac arrest and remain stable in survivors in the subacute phase. *Lipids,* 46, 151–161.

Agren, J.J., Hänninen, O., Julkunen, A., Fogelholm, L., Vidgren, H., Schwab, U., Pynnönen, O., & Uusitupa, M., (1996). Fish diet, fish oil and docosahexaenoic acid rich oil lower fasting and postprandial plasma lipid levels. *Eur. J. Clin. Nutr.,* 50, 765-71.

Albert, C.M., Campos, H., Stampfer, M.J., Ridker, P.M., Manson, J.E., Willett, W.C., & Ma, J., (2002). Blood levels of long-chain n-3 fatty acids and the risk of sudden death. *N. Engl. J. Med.,* 346, 1113-8.

Albert, C.M., Hennekens, C.H., O'Donnell, C.J., Ajani, U.A., Carey, V.J., Willett, W.C., Ruskin, J.N., & Manson, J.E., (1998). Fish consumption and risk of sudden cardiac death. *JAMA,* 279, 23-8.

Andrews, T.C., Reimold, S.C., Berlin, J.A., & Antman, E.M., (1991). Prevention of supraventricular arrhythmias after coronary artery bypass surgery: A meta-analysis of randomized control trials. *Circulation,* 84, 236–244.

Aranki, S.F., Shaw, D.P., Adams, D.H., Rizzo, R.J., Couper, G.S., VanderVliet, M., Collins, J.J., Cohn, L.H., & Burstin, H.R., (1996). Predictors of atrial fibrillation after coronary artery surgery: Current trends and impact on hospital resources. Circulation, 94, 390–397.

Bang, H.O., Dyerberg, J., & Nielsen, A.B., (1971). Plasma lipid and lipoprotein pattern in Greenlandic West-coast Eskimos. *Lancet,* 1, 1143-5.

Bays, H.E., Tighe, A.P., Sadovsky, R., & Davidson, M.H., (2008). Prescription omega-3 fatty acids and their lipid effects: physiologic mechanisms of action and clinical implications. Expert *Rev Cardiovasc. Ther.,* 6, 391-409.

Bei, R., Frigiola, A., & Masuelli, L., et al., (2011). Effects of omega-3 polyunsaturated fatty acids on cardiac myocyte protection. *Front. Biosci.,* 16, 1833-43.

Bianconi, L., Calò, L., Mennuni, M., Santini, L., Morosetti, P., Azzolini, P., Barbato, G., Biscione, F., Romano, P., & Santini, M., (2011). n-3 polyunsaturated fatty acids for the prevention of arrhythmia recurrence after electrical cardioversion of chronic persistent atrial fibrillation: a randomized, double-blind, multicentre study. *Europace,* 13, 174–181.

Bosch, J., Gerstein, H.C., Dagenais, G.R., Díaz, R., Dyal, L., Jung, H., Maggiono, A.P., Probstfield, J., Ramachandran, A., Riddle, M.C., Rydén, L.E., & Yusuf, S., ORIGIN trial investigators., (2012). n-3 fatty acids and cardiovascular outcomes in patients with dysglycemia. *N. Engl. J. Med.,* 367, 309-18.

Brouwer, I.A., Raitt, M.H., Dullemeijer, C., Kraemer, D.F., Zock, P.L., Morris, C., Katan, M.B., Connor, W.E., Camm, J.A., Schouten, E.G., & McAnulty, J., (2009). Effect of fish oil on ventricular tachyarrhythmia in three studies in patients with implantable cardioverter defibrillators. *Eur. Heart J.,* 30, 820-6.

Brouwer, I.A., Zock, P.L., Camm, A.J., Böcker, D., Hauer, R.N., Wever, E.F., Dullemeijer, C., Ronden, J.E., Katan, M.B., Lubinski, A., Buschler, H., & Schouten, E.G., SOFA Study Group., (2006). Effect of fish oil on ventricular tachyarrhythmia and death in patients with implantable cardioverter defibrillators: the Study on Omega-3 Fatty Acids and Ventricular Arrhythmia (SOFA) randomized trial. *JAMA,* 295, 2613-9.

Burr, M.L., Ashfield-Watt, P.A., Dunstan, F.D., Fehily, A.M., Breay, P., Ashton, T., Zotos, P.C., Haboubi, N.A. & Elwood, P.C., (2003). Lack of benefit of dietary advice to men with angina: results of a controlled trial. *Eur. J. Clin. Nutr.,* 57, 193-200.

Burr, M.L., Dunstan, F.D. & George, C.H., (2005). Is fish oil good or bad for heart disease? Two trials with apparently conflicting results. *J. Membr. Biol.,* 206, 155-63.

Burr, M.L., Fehily, A.M., Gilbert, J.F., Rogers, S., Holliday, R.M., Sweetnam, P.M., & Deadman N.M.,(1989). Effects of changes in fat, fish and fibre intakes on death and myocardial reinfarction: diet and reinfarction trial (DART). *Lancet,,* 2, 757-61.

Calder, P.C., (2009). Polyunsaturated fatty acids and inflammatory processes: New twists in an old tale. *Biochimie,* 91, 791-5.

Calò, L., Bianconi, L., Colivicchi, F., Lamberti, F., Loricchio, M.L., de Ruvo, E., Meo, A., Pandozi, C., Staibano, M., &Santini, M., (2005). N-3 Fatty acids for the prevention of atrial fibrillation after coronary artery bypass surgery: A randomized, controlled trial. *J. Am. Coll Cardiol.,* 45, 1723–1728.

Capucci, A., & Guerra F., (2011). Late postoperative atrial fibrillation: minor surgery complication or major cardiovascular issue? *J. Cardiovasc. Med.*, 12, 381–382.

Cazzola, R., Russo-Volpe, S., Miles, E.A., Rees, D., Banerjee, T., Roynette, C.E., Wells, S.J., Goua, M., Wahle, K.W., Calder, P.C., & Cestaro, B., (2007). Age- and dose-dependent effects of an eicosapentaenoic acid-rich oil on cardiovascular risk factors in healthy male subjects. *Atherosclerosis,* 193, 159-67.

Chin, J.P., Gust, A.P., Nestel, P.J., & Dart, A.M., (1993). Marine oils dose-dependently inhibit vasoconstriction of forearm resistance vessels in humans. *Hypertension,* 21, 22-8.

Christensen, J.H., Riahi, S., Schmidt, E.B., Mølgaard, H, Kirstein Pedersen, A., Heath, F., Cosedis Nielsen, J., & Toft, E., (2005). n-3 Fatty acids and ventricular arrhythmias in patients with ischaemic heart disease and implantable cardioverter defibrillators. *Europace,* 7, 338-44.

Conquer, J.A., & Holub, B.J., (1996). Supplementation with an algae source of docosahexaenoic acid increases (n-3) fatty acid status and alters selected risk factors for heart disease in vegetarian subjects. *J. Nutr.,* 126, 3032-9.

Dangardt, F., Osika, W., Chen, Y., Nilsson, U., Gan, L.M., Gronowitz, E., Strandvik, B., & Friberg, P., (2010). Omega-3 fatty acid supplementation improves vascular function and reduces inflammation in obese adolescents. *Atherosclerosis,* 212, 580-5.

Daviglus, M.L., Stamler, J., Orencia, A.J., Dyer, A.R., Liu, K., Greenland, P., Walsh, M.K., Morris, D., & Shekelle, R.B., (1997). Fish consumption and the 30-year risk of fatal myocardial infarction. *N. Engl. J. Med.,* 336, 1046-53.

De Caterina, R., (2011). n–3 Fatty Acids in Cardiovascular Disease. *N. Engl. J. Med.,* 364, 2439-50.

De Caterina, R., & Massaro, M., (2005). Omega-3 fatty acids and the regulation of expression of endothelial pro-atherogenic and pro-inflammatory genes. *J. Membr. Biol.,* 206, 103-16.

De Caterina, R., Cybulsky, M.I., Clinton, S.K., Gimbrone, M.A. Jr, & Libby P., (1994). The omega-3 fatty acid docosahexaenoate reduces cytokine-induced expression of proatherogenic and proinflammatory proteins in human endothelial cells. *Arterioscler. Thromb.,* 14, 1829-36.

De Lorgeril, M., Renaud, S., Mamelle, N., Salen, P., Martin, J.L., Monjaud, I., Guidollet, J., Touboul, P., & Delaye, J., (1994). Mediterranean alpha-linolenic acid-rich diet in secondary prevention of coronary heart disease. *Lancet,* 343, 1454-9.

Dijkstra, S.C., Brouwer, I.A., van Rooij, F.J., Hofman, A., Witteman, J.C., & Geleijnse, J.M., (2009). Intake of very long chain n-3 fatty acids from fish and the incidence of heart failure: the Rotterdam Study. *Eur. J. Heart Fail.,* 11, 922–928.

Dyerberg, J., Bang, H.O., & Aagaard, O., (1980). Alpha-linolenic acid and eicosapentaenoic acid. *Lancet,* 1, 199.

EFSA Panel on Dietetic Products, Nutrition and Allergies (NDA), (2012). Scientific Opinion on the Tolerable Upper Intake Level of eicosapentaenoic acid (EPA), docosahexaenoic acid (DHA) and docosapentaenoic acid (DPA). *EFSA Journal,* 10, 2815-63.

Egert, S., & Stehle, P., (2011). Impact of n-3 fatty acids on endothelial function: results from human interventions studies. *Curr. Opin. Clin. Nutr. Metab. Care.*, 14, 121-31.

Endres, S., Ghorbani, R., Kelley, V.E., Georgilis, K., Lonnemann, G., van der Meer, J.W., Cannon, J.G., Rogers, T.S., Klempner, M.S., Weber, P.C., Schaefer, E.J., Sheldon M., & Dinarello C.A., (1989). The effect of dietary supplementation with n-3 polyunsaturated fatty acids on the synthesis of interleukin-1 and tumor necrosis factor by mononuclear cells. *N. Engl. J. Med.,* 320, 265-71.

Farquharson, A.L., Metcalf, R.G., Sanders, P., Stuklis, R., Edwards, J.R., Gibson, R.A., Cleland, L.G., Sullivan, T.R., James, M.J., & Young, G.D., (2011). Effect of dietary fish oil on atrial fibrillation after cardiac surgery. *Am. J. Cardiol.,* 108, 851–856.

Fox, P.L., & DiCorleto, P.E., (1988). Fish oils inhibit endothelial cell production of platelet-derived growth factor-like protein. *Science,* 241, 453-6.

Gajos, G., Rostoff, P., Undas, A., & Piwowarska, W., (2010). Effects of polyunsaturated omega-3 fatty acids on responsiveness to dual antiplatelet therapy in patients undergoing percutaneous coronary intervention: the OMEGA-PCI (OMEGA-3 fatty acids after pci to modify responsiveness to dual antiplatelet therapy) study. *J. Am. Coll Cardiol.*, 55, 1671-8.

Gajos, G., Zalewski, J., Rostoff, P., Nessler, J., Piwowarska, W., & Undas, A., (2011). Reduced thrombin formation and altered fibrin clot properties induced by polyunsaturated omega-3 fatty acids on top of dual antiplatelet therapy in patients undergoing percutaneous coronary intervention (OMEGA-PCI clot). *Arterioscler. Thromb. Vasc. Biol.,* 31, 1696-702.

Galan, P., Kesse-Guyot, E., Czernichow, S., Briancon, S., Blacher, J., & Hercberg, S., and SU.FOL.OM3 Collaborative Group. (2010). Effects of B vitamins and omega 3 fatty acids on cardiovascular diseases: a randomised placebo controlled trial. *BMJ,* 341, c6273.

Geleijnse, J.M., Giltay, E.J., Grobbee, D.E., Donders, A.R., & Kok, F.J., (2002). Blood pressure response to fish oil supplementation: metaregression analysis of randomized trials. *J. Hypertens,* 20, 1493-9.

Ghio, S., Scelsi, L., Latini, R., Masson, S., Eleuteri, E., Palvarini, M., Vriz, O., Pasotti, M., Gorini, M., Marchioli, R., Maggioni, A., &Tavazzi, L.; GISSI-HF investigators. (2010). Effects of n-3 polyunsaturated fatty acids and of rosuvastatin on left ventricular function in chronic heart failure: a substudy of GISSI-HF trial. *Eur. J. Heart Fail,* 12, 1345-53.

Grimsgaard, S., Bonaa, K.H., Hansen, J.B., & Nordøy, A., (1997). Highly purified eicosapentaenoic acid and docosahexaenoic acid in humans have similar triacylglycerol-lowering effects but divergent effects on serum fatty acids. *Am. J. Clin. Nutr.,* 66, 649-59.

Gruppo Italiano per lo Studio della Sopravvivenza nell'Infarto Miocardico, (1999). Dietary supplementation with n-3 polyunsaturated fatty acids and vitamin E after myocardial infarction: results of the GISSI-Prevenzione trial. *Lancet,* 354, 447-55.

Guerra, F., Shkoza, M., Scappini, L., Roberti, L.,& Capucci, A., (2013). Omega-3 PUFAs and atrial fibrillation: have we made up our mind yet? *Ann. Noninvasive Electrocardiol.,* 18, 12–20.

Harris, W.S., Rambjør, G.S., Windsor, S.L., & Diederich, D., (1997). n-3 fatty acids and urinary excretion of nitric oxide metabolites in humans. *Am. J. Clin. Nutr.,* 65, 459-64.

Heidarsdottir, R., Arnar, D.O., Skuladottir, G.V., Torfason, B., Edvardsson, V., Gottskalksson, G., Palsson, R., & Indridason, O.S., (2010). Does treatment with n-3 polyunsaturated fatty acids prevent atrial fibrillation after open heart surgery? *Europace,* 12, 356–363.

Heidt, M.C., Vician, M., Stracke, S.K., Stadlbauer, T., Grebe, M.T., Boening, A., Vogt, P.R., & Erdogan, A., (2009). Beneficial effects of intravenously administered N-3 fatty acids for the prevention of atrial fibrillation after coronary artery bypass surgery: a prospective randomized study. *Thorac. Cardiovasc. Surg.,* 57, 276–280.

Hooper, L., Thompson, R.L., Harrison, R.A., Summerbell, C.D., Ness, A.S., Moore, H.J., Worthington, H.V., Durrington, P.N., Higgins, J.P., Capps, N.E., Riemersma, R.A., Ebrahim, S.B., & Smith, G.D., (2006). Risks and benefits of omega 3 fats for mortality, cardiovascular disease, and cancer: systematic review. *BMJ,* 332, 752-60.

Hu, F.B., Cho, E., Rexrode, K.M., Albert, C.M., & Manson, J.E., (2002). Fish and omega 3 fatty acid intake and risk of coronary heart disease in women. *JAMA,* 287, 1815-21.

Jacobson, T.A., Glickstein, S.B., Rowe, J.D., & Soni, P.N., (2012). Effects of eicosapentaenoic acid and docosahexaenoic acid on low-density lipoprotein cholesterol and other lipids: a review. *J. Clin. Lipidol.,* 6, 5-18.

Jacobson, T.A., Miller, M., & Schaefer, E.J., (2007). Hypertriglyceridemia and cardiovascular risk reduction. *Clin. Ther.,* 29, 763-77.

Kaminski, W.E., Jendraschak, E., Kiefl, R., & von Schacky, C., (1993). Dietary omega-3 fatty acids lower levels of platelet-derived growth factor mRNA in human mononuclear cells. *Blood,* 81, 1871-9.

Kelley, D.S., Siegel, D., Vemuri, M., & Mackey, B.E., (2007). Docosahexaenoic acid supplementation improves fasting and postprandial lipid profiles in hypertriglyceridemic men. *Am. J. Clin. Nutr.,* 86, 324-33.

Kenny, D., Warltier, D.C., Pleuss, J.A., Hoffmann, R.G., Goodfriend, T.L., & Egan, B.M., (1992). Effect of omega-3 fatty acids on the vascular response to angiotensin in normotensive men. *Am. J. Cardiol.,* 70, 1347-52.

Kromhout, D., Giltay, E.J., & Geleijnse, J.M., (2010). n–3 Fatty acids and cardiovascular events after myocardial infarction. *N. Engl. J. Med.,* 363, 2015-26.

Kromhout, D., Giltay, E.J., & Geleijnse, J.M., for the Alpha Omega Trial Group. (2010). n-3 Fatty Acids and Cardiovascular Events after Myocardial Infarction. *N. Engl. J. Med.,* 363, 2015-26.

Leaf, A., Albert, C.M., Josephson, M., Steinhaus, D., Kluger, J., Kang, J.X., Cox, B., Zhang, H., & Schoenfeld, D., Fatty Acid Antiarrhythmia Trial Investigators, (2005). Prevention of fatal arrhythmias in high-risk subjects by fish oil n-3 fatty acid intake. *Circulation,* 112, 2762-8.

Leaf, A., & Weber, P.C., (1988) Cardiovascular effects of n–3 fatty acids. *N. Engl. J. Med.*, 318, 549-57.

Lev, E.I., Solodky, A., Harel, N., Mager, A., Brosh, D., Assali, A., Roller, M., Battler, A., Kleiman, N.S., & Kornowski, R., (2010). Treatment of aspirin resistant patients with omega-3 fatty acids versus aspirin dose escalation. *J. Am. Coll. Cardiol.,* 55, 114-21.

Libby, P., (2013) Mechanisms of Disease: Mechanisms of Acute Coronary Syndromes and Their Implications for Therapy. *N. Engl. J. Med.*, 368, 2004-2013.

Liu, T., Korantzopoulos, P., Shehata, M., Li, G., Wang, X., & Kaul, S., (2007). Prevention of atrial fibrillation with omega-3 fatty acids: a meta-analysis of randomised clinical trials. *Heart,* 97, 1034–1040.

López, D., Möller, M., Denicola, A., Casós, K., Rubbo, H., Ruiz-Sanz, J.I., & Mitjavila, M.T., (2008). Long-chain n-3 polyunsaturated fatty acid from fish oil modulates aortic nitric oxide and tocopherol status in the rat. *Br. J. Nutr.,* 100, 767-75.

Macchia, A., Grancelli, H., Varini, S., Nul, D., Laffaye, N., Mariani, J., Ferrante, D., Badra, R., Figal, J., Ramos, S., Tognoni, G.,& Doval, H.C.; GESICA Investigators. (2013). Omega-3 Fatty Acids for the Prevention of Recurrent Symptomatic Atrial Fibrillation: Results of the FORWARD (Randomized Trial to Assess Efficacy of PUFA for the Maintenance of Sinus Rhythm in Persistent Atrial Fibrillation) Trial. *J. Am. Coll Cardiol.,* 61, 463-8.

Macchia, A., Monte, S., Pellegrini, F., Romero, M., Ferrante, D., Doval, H., D'Ettorre, A., Maggioni, A.P., & Tognoni, G., (2004). Omega-3 fatty acid supplementation reduces one-year risk of atrial fibrillation in patients hospitalized with myocardial infarction. *Eur. J. Clin. Pharmacol.,* 64, 627–634.

Madsen, T., Christensen, J.H., Thøgersen, A.M., Schmidt, E.B., & Toft, E., (2010). Intravenous infusion ofn-3polyunsaturated fatty acids and inducibility of ventricular tachycardia in patients with implantable cardioverter defibrillator. *Europace,* 12, 941–946.

Maki, K.C., Van Elswyk, M.E., McCarthy, D., Hess, S.P., Veith, P.E., Bell, M., Subbaiah, P., & Davidson, M.H., (2005) Lipid responses to a dietary docosahexaenoic acid supplement in men and women with below average levels of high density lipoprotein cholesterol. *J. Am. Coll. Nutr.,* 24, 189-99.

Mann, N.J., O'Connell, S.L., Baldwin, K.M., Singh, I., & Meyer, B.J., (2010). Effects of seal oil and tuna-fish oil on platelet parameters and plasma lipid levels in healthy subjects. *Lipids,* 45, 669-81.

Marchioli, R., Barzi, F., Bomba, E., Chieffo, C., Di Gregorio, D., Di Mascio, R., Franzosi M.G., Geraci E., Levantesi G., Maggioni A.P., Mantini L., Marfisi R.M., Mastrogiuseppe, G., Mininni, N., Nicolosi G.L., Santini M., Schweiger C., Tavazzi L., Tognoni G., Tucci C., & Valagussa F., GISSI-Prevenzione Investigators., (2002). Early protection against sudden death by n-3 polyunsaturated fatty acids after myocardial infarction: time-course analysis of the results of the Gruppo Italiano per lo Studio della Sopravvivenza nell'Infarto Miocardico (GISSI)- Prevenzione. *Circulation,* 105, 1897-903.

Mathew, J.P., Parks, R., Savino, J.S., Mathew, J.P., Parks, R., Savino, J.S., Friedman, A.S., Koch, C., Mangano, D.T., & Browner, W.S., (1996). Atrial fibrillation following coronary artery bypass graft surgery: Predictors, outcomes, and resource utilization: Multicenter Study of Perioperative Ischemia Research Group. *JAMA,* 276, 300– 306.

McMurray, J.J., Adamopoulos, S., Anker, S.D., Auricchio, A., Böhm, M., Dickstein, K., Falk, V., Filippatos, G., Fonseca, C., Gomez-Sanchez, M.A., Jaarsma, T., Køber, L., Lip, G.Y., Maggioni, A.P., Parkhomenko, A., Pieske, B.M., Popescu, B.A., Rønnevik, P.K., Rutten, F.H., Schwitter, J., Seferovic, P., Stepinska, J., Trindade, P.T., Voors, A.A., Zannad, F., & Zeiher, A.; ESC Committee for Practice Guidelines., (2012). ESC Guidelines for the diagnosis and treatment of acute and chronic heart failure 2012: The Task Force for the Diagnosis and Treatment of Acute and Chronic Heart Failure 2012 of the European Society of Cardiology. Developed in collaboration with the Heart Failure Association (HFA) of the ESC. *Eur. Heart J.,* 33, 1787-847.

Metcalf, R.G., Sanders, P., James, M.J., Cleland, L.G., & Young, G.D., (2008). Effectofdietaryn-3 polyunsaturated fatty acids on the inducibility of ventricular tachycardia in patients with ischemic cardiomyopathy. *Am. J. Cardiol.,* 101, 758–761.

Moertl, D., Hammer, A., Steiner, S., Hutuleac, R., Vonbank, K., & Berger, R., (2011). Dose-dependent effects of omega-3-polyunsaturated fatty acids on systolic left ventricular function, endothelial function, and markers of inflammation in chronic heart failure of nonischemic origin: a double-blind, placebo-controlled, 3-arm study. *Am. Heart J.,* 161, 915.e1-9.

Mori, T.A., Bao, D.Q., Burke, V., Puddey, I.B., & Beilin, L.J., (1999). Docosahexaenoic acid but not eicosapentaenoic acid lowers ambulatory blood pressure and heart rate in humans. *Hypertension,* 34, 253–60.

Mori, T.A., Burke, V., Puddey, I.B, Watts, G.F., O'Neal, D.N., Best, J.D., & Beilin, L.J., (2000). Purified eicosapentaenoic and docosahexaenoic acids have differential effects on serum lipids and lipoproteins, LDL particle size, glucose, and insulin in mildly hyperlipidemic men. *Am. J. Clin. Nutr.,* 71, 1085-94.

Mori, T.A., Watts, G.F., Burke, V., Hilme, E., Puddey, I.B., & Beilin, L.J., (2000). Differential effects of eicosapentaenoic acid and docosahexaenoic acid on vascular reactivity of the forearm microcirculation in hyperlipidemic, overweight men. *Circulation,* 102, 1264-9.

Mori, T.A., & Woodman, R.J., (2006). The independent effects of eicosapentaenoic acid and docosahexaenoic acid on cardiovascular risk factors in humans. *Curr. Opin. Clin. Nutr. Metab. Care,* 9, 95-104.

Mori, T.A., Woodman, R.J., Burke, V., Puddey, I.B., Croft, K.D., & Beilin, L.J., (2003). Effect of eicosapentaenoic acid and docosahexaenoic acid on oxidative stress and inflammatory markers in treated-hypertensive type 2 diabetic subjects. *Free Radic. Biol. Med.,* 35, 772-81.

Mozaffarian, D., Bryson, C.L., Lemaitre, R.N., & Siscovick, D.S., (2005). Fish intake and risk of incident heart failure. *J. Am. Coll. Cardiol.,* 45, 2015–2021.

Mozaffarian, D., Lemaitre, R.N., King, I.B., Song, X., Spiegelman, D., Sacks, F.M., Rimm, E.B., & Siscovick, D.S., (2011). Circulating long-chain ω-3 fatty acids and incidence of congestive heart failure in older adults: the cardiovascular health study: a cohort study. *Ann. Intern. Med.,* 155, 160-70.

Mozaffarian, D., Marchioli, R., Macchia, A., Silletta, M.G., Ferrazzi, P., Gardner, T.J., Latini,R., Libby, P., Lombardi, F., O'Gara, P.T., Page, R.L., Tavazzi, L.,& Tognoni, G.; OPERA Investigators., (2012). Fish oil and postoperative atrial fibrillation: the Omega-3 Fatty Acids for Prevention of Post-operative Atrial Fibrillation (OPERA) randomized trial. *JAMA,* 308, 2001-11.

Mozaffarian, D., & Wu, J.H., (2012). (n-3) fatty acids and cardiovascular health: are effects of EPA and DHA shared or complementary? *J. Nutr.,* 142, 614S-625S.

Mozaffarian, D.. Psaty, B.M., Rimm, E.B., Lemaitre, R.N., Burke, G.L., Lyles, M.F., Lefkowitz, D., & Siscovick, D.S., (2004). Fish intake and risk of incident atrial fibrillation. *Circulation,* 110, 368-73.

Murad, M.H., Hazem, A., Coto-Yglesias, F., Dzyubak, S., Gupta, S., Bancos, I., Lane, M.A., Erwin, P.J., Berglund, L., Elraiyah, T., & Montori, V.M., (2012). The association of hypertriglyceridemia with cardiovascular events and pancreatitis: a systematic review and meta-analysis. *BMC Endocr. Disord.,* 31, 12-2.

Neff, L.M., Culiner, J., Cunningham-Rundles, S., Seidman, C., Meehan, D., Maturi, J., Wittkowski, K.M., Levine, B., & Breslow, J.L., (2011). Algal docosahexaenoic acid affects plasma lipoprotein particle size distribution in overweight and obese adults. *J. Nutr.,* 141, 207-13.

Ness, A.R., Ashfield-Watt, P.A., Whiting, J.M., Smith, G.D., Hughes, J. & Burr, M.L., (2004). The long-term effect of dietary advice on the diet of men with angina: the diet and angina randomized trial. *J. Hum. Nutr. Diet.,* 17, 117-9.

Nestel, P., Shige, H., Pomeroy, S., Cehun, M., Abbey, M., & Raederstorff, D., (2002). The n-3 fatty acids eicosapentaenoic acid and docosahexaenoic acid increase systemic arterial compliance in humans. *Am. J. Clin. Nutr.,* 76, 326-30.

Newman, W.P., Middaugh, J.P., Propst, M.T., & Rogers, D.R., (1993). Atherosclerosis in Alaska Natives and non-natives. *Lancet,* 341, 1056-7.

Nodari, S., Triggiani, M., Campia, U., Manerba, A., Milesi, G., Cesana, B.M., Gheorghiade, M.,& Dei Cas, L., (2011). Effects of n-3 polyunsaturated fatty acids on left ventricular function and functional capacity in patients with dilated cardiomyopathy. *J. Am. Coll. Cardiol.,* 57, 870-9.

Nodari, S., Triggiani, M., Campia, U., Manerba, A., Milesi, G., Cesana, B.M., Gheorghiade, M., DeiCas, L., (2011). n-3 polyunsaturatedfatty acids in the prevention of atrial fibrillation recurrences after electrical cardioversion: a prospective, randomized study. *Circulation,* 124, 1100–1106.

O'Keefe, J.H. Jr, Abuissa, H., Sastre, A., Steinhaus, D.M., & Harris, W.S., (2006). Effects of omega-3 fatty acids on resting heart rate, heart rate recovery after exercise, and heart rate variability in men with healed myocardial infarctions and depressed ejection fractions. *Am. J. Cardiol.,* 97, 1127-30.

Okpala, I., Ibegbulam, O., Duru, A., Ocheni S., Emodi I., Ikefuna A., Umar G., Asinobi I., Madu, A., Okoye, A., Nwagha, T., Oquonu U., Uamai, I., Aqwu, O., nonyelu, C., Anike, U., Agu, K., Anigbo, C., Chukwura A., Ugwu, O., & Herrada, S., (2011). Pilot study of omega-3 fatty acid supplements in sickle cell disease. *APMIS,* 119, 442-8.

Patel, D., Shaheen, M., Venkatraman, P., Armaganijan, L., Sanchez, J., Horton, R.P., DiBiase L, Mohanty, P., Canby, R., Bailey, S.M., Burkhardt, J.D., Gallinghouse, G.J., Zagrodzky, J.D., Kozeluhova, M., & Natale, A., (2010). Omega-3polyunsaturated fatty acid supplementation reduced atrial fibrillation recurrence after pulmonary vein antrum isolation. *Indian. Pacing. Electrophysiol. J.,* 9, 292–298.

Raitt, M.H., Connor, W.E., Morris, C., Kron, J., Halperin, B., Chugh, S.S., McClelland, J., Cook, J., MacMurdy, K., Swenson, R., Connor, S.L., Gerhard, G., Kraemer, D.F., Oseran, D., Marchant, C., Calhoun, D., Shnider, R., & McAnulty, J., (2005). Fish oil supplementation and risk of ventricular tachycardia and ventricular fibrillation in patients with implantable defibrillators: a randomized controlled trial. *JAMA,* 293, 2884-91.

Rauch, B., Schiele, R., Schneider, S., Diller, F., Victor, N., Gohlke, H., Gottwik, M., Steinbeck, G., Del Castillo, U., Sack, R., Worth, H., Katus, H., Spitzer, W., Sabin, G., & Senges, J.; OMEGA Study Group., (2010). OMEGA, a Randomized, Placebo-Controlled Trial to Test the Effect of Highly Purified Omega-3 Fatty Acids on Top of Modern Guideline-Adjusted Therapy After Myocardial Infarction. *Circulation,* 122, 2152-2159.

Rees, D., Miles, E.A., Banerjee, T., Wells, S.J., Roynette, C.E., Wahle, K.W., & Calder, P.C., (2006). Dose-related effects of eicosapentaenoic acid on innate immune function in healthy humans: a comparison of young and older men. *Am. J. Clin. Nutr.,* 83, 331-42.

Reiner, Z., Catapano, A.L., De Backer, G., Graham, I., Taskinen, M.R., Wiklund, O., Agewall, S., Alegria, E., Chapman, M.J., Durrington, P., Erdine, S., Halcox, J., Hobbs, R., Kjekshus, J., Filardi, P.P., Riccardi, G., Storey, R.F., & Wood, D.; ESC Committee for Practice Guidelines (CPG) 2008-2010 and 2010-2012 Committees., (2011). ESC/EAS Guidelines for the management of dyslipidaemias: the Task Force for the management of dyslipidaemias of the European Society of Cardiology (ESC) and the European Atherosclerosis Society (EAS). *Eur. Heart J.,* 32, 1769-818.

Rizos, E.C., Ntzani, E.E., Bika, E., Kostapanos, M.S., & Elisaf, M.S., (2012). Association Between Omega-3 Fatty Acid Supplementation and Risk of Major Cardiovascular Disease Events A Systematic Review and Meta-analysis. *JAMA,* 308, 1024-1033.

Saravanan, P., Bridgewater, B., West, A.L., O'Neill, S.C., Calder, P.C., & Davidson, N.C., (2010). Omega- 3 fatty acid supplementation does not reduce risk of atrial fibrillation after coronary artery bypass surgery. *Circ. Arrhythm. Electrophysiol.*, 3, 46–53.

Sekikawa, A., Curb, J.D., Ueshima, H., El-Saed, A., Kadowaki, T., Abbott, R.D., Evans, R.W., Rodriguez, B.L., Okamura, T., Sutton-Tyrrell, K., Nakamura, Y., Masaki, K., Edmundowicz, D., Kashiwagi, A., Willcox, B.J., Takamiya, T., Mitsunami, K., Seto, T.B., Murata, K., White, R.L., & Kuller, L.H.; ERA JUMP (Electron-Beam Tomography, Risk Factor Assessment Among Japanese and U.S. Men in the Post-World War II Birth Cohort) Study Group., (2008). Marine-derived n-3 fatty acids and atherosclerosis in Japanese, Japanese-American, and white men: a cross-sectional study. *J. Am. Coll. Cardiol.,* 52, 417-24.

Sekikawa, A., Ueshima, H., Kadowaki, T., El-Saed, A., Okamura, T., Takamiya, T., Kashiwagi, A., Edmundowicz, D., Murata, K., Sutton-Tyrrell, K., Maegawa, H., Evans, R.W., Kita, Y., & Kuller, L.H., (2007). Less subclinical atherosclerosis in Japanese men in Japan than in White men in the United States in the post-World War II birth cohort. *Am. J. Epidemiol.,* 165, 617-24.

Serhan, C.N., (2010). Novel lipid mediators and resolution mechanisms in acute inflammation: to resolve or not? *Am. J. Pathol.,* 177, 1576-91.

Serhan, C.N., Clish, C.B., Brannon, J., Colgan, S.P., Chiang, N., & Gronert, K., (2000). Novel functional sets of lipid-derived mediators with antiinflammatory actions generated from omega-3 fatty acids via cyclooxygenase 2-nonsteroidal antiinflammatory drugs and transcellular processing. *J. Exp. Med.,* 192, 1197-204.

Serhan, C.N., Clish, C.B., Brannon, J., Colgan, S.P., Gronert, K., & Chiang, N., (2000). Anti-microinflammatory lipid signals generated from dietary N-3 fatty acids via cyclooxygenase-2 and transcellular processing: a novel mechanism for NSAID and N-3 PUFA therapeutic actions. *J. Physiol. Pharmacol.,* 51, 643-54.

Skuladottir, G.V., Heidarsdottir, R., Arnar, D.O., Torfason, B., Edvardsson, V., Gottskalksson, G., Palsson, R., & Indridason, O.S., (2011). Plasma n-3andn-6fattyacidsand the incidence of atrial fibrillation following coronary artery bypass graft surgery. *Eur. J. Clin. Invest.,* 41, 995–1003.

Smith, P.J., Blumenthal, J.A., Babyak, M.A, Georgiades, A., Sherwood, A., Sketch, M.H., & Watkins, L.L., (2009). Association between n-3 fatty acid consumption and ventricular ectopy after myocardial infarction. *Am. J. Clin. Nutr.,* 89, 1315–1320.

Stirban, A., Nandrean, S., Götting, C., Tamler, R., Pop, A., Negrean, M., Gawlowski, T., Stratmann, B., & Tschoepe, D., (2010). Effects of n-3 fatty acids on macro- and microvascular function in subjects with type 2 diabetes mellitus. *Am. J. Clin. Nutr.,* 91, 808-13.

Theobald, H.E., Goodall, A.H., Sattar, N., Talbot, D.C., Chowienczyk, P.J., & Sanders, T.A., (2007). Low-dose docosahexaenoic acid lowers diastolic blood pressure in middle-aged men and women. *J. Nutr.,* 137, 973-8.

Villareal, R.P., Hariharan, R., Liu, B.C., Kar, B., Lee, V.V., Elayda, M., Lopez, J.A., Rasekh, A., Wilson, J.M., & Massumi, A., (2004). Postoperative atrial fibrillation and mortality after coronary artery bypass surgery. *J. Am. Coll. Cardiol.,* 43, 742–748.

Wang, Q., Liang, X., Wang, L., Lu, X., Huang, J., Cao, J., Li, H., & Gu, D., (2012). Effect of omega-3 fatty acids supplementation on endothelial function: a meta-analysis of randomized controlled trials. *Atherosclerosis,* 221, 536-43.

Woodman, R.J., Mori, T.A., Burke, V., Puddey, I.B., Watts, G.F., & Beilin, L.J., (2002). Effects of purified eicosapentaenoic and docosahexaenoic acids on glycemic control, blood pressure, and serum lipids in type 2 diabetic patients with treated hypertension. *Am J. Clin. Nutr.,* 76, 1007-15.

Yamamoto, H., Yoshimura, H., Noma, M., Suzuki, S., Kai, H., Tajimi, T., Sugihara, M., & Kikuchi, Y., (1995). Improvement of coronary vasomotion with eicosapentaenoic acid does not inhibit acetylcholine-induced coronary vasospasm in patients with variant angina. *Circ. J.,* 59, 608-16.

Yokoyama, M., Origasa, H., Matsuzaki, M., Matsuzawa, Y., Saito, Y., Ishikawa, Y., Oikawa, S., Sasaki, J., Hishida, H., Itakura, H., Kita, T., Kitabatake, A., Nakaya, N., Sakata, T., Shimada, K., & Shirato, K., Japan EPA lipid intervention study (JELIS) Investigators, (2007). Effects of eicosapentaenoic acid on major coronary events in hypercholesterolaemic patients (JELIS): a randomised open-label, blinded endpoint analysis. *Lancet*, 369, 1090-8.

In: Omega-3 Fatty Acids
Editor: Md. Wasim Khan
ISBN: 978-1-62948-516-4

Chapter 4

Fish Oil in Colon Cancer Prevention: Can Biomarkers Tell us How Much is Enough?

***Yan Jiang*[1], *Dean E. Brenner*[1,2,3], *William L. Smith*[4] *and Zora Djuric*[5*]**

[1]Dept. of Internal Medicine,
[2]Dept of Pharmacology, University of Michigan Medical School, University of Michigan, US
[3]VA Medical Center, Ann Arbor, MI, US
[4]Dept. of Biological Chemistry, [5]Dept. of Family Medicine
University of Michigan Medical School, University of Michigan, Ann Arbor, MI, US

Abstract

Experimental studies, performed *in vivo* and *in vitro*, have provided strong indications for the anti-tumor actions of omega (ω) 3 fatty acids from fish oil against colon cancer. The molecular mechanisms of action have also been investigated extensively. However, the heterogeneity of the models utilized in these studies do not allow for prediction of the optimal dose of fish oil for healthy individuals. Many nutrients or supplements have non-linear dose-response curves with regard to beneficial effects. Whether or not higher doses of ω-3 fatty acids will increase the efficacy of fish oil in terms of reducing the risk of colon cancer is still an unanswered question. In this chapter, the available evidence for the beneficial effects of ω-3 fatty acids from fish oil supplements for colon cancer prevention is presented with a view towards identifying possible minimum and maximum fish oil intakes that would lead to a preventative effect. It is also suggested that biomarkers, such as the ω-3/ω-6 fatty acid ratio, may serve to personalize the efficacy of fish oil supplementation for reducing the risk of colon cancer.

Keywords: Fish oil; colon; cancer prevention; biomarker

* Corresponding Author: Zora Djuric, Ph.D. Email: zoralong@med.umich.edu.

Introduction

A number of pharmacological and dietary approaches have been explored for the prevention of colorectal cancer. One major pathway that has been targeted with preventative approaches is cyclooxygenase (COX)-mediated formation of prostaglandin (PG)E_2 from the ω-6 fatty acid (FA) arachidonic acid (AA). PGE_2, which is formed by constitutive cyclooxygenase (COX-1) and inducible cyclooxygenase (COX-2) in the colonic mucosa, plays an important role in colonic crypt cellular expansion with subsequent adenoma formation (Eberhart et al. 1994; Kargman et al. 1995; Sano et al. 1995; Sonoshita et al. 2002; Takeda et al. 2003). COX inhibitors conversely block proliferation, induce apoptosis and inhibit angiogenesis in the colon (Chan 2003; Kitamura et al. 2004). Indeed, PGE_2 level appears to be an appropriate candidate as a prevention endpoint (Sanderson et al. 2004).

Unfortunately, nonsteroidal anti-inflammatory drugs (NSAIDs) that reduce colorectal PGE_2 concentrations through COX inhibition have unacceptable toxicity profiles for daily consumption in healthy populations. This renders this class of interventions ineffective as cancer preventative interventions in healthy individuals due to an unacceptable therapeutic index (Bresalier et al. 2005; Nussmeier et al. 2005; Solomon et al. 2005). An alternative approach to the classical drug - enzyme targeting strategy to reduce or inhibit a downstream product lies in altering substrate availability.

Changes in dietary fats can alter the types of PGs and other eicosanoids that are produced in cells. Importantly, the products derived from ω-3 fatty acids (FAs) appear to be less pro-inflammatory than the products from ω-6 FAs (Bortuzzo et al. 1996; Rao et al. 2001; Smith 2005; Wada et al. 2007).

The ω-3 FA eicosapentanoic acid (EPA) from fish oil is an alternate substrate for COXs resulting in production of PGE_3 (Smith 2005). Unlike PGE_2, PGE_3 appears to have anti-inflammatory properties in the colon (Smith 2005; Vanamala et al. 2008). In addition, COX-2 expression, PGE_2 production and cytokine production in the colon are inhibited by ω-3 fatty acids (Gravaghi et al. 2011). Both EPA and docosahexaenoic acids (DHA) from fish oil, inhibited the growth of experimental tumors in tissue culture and in animals, including xenografts of human tumors in nude mice (Broughton and Wade 2002; Tsuzuki et al. 2004).

In epidemiological studies, a significant inverse relationship between ω-3 FA intake and reduced risk of colorectal neoplasia has been demonstrated in a meta-analysis of studies (Geelen et al. 2007). However, there is wide variation among fish oil dosages in these trials, and few trials have determined the minimally-effective dose. More data are therefore needed to examine dose/response profiles for fish oil in humans (Akabas and Deckelbaum 2006). Burns *et al.* determined that the maximally tolerated dose of ω-3 FAs from fish oil was 13.1 grams, which is a much higher dose than that anticipated from published studies of many human diseases (Burns et al. 1999). Whether a high dose of fish oil provides a maximal benefit for cancer prevention is unknown.

Metabolism of Omega-3 and Omega-6 Fatty Acids

Mammalian cells contain saturated fatty acids (SFA) (no double bonds), monounsaturated FA (MUFA) (one double bond), and polyunsaturated FA (PUFAs) (multiple double bonds). Mammals are able to add ω-9 double bonds through the action of Δ9-

desaturase, but are unable to introduce double bonds nearer the methyl (ω) end. Thus, ω-3 and ω-6 FA, are essential nutrients, and must be obtained from external sources. The shorter-chain linoleic acid (LA, ω-6, 18:2) and α-linolenic acid (ALA, ω-3, 18:3) are the precursors of AA (ω-6, 20:4) and EPA (ω-3, 20:5) (Marszalek and Lodish 2005).

LA, the major dietary source of ω-6 FA is found in vegetable oils such as corn, soybean, rapeseed, palm and sunflower oil (Willett 2007). LA is efficiently incorporated into membrane phospholipids and can be efficiently converted to AA. On the other hand, ALA is found in diets high in grains, nuts and certain fruits. The highest concentrations of EPA and DHA are found in cold water fish such as salmon, tuna, and herring. Although ALA can serve as the precursor for EPA and DHA synthesis in humans, the fractional conversion of ALA to EPA is low: 0.8% to 8% in men; up to 21% in women (Arterburn et al. 2006; Marszalek and Lodish 2005). Moreover, ALA is not efficiently incorporated into cellular membrane phospholipids, so that dietary intake of EPA, not ALA metabolic conversion, is the major contributor of EPA in cellular membrane phospholipids. Therefore, direct dietary consumption of omega-3 fats rich in EPA is of the most benefit clinically. The high amount of ω-6 FAs in the Western diet reduces EPA incorporation into membrane phospholipids, making the ω-6 to ω-3 ratio in the diet a key consideration.

Certain polyunsaturated FAs containing 20 and 22 carbons can be used as substrates for PG production. PGs are synthesized from ω-6 and ω-3 fatty acids in response to various hormones and physical stimuli (Smith and Murphy 2002). Both pathways involve three stages (Figure 1).

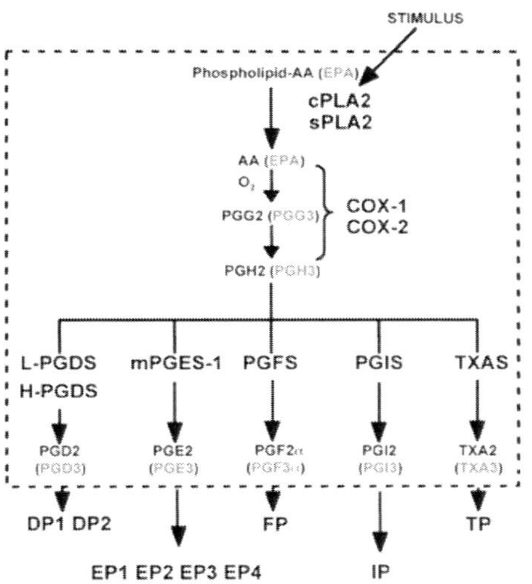

Figure 1. Effectiveness of AA-derived versus EPA-derived substrates and products with enzymes and receptors of the PG pathway. Adapted from (Wada et al. 2007). Abbreviations: TXAS, thromboxane A synthase; DP, EP, FP, IP, and TP are receptors for PGD, PGE, PGF, PGI, and TxA/PGH, respectively.

These are (a) mobilization of AA from membrane phospholipids by cytosolic phospholipase $A_2\alpha$ ($cPLA_2\alpha$) sometimes in conjunction with secretory $sPLA_2$s; (b) conversion of AA to the prostaglandin endoperoxide PGH_2 by COX-1 or 2; and (c) conversion of PGH_2 to a "2-series" product—PGD_2, PGE_2, $PGF_{2\alpha}$, PGI_2, or thromboxane A_2 (TxA_2) by specific synthases. PGs act at or near their sites of synthesis and are rapidly metabolized. Newly formed PGs exit cells and function primarily through G-protein coupled receptors on neighboring or parent cells to elicit responses. For modulation of cancer risk, most information available has implicated a role for PGE_2 (Backlund et al. 2005).

EPA can serve as an alternative substrate for PG formation generating "3-series" PG products including PGD_3, PGE_3, $PGF_{3\alpha}$, PGI_3 and TxA_3 (Figure 1). These products are generally less pro-inflammatory than the homologous 2-series (Bagga et al. 2003; Wada et al. 2007; Yang et al. 2004). While EPA- and AA-containing phospholipids are both substrates for $cPLA_2$, DHA-containing phospholipids are essentially inactive with $cPLA_2$ (Kramer and Sharp 1997; Shikano et al. 1994). Therefore, it is reasonable to focus on dietary EPA and AA in examining effects of ω-3 and ω-6 FAs on PG formation.

Typical North American and Western European diets contain high levels of ω-6 FA from plant oils. However, diets enriched in ω-3 FAs are associated with reduced risk for colorectal adenocarcinoma (Busstra et al. 2003; Kojima et al. 2005; Nkondjock et al. 2003; Rose and Connolly 1999; Slattery et al. 1997) as well as other epithelial malignancies (prostate, breast) (Bartsch et al. 1999). The ratio of ω-3/ω-6 FA in tissue phospholipids from Western populations averages less than 0.1 (Marszalek and Lodish 2005). Raising the dietary ratio of ω-3/ω-6 FA in humans or in laboratory animals causes corresponding increases in tissue phospholipid ω-3/ω-6 FA ratios (Anti et al. 1994; Anti et al. 1992; Deschner et al. 1990; Huang et al. 1996). Thus, dietary changes can potentially modulate colonic PGE_3/PGE_2 ratios.

Preventative Effects of Omega-3 Fatty Acids in Colon Cancer

Most data available for evaluation of ω-3 and ω-6 FA in colon cancer are from animal studies. The common rodent model used in these studies utilizes carcinogens to induce colonic tumors, namely either azoxymethane (AOM) or 1,2-dimethylhydrazine (DMH). This has been shown to cause development of colorectal aberrant crypt foci (ACF) after 6-8 weeks, and colonic tumors 20 weeks after injection of the carcinogen (De Robertis et al. 2011). Use of dextran sodium sulfate (DSS) in the drinking water augments the process of carcinogenesis induced by AOM via an inflammatory stress in the colon that results in tumor promotion (De Robertis et al. 2011). On the other hand, the effect of dietary ω-3 and ω-6 PUFAs on adenoma formation has also been investigated in genetic mouse models, including the $Apc^{Min/+}$ mouse model of familial adenomatous polyposis (FAP) and the $Apc^{\Delta 716}$ truncation model. These genetic models are characterized by impairments of the adenomatous polyposis coli (APC) protein, a product of the tumor suppressor gene *APC* (Moser et al. 1990; Oshima et al. 1995a).

Tumor incidence and multiplicity in chemically-induced colon cancer in rodents. Data from many laboratories have demonstrated that dietary ω-6 FAs (Bull et al. 1979) and SFA have colon tumor-promoting effects whereas ω-3 FAs lack promoting effects (Reddy and Maeura 1984; Reddy and Sugie 1988; Wargovich and Felkner 1982). In these studies, menhaden oil has been a common source of fish oil ω-3 FAs ranging from 4% to 20% by weight in the diet. It caused consistent reductions in both tumor incidence and multiplicity. Within a low fat intake background, studies showed that 4-5% fish oil reduced colonic tumor incidence and tumor yields significantly compared to rats on the control diet (5% fat from corn oil or coconut oil) (Minoura et al. 1988; Monk et al. 2012; Reddy and Sugie 1988). On the other hand, when animals were fed a high total fat diet supplemented with fish oil (10-12%), the results are more variable. The data seem to show a consistent reduction of both tumor incidence and multiplicity by fish oil but the percentage decrease depends on the type of fat given to the control animal (Chang et al. 1998; Deschner et al. 1990; Reddy and Sugie 1988). Two studies have used as high as 20% fish oil feeding and showed about a 40% reduction in tumor incidence and a 70% reduction in tumor multiplicity (Hendrickse et al. 1995; Singh et al. 1998). Therefore, in rodents, in terms of tumor outcome, a diet containing 4-20% fish oil has shown a preventive effect on colon cancer.

Numbers of polyps in APC mice. FAP patients carry mutations in the *APC* gene. The first genetic mouse model containing a mutation in the *APC* gene was designated *Min* for multiple intestinal neoplasia (Moser et al. 1990). Similar to $Apc^{Min/+}$ mice, mice carrying an allele of Apc ($Apc^{\Delta 716}$) that causes APC protein truncation were described by Oshima *et al.* (Oshima et al. 1995a). However, one needs to be aware of the limitations of these mouse models because the majority tumors appear to be in the small intestine rather than the colon (Heyer et al. 1999).

In genetic colon cancer rodent models such as Min mice (Rose and Connolly 1999) and the Oshima $Apc^{\Delta 716}$ truncation model (Oshima et al. 1995b), a higher dietary intake of fish oil showed potential colon cancer preventative effects. A high fat diet (10% or 15% fat by weight) itself induces polyps in Min mice compared to low-fat fed mice (Wasan et al. 1997). When fed with a high fat diet containing 2.3% ω-3 PUFAs, Min mice had reduced numbers of tumors and reduced tumor incidence in the colon compared to controls fed with corn oil (Paulsen et al. 1997). However, neither of the reductions were significant due to the small sample size (n=6 or less) and small numbers of colonic tumor developed. Similar problem was seen in several other studies. For example, Petrick et al. observed no statistical difference in average numbers of tumors per mouse between Min mice feed with a control (12% fat by weight) versus an EPA diet (3% EPA+9% control oil) after 7 weeks (Petrik et al. 2000b); moreover, both an oleic acid-enriched diet (10% fat by weight) and an EPA-enriched diet (10% fat by weight containing 1.5 % EPA) resulted in similar numbers of colonic tumors (Petrik et al. 2000a).

A recent report by Fini and colleagues demonstrated that a diet containing 2.5% free EPA plus 4.5% corn oil, reduced polyp number significantly compared to the controls fed 7% corn oil in the Min mice (Fini et al. 2010). Interestingly, they found that 5% free EPA feeding resulted in slightly more tumors, which might suggest that an overdose of EPA could cause an adverse effect in these mice.

On the other hand, in $Apc^{\Delta 716}$ heterozygous mice fed a purified AIN-76A diet with or without 3% DHA for 7 weeks showed that DHA-treated females had a 69% reduction in polyp numbers compared with the control animals, whereas DHA-fed males showed no significant decrease in total polyp number (Oshima et al. 1995b). Unfortunately, they did not report on colonic polyps.

Table 1. Tumor incidence and multiplicity in chemically induced colon cancer rodent models

Model	ω-3 FA dose	Control	Tumor incidence	Tumor multiplicity	Reference
Sprague-Dawley rat	17% MO	17% CO	- negative	-36% reduction	(Nelson et al. 1988)
Donryu rat	4.7% EPA+0.3% LA	5% LA	-52% reduction	-75% reduction	(Minoura et al. 1988)
F344 rat	4% MO + 1% CO 5.9% MO+17.6% CO 11.8% MO+11.8%CO 17.6%MO+5.9% CO	5% CO and 23.5% CO	-4% MO 16% reduction compared to 5% CO; 37% reduction compared to 23.5% CO -5.9% MO 25% reduction -11.8% MO 32% reduction -17.6%MO 32% reduction compared to 23.5% CO	-4% MO 26% reduction compared to 5% CO; 59% reduction compared to 23.5% CO -5.9% MO 32% reduction -11.8% MO 33% reduction -17.6% MO 45% reduction compared to 23.5% CO	(Reddy and Sugie 1988)
CF1 mouse Female	4.4% MaxEPA+16% CO 10.2% MaxEPA+10.2% CO 16%MaxEPA+4.4%CO	4% CO and 20.4% CO	-4.4% MaxEPA negative -10.2% MaxEPA 25% reduction compared to 4.4% CO; 52% reduction compared to 20.4% CO -16% MaxEPA 26% reduction compared to 20.4% CO	-4.4% MaxEPA negative -10.2% MaxEPA 58% reduction compared to 4% CO; 61% reduction compared to 20.4% CO -16% MaxEPA negative	(Deschner et al. 1990)
F344 rat	18.5% MO+5%CO	5% CO and 23.5% CO	-51%, 76% reduction compared to 5% CO, 23.5% CO	-71% reduction compared to 23.5% CO	(Reddy et al. 1991)
Wistar rat anastomosis or sham	20% MaxEPA	20% coconut oil	-40% reduction	-70% reduction	(Hendrickse et al. 1995)
R344 rat	1 ml DHA	1mL water 5 times a week	-negative	-35% reduction	(Takahashi et al. 1997)
Sprague-Dawley rat	11.5% MO+3.5% CO	15% CO	-20% reduction	-Not reported	(Chang et al. 1998)
F344 rat	18% MO+5% CO	5% CO and 23% CO	-28% reduction compared to 23% CO	-Negative	(Good et al. 1998)
F344 rat	20.5% MO+3% CO	5% CO and 23.5% CO	-30% , 47% reduction compared to 5% CO, 23.5% CO	-38%, 67%, reduction compared to 5% CO, 23.5% CO	(Singh et al. 1998)
F344 rat	17% MO+3% CO	5% CO and 20% mixed lipids	-30% reduction compared to high fat group	-17% reduction compared to 5% CO; 60% reduction compared to high fat group	(Rao et al. 2001)
Wistar rat	18% FO (24% EPA+20% DHA)	18% soybean oil	-67% reduction	-Not reported	(Moreira et al. 2009)
Wistar rat	15% canola oil rich in ω-3 PUFA	15% corn oil	-20% reduction	-58% reduction	(Bhatia et al. 2011)
C57BL/6 mice mixed with $PPAR\delta^{\Delta IEpC}$	4% fish oil + 1% corn oil	5% corn oil	-30% reduction	-59% reduction	(Monk et al. 2012)

FO: fish oil; MO: menhaden oil (2.4% EPA+11% DHA); CO: corn oil; MaxEPA:18% EPA+12% DHA.

Available data suggest protective effects of ω-3 FAs on the numbers of polyps in these mice having a genetic APC alteration; however, factors such as fat content in the diet, sample sizes and the type of ω-3 FAs could lead to variations in the outcomes. It is hard to determine the minimum dose of ω-3 FAs for colon cancer prevention in these rodents based on the tumor endpoints.

Aberrant crypt foci (ACF) in rodents. In rodents, ACFs are one of the earliest changes seen in the colon that may lead to formation of colorectal polyps. The relevance of ACF as a surrogate endpoint biomarker for carcinogenesis is still controversial because the clinical importance of ACFs in humans is unknown (Lopez-Ceron and Pellise 2012). In animals, ACF formation has been extensively studied in chemically induced colon cancer animal models. Takahashi published two studies showing that the numbers of ACF per colon were statistically significantly reduced by intragastric gavage of 1 mL of DHA in rats injected with DMH or AOM as early as four weeks post injection (Takahashi et al. 1997; Takahashi et al. 1993). Latham and colleagues reported that ACF in the distal colon was decreased by 47% in DMH induced rats fed fish oil (19% EPA plus 8% DHA) compared to those fed corn oil (Latham et al. 1999). A similar study performed in AOM treated rats demonstrated that a high fish oil diet (17% menhaden oil plus 3% corn oil) decreased total colonic ACF by 50% compared to a high fat corn oil diet (20%) after eight weeks (Rao et al. 2001). Lesions with four or more aberrant crypts were also reduced by the high fish oil diet. Additionally, Moreira *et al.* demonstrated that a diet containing 18% fish oil induced about 50% smaller ACFs in the colon (Moreira et al. 2009). The narrow range of doses (17-19% menhaden oil) makes it difficult to compare the efficacy of different dosages of fish oil but do suggest a reduction in ACF by fish oil in chemically induced colon cancer models. However, it is interesting to recognize that one study suggested an increase of ACFs by 18% menhaden oil, yet the overall tumor numbers was significantly lower in these rats fed with high fat menhaden oil compared to the ones that were fed with high fat corn oil (Good et al. 1998). Therefore, the high numbers of ACFs in the fish oil fed animals could result from fish oil slowing development of the carcinogenic process and prevented the ACFs from progressing.

The colon of Min mice (Jacoby et al. 1996) or $Apc^{\Delta 716}$ mice (Oshima et al. 1997) do not form spontaneous 'classical' ACFs. Paulsen *et al.* discovered that $APC^{min/+}$ mice have a different ACF phenotype, called ACF_{min}, which were not elevated above the surrounding mucosa (Paulsen et al. 2000). The detection of ACF_{min} varies depending on methylene blue staining and transillumination. Several years later, the same group reported that dietary fish oil significantly reduced incidence and multiplicity of ACF_{min} in female but not male Min mice (Paulsen et al. 1997). The relevance of ACF_{min} is questionable since it might have completely different genetic or molecular mechanisms of development than the common type of ACF.

Polyp numbers in colon cancer patients. Although animal studies have strongly suggested a chemoprotective effect of dietary fish oil for colon cancer, examining the potential of any dietary agent on risk reduction in humans requires a large number of subjects and costly, long-term follow-up. To minimize the time needed to a cancer or adenoma endpoint, this type of clinical trial is frequently designed using persons at high risk of colorectal cancer, which could be difficult to translate to low risk individuals. In addition, inter-individual differences in metabolism might require individualized dosing. Identification of biomarkers would be useful to allow predication of the optimum daily intake of fish oil for prevention. Studies of potential chemopreventative agents in patients with FAP allow polyp size and number to be used as end points over a relatively short time (6 to 12 months). A recent phase III randomized trial investigated treatment with 2 g of EPA free fatty acid daily for 6 months in FAP patients (West et al. 2010). Rectal polyp multiplicity and size were measured, and a 22% reduction in polyp number in the EPA group was found compared with placebo. In an earlier study, five FAP patients were given a higher dose (2.2 g DHA+0.6 g EPA daily) for 1 to 2 years, however, no significant change in polyp number was seen (Akedo et al. 1998). Murff *et al.* performed a case-control study that included polyp-free control subjects, adenomatous polyp patients, and hyperplastic polyp cases (Murff et al. 2012). They found that there were no associations between the intake of marine-derived ω-3 PUFAs and adenomatous or hyperplastic polyp risk in men. Compared with women who reported the lowest intake of marine-derived ω-3 PUFAs (0-0.023 g/day), women who reported the highest intake of marine derived ω-3 PUFAs (0.14-1.01 g/day) had 33% lower risk for adenomatous polyp risk. There was a significant correlation between increasing marine-derived ω-3 PUFA intake and decreased adenomatous polyp risk in women. These published reports, although the data is limited; do indicate that up to 2 g of fish oil could be associated with a lower risk of polyp development. However, it is unclear whether this is limited to FAP patients or is applicable to a general population.

Cell proliferation and apoptosis. Some animal data suggest that measurements of differentiation and apoptosis may have greater prognostic value than measurements of cell proliferation in detecting dietary effects on tumor incidence (Chang et al. 1997; Chang et al. 1998). However, several studies have demonstrated that ω-3 PUFAs reduce human colorectal proliferation (Table 2). Countey et al. investigated mucosal cell proliferation and apoptosis by scoring and analyzing Ki67 (proliferation) or M30 (apoptosis) labeling in the colonic mucosa upon an intake of 2 g of EPA in patients who had colorectal adenomas (Courtney et al. 2007). They found that labeling index curves of Ki67 cells showed a significant downwards shift (20% reduction in total crypt Ki67 label index) after EPA treatment compared to no treatment groups. Both the apoptosis grade and the number of apoptotic cells increased significantly (4-fold, 75-fold, respectively) after EPA supplementation. After administration for one month, 7.7 g of EPA+DHA per day demonstrated a slightly larger reduction in the mean labeled index (Ki67) for high crypt region compared to dosages of 2.5 g and 5.1 g per day; however, the 2.5 g/d dose had highest total label index in rectal epithelial cells (Bartoli et al. 1993). Larger doses of fish oil--more than 7 g per day--reduced the Ki67 labeling index in the apex of the colonic crypts (Anti et al. 1992). In a short term dose response study, the authors did not find any reduction of proliferation with fish oil consumption except in the highest dose group (7.7 g per day) (Anti et al. 1994). This decrease in proliferative indices was maintained with extended fish oil intake up to 6 month. Interestingly, one group reported no evidence of any effect of supplementation with 1.4 g of EPA for 12 days on mitotic cells within the crypt, despite clear evidence of incorporation of EPA into epithelial cells (Gee et al. 1999). One important difference in this particular study was that whereas other groups measured incorporation of label into the DNA of crypt cells incubated *ex vivo*, they measured the frequency and position of mitotic profiles in the mucosa immediately after recovery of the biopsy. The authors speculated that the distribution of DNA synthesis may not necessarily reflect the true distribution of cells which enter mitosis in the native colorectal mucosa. Most of data available to date show a trend of decreasing colonic cell proliferation upon dietary fish oil intake. However, to determine if higher doses provide any extra benefit, will require analysis of markers of both proliferation and apoptosis.

Table 2.1. Fish Oil Administration and ω-3/ω-6 Response in Human Blood Phospholipids

EPA+DHA g/day	Time (months)	Pre Intervention						Post Intervention						Reduction of Proliferation	References
		ω-3 (%)	ω-6 (%)	ω-3/ω-6 Ratio	EPA (%)	AA (%)	EPA/AA Ratio	ω-3 (%)	ω-6 (%)	ω-3/ω-6 Ratio	EPA (%)	AA (%)	EPA/AA Ratio		
7.7	1	3.2	29.4	0.11	1.0	7.1	0.14	11.4	25.7	0.44	6.7	6.5	0.97	Yes	(Bartoli et al. 1993)
5.1	1	3.2	28.6	0.11	0.8	6.7	0.12	7.6	27	0.28	3.9	6.6	0.59	ND	(Bartoli et al. 1993)
3	1	ND	ND	ND	ND	ND	0.025	ND	ND	ND	ND	ND	0.20	ND	(Burns et al. 2007)
2.5	1	3.1	26.8	0.12	1.0	6.0	0.17	8	25	0.32	3.9	6.1	0.64	Yes	(Bartoli et al. 1993)
2.28	3	5.6	32.6	0.17	1.1	6.6	0.17	11.8	26.9	0.44	5	6	0.83	ND	(Vidgren et al. 1997)
1.5	1	ND	ND	ND	ND	ND	0.025	ND	ND	ND	ND	ND	0.11	ND	(Burns et al. 2007)
1.2	2	4.26	35.3	0.12	0.73	9.54	0.076	8.86	29.61	0.3	2.38	8.06	0.3	ND	(Young et al. 2011)

Table 2.2. Fish Oil Administration and ω-3/ω-6 Response in Human Colorectal Mucosa

EPA+DHA g/day	Time (months)	Pre Intervention						Post Intervention						Reduction of Proliferation	References
		ω-3 (%)	ω-6 (%)	ω-3/ω-6 Ratio	EPA (%)	AA (%)	EPA/AA Ratio	ω-3 (%)	ω-6 (%)	ω-3/ω-6 Ratio	EPA (%)	AA (%)	EPA/AA Ratio		
7.7	3	3.3	18.3	0.18	1.3	5.7	0.23	5.7	13.8	0.41	3.3	4.4	0.75	Yes	(Anti et al. 1992)
7.7	1	3.1	16.1	0.19	1.1	5.8	0.19	7.2	13.6	0.53	2.6	4.2	0.62	Yes	(Bartoli et al. 1993)
7.2	6	7	29	0.24	2	11	0.18	12	23	0.52	6	7	0.88	Yes	(Huang et al. 1996)
5.4	3	4.3	21.9	0.20	0.44	10.0	0.044	7.2	17.5	0.41	3.1	5.6	0.55	ND	(Hillier et al. 1991)
5.1	1	2.26	18.7	0.12	0.85	5.69	0.15	4.54	15.24	0.3	1.93	3.64	0.53	ND	(Anti et al. 1994)
5.1	1	2.3	18.4	0.125	0.8	5.5	0.15	4.4	16	0.28	1.7	4.2	0.24	ND	(Bartoli et al. 1993)
2.5	6	2.7	18.4	0.15	0.6	6.3	0.10	4.3	13.8	0.31	1.5	3.6	0.42	Yes	(Anti et al. 1994)
2.5	1	2	17.7	0.11	0.7	6.1	0.11	2.9	16.1	0.18	1.5	5.5	0.27	Yes	(Bartoli et al. 1993)
2	3	4.9	24.7	0.20	1.15	9.4	0.12	6.83	22.01	0.31	2.65	7.99	0.33	Yes	(Courtney et al. 2007)

ND= Not done; Hillier et al (Hillier et al. 1991) studied patients with inflammatory bowel disease.

Dose-Limiting Adverse Effects in Humans

The ω-3 FA formulations appear to be well tolerated by humans, although at higher doses, some subjects complained of a 'fishy' smell. In one prospective study where the effect of ω-3 PUFA in cardiovascular disease was studied, 2836 patients were given 0.85 g ω-3 PUFAs for 3.5 years and only a 3.8% dropout rate was observed with the common side effects being gastrointestinal disturbances (4.9%) and nausea (1.4%) (GISSI 1999). A recent EPA phase III trial using a dose as high as 2 g reported only 3.4% dropout from the study (West et al. 2010). Other typical gastrointestinal symptoms were mild diarrhea, indigestion, belching and a 'fishy' taste (Anti et al. 1994; Bartram et al. 1993; Bays 2006; Burns et al. 2007; Courtney et al. 2007; Kris-Etherton et al. 2002). With a high dose of fish oil (over 7 g per day), some studies reported that no participants discontinued taking fish oil prior to termination of the studies (Anti et al. 1992; Bartoli et al. 1993; Huang et al. 1996), while others observed a 13% dropout (Anti et al. 1994).

Omega-3 PUFAs cause a reduction in thromboxane A_2 formation by platelets, and thromboxane A_2 is involved in platelet activation and aggregation (Dyerberg et al. 1978). However, no signs of abnormal bleeding, hyperglycemia, or abnormalities in kidney function were reported in several trials when patients were prescribed 4 capsules/d of fish oil containing a total of 1.86 g of EPA and 1.5 g of DHA (Bays 2006). Some studies with ω-3 fatty acid treatment in cardiovascular disease demonstrated prolongation of bleeding time, but it still did not exceed the normal limits of bleeding time and did not produce clinically significant bleeding episodes. One other study reported no bleeding complications when fish oil capsules were taken with aspirin (Eritsland et al. 1996). Further, no GI bleeding with ω-3 FAs was reported in any of the colorectal cancer studies.

Biomarkers for Fish Oil Intervention in Colorectal Cancer Prevention

FA profiles in rodents upon fish oil feeding. One potential biomarker for dietary fish oil efficacy may be FAs. The levels of lipids in membranes reflect the net outcome of dietary intake, absorption, transport and metabolism of fats. *In vitro* data indicate that EPA can compete with AA as a substrate for COXs (Wada et al. 2007). Takahashi *et al.* measured blood lipids from DMH induced rats given DHA by gastric intubation and found that plasma EPA/AA ratio increased from 0.08 (control group) to 0.25 (DHA group) after 4 weeks; this value further increased to 0.42 after 12 weeks (Takahashi et al. 1993). A similar finding was reported by the same research group a few years later in AOM treated rats, where the plasma EPA/AA ratio was 0.23 upon DHA gastric intubation for four weeks compared to 0.003 in the control rats; the EPA/AA ratio increased to 0.54 after 36 weeks. In these studies, the authors used low fat diets and a pure EPA diet or DHA intubation that would not be directly relevant for comparison to human dietary fish oil intake.

In rats, dietary fish oil induced higher amounts of EPA in both phospholipid and neutral lipids in both colonic tumors and normal flat mucosa. This indicates that changes in dietary intake of ω-3 and ω-6 FAs can modify tissue fatty acid pools that would cause a shift of fatty acid metabolites between 2-series and 3-series PG products (Minoura et al. 1988). The same study reported that the ratio of EPA/AA in the normal mucosa was 1.7 upon feeding the rats with a 4.7% EPA diet compared to a control LA (5%) diet (Minoura et al. 1988). Reddy *et al.* demonstrated that a low fat corn oil (5%) diet resulted in a small EPA/AA ratio (0.012) in the colonic mucosa, while this ratio increased to 5.41 when using 4% menhaden oil (Reddy and Sugie 1988). When switched to a high fat background (23.5% fat in diet), the ratio of EPA/AA in the control did not change (0.015). However, the highest percentage of dietary menhaden oil (17.6%) resulted in an EPA/AA ratio of only 2.09 (Reddy and Sugie 1988). Another study also showed that a high ω-3 diet (20%) induced an increase in EPA/AA ratio of 8.7-fold and 6.5-fold in the colonic mucosa and tumors, respectively, versus a control coconut oil diet (Hendrickse et al. 1995). A recent study looking at the lipid profiles of colons from rats treated with DMH showed that 18% fish oil in the diet lead to an EPA/AA ratio of 1.58 in the colon (Moreira et al. 2009). Additionally, there is one other study that has reported a dose response of dietary fish oil and colonic EPA/AA ratios (Reddy and Sugie 1988). The data available from these carcinogen treated rodents suggest that a high ω-3 FA diet increases both blood and colonic tissue leve EPA/AA ratios regardless of the low fat (5% by weight) or high fat (> 20% by weight) background.

Several studies reported the fatty acid composition upon dietary fish oil feeding in $APC^{min/+}$ mice. In testing the phospholipids from small intestines of mice fed control (12% fat wt/wt in diet) vs. EPA-enriched (3% EPA+9% control oil in diet) diets, the EPA/AA ratio of 0.015 in the control was increased to 1.7 by dietary EPA (Petrik et al. 2000b). Similarly, the mean EPA/AA ratio in the small intestine was 0.017 in a group of oleic acid fed mice, while it was 1.14 in EPA fed mice (Petrik et al. 2000a). Further, Finin *et al.* evaluated the fatty acid composition of nonpolypoid small intestinal tissue from Min mice fed either a control diet (7% corn oil) or EPA diet (2.5% free EPA or 5% free EPA) (Fini et al. 2010). They found a higher ratio of EPA/AA in the small intestine: 1.95 and 1.25 in mice fed 2.5% and 5% EPA diets, respectively, versus 0.14 in the control mice. Interestingly, 5% EPA caused a smaller increase of EPA/AA ratio in the small intestine than 2.5% EPA, which was consistent with the fact that 5% EPA diet was less effective in decreasing polyp numbers.

The preponderance of the evidence from rodents has shown that dietary serum and tissue EPA/AA ratios can potentially be a biomarker for investigating the preventive effects of ω-3 FAs on development of colon cancer. Determination of the optimum effective dose of fish oil using this marker, however, is difficult, and is affected by individual metabolic factors.

Effects of fish oil supplementation on fatty acid compositions in humans. Intakes of ω-6 FAs are very high in the American diet with an average ω-6/ω-3 ratio of about 11. This is notable because humans likely evolved eating a diet with a ω-6/ω-3 ratio of close to one, and a ratio of 2.3 has been recommended (Cordain et al. 2005; Kris-Etherton et al. 2000; Simopoulos 1999; Tappel 2007). It is important to remember that this is a ratio. Lower intakes of fish oil fatty acids are required to achieve the same ω-6/ω-3 dietary ratios when the ω-6 intake is relatively low. Dietary ratios of ω-6/ω-3 (as percent of energy intake) are a driving force in equations that predict the relative amounts of ω-3 and ω-6 20-carbon highly unsaturated fatty acids (HUFAs) in membranes (Lands 2007). For example, a fish oil dose of 3.5 g/day has been recommended for increasing ω-3 HUFA in membranes, but 0.35 g/day may be sufficient if ω-6 intake is limited (Hibbeln et al. 2006). When consumption of ω-6 FA was low, tissue levels of EPA were much higher after ω-3 FA supplementation versus tissue EPA levels accompanied by a high ω-6 FA diet (James et al. 2000).

High ω-6/ω-3 ratios retard ω-3 incorporation into membranes, even though ω-6 FAs are preferred substrates for elongation and desaturation relative to ω-3 FAs (Lands et al. 1992; Pan et al. 1994). In humans, conversion of linolenic acid (18:3, ω-3) to longer chain ω-3 FAs occurs with an efficiency of about 5-10%, and this is inhibited with high ω-6 FA intakes (Raatz et al. 2001; Rose and Connolly 1999). It is interesting to recognize that a low-fat diet alone can increase ω-3 and decrease ω-6 FAs in phospholipids somewhat (Raatz et al. 2001).

Many studies have shown that supplementation with 1-4 g/day of fish oil can increase blood ratios of EPA/AA ranging from 0.025~0.17 at the baseline to 0.11-1.0 after fish oil consumption in humans. These studies are presented in Table 2. Serum fatty acids have quicker responses which can be identified within one month of treatment, but slower incorporation into colon cell membrane. Supplementation of a purified fish oil preparation enhanced tissue colorectal mucosal EPA and ω-3 concentrations, and increased the EPA/AA and ω-3/ω-6 PUFA ratio in multiple clinical trials (Table 2) (Anti et al. 1994; Anti et al. 1992; Huang et al. 1996). The maximum ω-3/ω-6 ratio achieved in any of these trials was 0.5, which was well below ratios of 1.0 used for anti-carcinogenic effects in rodent models (Deschner et al. 1990).

Huang and colleagues performed a 6 month study using 7.2 g of fish oil per day, one of the highest doses used, and demonstrated the highest EPA/AA ratio and ω-3/ω-6 FA ratio achieved in the colonic mucosa among all the human trials to date. Hence, there seems to be a high dose response with fish oil intake and tissue fatty acid profiles. Though Bartram *et al.* did not report FA levels before the intervention in their trial, they showed that fish oil intake caused a slight reduction of ω-3/ω-6 FA ratio in the colonic mucosa compared to that in the control subjects after 4 weeks of fish oil treatment (Bartram et al. 1993).

The levels of AA in blood are fairly tightly regulated in humans and alteration of dietary ω-6 FAs, with the exception of deficiency, does not modulate levels of AA in phospholipids greatly (Lands 1995). Consistent with this, most studies showed fish oil supplementation caused greater EPA/AA ratios due to the increase in EPA and small or no reduction in AA. King *et al.* examined changes in blood FAs with a low-fat diet. There were decreases in both 18:2 ω-6 and trans 18:1 while there was an increase in 18:1 ω-7 and 16:0. Interestingly, AA levels increased slightly (King et al. 2006). It is encouraging that colon fatty acid ratios can be modulated by fish oil supplementation. The relationships between serum fatty acid ratios and colonic fatty acid ratios have not been established, however, and it would be important to determine if serum levels can be used to predict colon levels.

Colonic eicosanoids in rodents with fish oil feeding. Dietary ω-3 FAs cause a shift in eicosanoid metabolism from 2-series PG products derived from AA to 3-series PGs from EPA. Levels of PGE_2 can be assayed relatively easily by ELISA or LC/MS-MS. The literature indicates that colonic PGE_2 can be consistently reduced by dietary intake of fish oil. In a feeding experiment by Minoura *et al.*, PGE_2 was measured and showed, as expected, five-fold higher levels in the tumors from rats fed with LA than tumors from EPA fed rats when using a diet low in total fat. Colonic mucosa PGE_2 in the normal colon tissue was decreased about 30% by dietary EPA (Minoura et al. 1988). Feeding C57/Bl6 mice with a ω-3/ω-6 ratio of 1:1 in the context of 5-10% fat by weight caused a 50% to 60% reduction in PGE_2 concentrations in stimulated peritoneal cells compared to mice fed a ω-3/ω-6 ratio of 0:1 with the same, low fat diet. There was a reduced impact of increased ω-3 FA intake on PGE_2 concentrations in stimulated peritoneal cells from mice fed high fat (15-20% by weight) compared to mice fed the low fat diets (Broughton and Wade 2002). One piece of information lacking in these studies was PGE_3 levels. Production of PGE_3 was reported in a recently study by Vanamala *et al.* but the apparent level of PGE_3 was surprisingly high. In rats treated with AOM, a 78% reduction in PGE_2 and production of PGE_3 was reached by substituting dietary LA with EPA and DHA in a background of a 15% fat by weight in diet (Vanamala et al. 2008). More data are necessary to answer the question of dose-dependent changes in PGE_2 and PGE_3 and to clarify the issue of the optimal dietary dose of ω-3 for controlling tissue levels of PGE_2 and preventing colon cancer.

In Min mice, tissue production of PGE_2 was reduced by dietary ω-3 FAs. Petrick *et al.* observed a 56% decrease in PGE_2 in normal-appearing small intestines upon consumption of diets supplemented with 3% EPA. Another AA-derived tissue metabolite, 6-keto-$PGF_{1\alpha}$, was also reduced by 54% by EPA in the diet (Paulsen et al. 1997). The same research group reported that production of PGE_2 and 6-keto-$PGF_{1\alpha}$ were correlated with the arachidonic acid content in the tissue phospholipids in normal small intestine (Petrik et al. 2000a). This information supports the concept that the tissue EPA/AA ratios change upon increasing dietary ω-3 FA, which, in turn, relates to differences in PG production. Additionally, in small intestinal tumor tissue, a 90% reduction in PGE_2 levels was achieved when Min mice were fed a high fat fish oil diet (12% menhaden oil and 8% other fat) compared to a control diet with 20% fat (Bose et al. 2007). A bigger reduction of PGE_2 was observed in the tumors compared to normal intestine tissue, which could indicate that dietary ω-3 FAs may target PGE_2 in tumor tissue. Further studies are needed to investigate the response of PGE_2 in different stages of carcinogenesis.

Eicosanoid levels in humans. Only a few human studies have reported blood eicosanoid production after dietary ω-3 FA intervention. The data were limited mainly to PGE_2 in *ex vivo* lipopolysaccharide stimulated peripheral blood mononuclear cells (Gallai et al. 1995; Swails et al. 1997) and whole blood (Faber et al. 2013; Warstedt et al. 2009). The available data suggested a trend toward reduction of PGE_2 production by fish oil.

There are few data on the effects of fish oils on colonic tissue concentrations of eicosanoids with varying ω-6 intakes, and the literature is conflicting with regard to whether or not ω-6 intake needs to be reduced to maximize the effect of fish oils on disease prevention (Harris et al. 2006; Hwang et al. 1997; Moore et al. 2006). A low AA diet (<90 mg/day, which can be achieved with a low-fat diet) resulted in decreased PGE_2 in the blood of patients with rheumatoid arthritis (Adam et al. 2003). PGE_2 levels in mononuclear cells of patients with fatty liver disease were directly responsive to dietary linoleic (18:2 ω-6) and eicosatrienoic acid (20:3 ω-6) levels, which are precursors of AA (Okita et al. 2001).

In rats, an increased ω-3/ω-6 dietary ratio was more important than the total amount of ω-3 FA intake in altering eicosanoid production (Boudreau et al. 1991) although this is still controversial in humans (Hwang et al. 1997). Bartram *et al.* found that 4.4 g/day ω-3 fatty acid supplementation decreased PGE_2 levels in rectal mucosa when subjects were consuming a low-fat diet, but there was no effect of fish oil on PGE_2 levels in mucosa when the background diet was high in fat (Bartram et al. 1995; Bartram et al. 1993). However, another study demonstrated that fish oil (MaxEPA) with 5.4g EPA plus DHA was able to decrease PGE_2, thromboxane B_2 and 6-keto-$PGF_{1\alpha}$ within three weeks in patients with inflammatory bowel disease (Hillier et al. 1991). ω-3 fatty acids reduced PGE_2 release and ornithine decarboxylase activity, and increased PGE_3 formation in human colorectal mucosa (Bartram et al. 1993; Ferretti et al. 1988; Fischer et al. 1988; Kobayashi et al. 2006; La Guardia et al. 2005; Rees et al. 2006). EPA incorporation into cell membranes was linear in response to dietary EPA dose (1.35, 2.7 or 4.5 g/day for 12 weeks), with concomitant and significant decreases in PGE_2 (but not cytokines) using the two higher doses (Rees et al. 2006). Dosing with 0.3, 1 or 2 g/d EPA plus DHA decreased PGE_2 in stimulated peripheral blood mononuclear cells with the biggest decrease being observed at 1 g/d (96% decrease) and little further decrease by going to 2 g/d (Trebble et al. 2003). Others have shown that increased ω-3 intakes decrease eicosanoids and triglycerides without detrimental effects on insulin sensitivity or immune function (Griffin et al. 2006; Rees et al. 2006). There is appear to be strong mechanistic and epidemiological data that support a maximum of 2 g of dietary fish oil (EPA+DHA) and colorectal PGE_2 reduction; however, a higher dose might lead to further impact on eicosanoids depending on the eating habits and health status of individuals.

Human urinary PG metabolites. In addition to measurements of FAs or PGs in colon tissue, it is also possible to quantify the major PGE_2 metabolite termed PGE-M (Murphey et al. 2004) and the major urinary metabolites of AA-derived 6-keto-$PGF_{1\alpha}$ (Ward et al. 1991). Two related 2,3-dinor metabolites of EPA-derived 6-keto-$PGF_{2\alpha}$ were recently shown to be major urinary metabolites of 6-keto-$PGF_{2\alpha}$ in mice (Dmitry V. Kuklev et al. 2013). This information may be useful in developing a method to measure urinary metabolites of 3-series PGs. Urinary metabolite levels are markers of the total body production of the PGs, which may be altered in a different manner than specific tissue levels of PG (Yin et al. 2009). More research is needed to determine if systemic changes will mirror changes in the colon.

In a group of healthy male volunteers, low-fat diet (6% vs. 9% energy PUFA) decreased the urinary PGE-M by 14% (Ferretti et al. 1989). Interestingly, the authors also observed that body weight had a significant effect on the difference in PGE-M levels in the urine. The lighter weight individuals had greater reduction in urinary PGE-M levels. This same research group also reported that fish oil supplementation reduced urinary PGE-M by 14%, within a high fat intake background (Ferretti et al. 1991). Murff and colleagues demonstrated that the dietary intake of ω-3 PUFAs in women but not in men was negatively correlated to urinary PGE-M concentrations (Murff et al. 2012). The highest quintile of ω-3 PUFAs consumption was 0.14-1.01 g/day, which resulted in the lowest PGE-M in the urine. However, Young *et al.* performed a randomized, crossover study to compare the effects of diets containing varying levels of lipids and ω-3 FAs on plasma fatty acid profiles and urinary eicosanoid metabolites in healthy postmenopausal women (Young et al. 2011). They did not identify changes in PGE-M in urine samples from people who consumed 1.2 g of EPA and DHA per day for 8 weeks. These data indicate that PGE-M as a marker for estimation of dietary fish oil may be appropriate in certain populations, however, more studies with larger number and more diverse participants are necessary to verify this conclusion.

Conclusion

The studies reviewed here show that dietary fish oils can affect serum and colon levels of EPA and AA. This in turn may be responsible for decreases in PGE_2 production due to substrate competition for COX enzymes. Given the strong relationship of PGE_2 to colon cancer development, alteration of PGE_2 metabolism may be the key to the anti-carcinogenic mechanisms of fish oil. Further work is needed to determine if serum fatty acid ratios could be useful in predicting changes in PGE_2 in the colon of each individual. The changes in fatty acid will depend on the overall diet and metabolic factors as well as the fish oil dose. Accordingly, it should be possible to individualize fish oil dosing for cancer prevention.

References

Adam O., Beringer C., Kless T., Lemmen C., Adam A., Wiseman M., Adam P., Klimmek R., and Forth W. (2003). Anti-inflammatory effects of a low arachidonic acid diet and fish oil in patients with rheumatoid arthritis. *Rheumatol Int.* 23, 27-36.

Akabas S.R., and Deckelbaum R.J. (2006). Summary of a workshop on n-3 fatty acids: current status of recommendations and future directions. *Am. J. Clin. Nutr.* 83(6 Suppl):1536S-1538S.

Akedo I., Ishikawa H., Nakamura T., Kimura K., Takeyama I., Suzuki T., Kameyama M., Sato S., Matsuzawa Y., Kakizoe T. et al. (1998). Three cases with familial adenomatous polyposis diagnosed as having malignant lesions in the course of a long-term trial using docosahexanoic acid (DHA)-concentrated fish oil capsules. *Japanese journal of clinical oncology* 28(12):762-765.

Anti M., Armelao F., Marra G., Percesepe A., Bartoli G.M., Palozza P., Parrella P., Canetta C., Gentiloni N., De Vitis I. et al. (1994). Effects of different doses of fish oil on rectal cell proliferation in patients with sporadic colonic adenomas. *Gastroenterology* 107, 1709-1718.

Anti M., Marra G., Armelao F., Bartoli G.M., Ficarelli R., Percesepe A., De Vitis I., Maria G., Sofo L., Rapaccini G.L. et al. (1992). Effect of omega-3 fatty acids on rectal mucosal cell proliferation in subjects at risk for colon cancer. *Gastroenterology* 103, 883-891.

Arterburn L.M., Hall E.B., and Oken H. (2006). Distribution, interconversion, and dose response of n-3 fatty acids in humans. *Am. J. Clin. Nutr.* 83(6 Suppl):1467S-1476S.

Backlund M.G., Mann J.R., and Dubois R.N. (2005). Mechanisms for the prevention of gastrointestinal cancer: the role of prostaglandin E2. *Oncology* 1:28-32.

Bagga D., Wang L., Farias-Eisner R., Glaspy J.A., and Reddy S.T.. (2003). Differential effects of prostaglandin derived from omega-6 and omega-3 polyunsaturated fatty acids on COX-2 expression and IL-6 secretion. *Proc Natl Acad Sci U S A* 100, 1751-1756.

Bartoli G.M., Palozza P., Marra G., Armelao F., Franceschelli P., Luberto C., Sgarlata E., Piccioni E., and Anti M. (1993). n-3 PUFA and alpha-tocopherol control of tumor cell proliferation. *Molecular aspects of medicine* 14, 247-252.

Bartram H.P., Gostner A., Reddy B.S., Rao C.V., Scheppach W., Dusel G., Richter A., Richter F., and Kasper H. (1995). Missing anti-proliferative effect of fish oil on rectal epithelium in healthy volunteers consuming a high-fat diet: potential role of the n-3:n-6 fatty acid ratio. *Eur. J. Cancer. Prev.* 4, 231-237.

Bartram H.P., Gostner A., Scheppach W., Reddy B.S., Rao C.V., Dusel G., Richter F., Richter A., and Kasper H. (1993). Effects of fish oil on rectal cell proliferation, mucosal fatty acids, and prostaglandin E2 release in healthy subjects. *Gastroenterology.* 105, 1317-1322.

Bartsch H., Nair J., and Owen R.W. (1999). Dietary polyunsaturated fatty acids and cancers of the breast and colorectum: emerging evidence for their role as risk modifiers. Carcinogenesis 20, 2209-2218.

Bays H. (2006). Clinical overview of Omacor: a concentrated formulation of omega-3 polyunsaturated fatty acids. *Am. J. Cardiol.* 98, 71i-76i.

Bhatia E., Doddivenaka C., Zhang X., Bommareddy A., Krishnan P., Matthees D.P., and Dwivedi C. (2011). Chemopreventive effects of dietary canola oil on colon cancer development. *Nutr Cancer* 63, 242-247.

Bortuzzo C., Hanif R., Kashfi K., Staiano-Coico L., Shiff S.J., and Rigas B. (1996). The effect of leukotrienes B and selected HETEs on the proliferation of colon cancer cells. *Biochim Biophys Acta* 1300, 240-246.

Bose M., Hao X., Ju J., Husain A., Park S., Lambert J.D., and Yang C.S. (2007). Inhibition of tumorigenesis in ApcMin/+ mice by a combination of (-)-epigallocatechin-3-gallate and fish oil. *Journal of agricultural and food chemistry* 55, 7695-7700.

Boudreau M.D., Chanmugam P.S., Hart S.B., Lee S.H., and Hwang D.H. 1991. Lack of dose response by dietary n-3 fatty acids at a constant ratio of n-3 to n-6 fatty acids in suppressing eicosanoid biosynthesis from arachidonic acid. *Am. J. Clin. Nutr.* 54, 111-117.

Bresalier R.S., Sandler R.S., Quan H., Bolognese J.A., Oxenius B., Horgan K., Lines C., Riddell R., Morton D., Lanas A. et al. (2005). Cardiovascular events associated with rofecoxib in a colorectal adenoma chemoprevention trial. *N. Engl. J. Med.* 352, 1092-1102.

Broughton K.S., and Wade J.W. (2002). Total fat and (n-3):(n-6) fat ratios influence eicosanoid production in mice. *J. Nutr.* 132, 88-94.

Bull A.W., Soullier B.K., Wilson P.S., Hayden M.T., and Nigro N.D. (1979). Promotion of azoxymethane-induced intestinal cancer by high-fat diet in rats. *Cancer Res.* 39, 4956-4959.

Burns C.P., Halabi S., Clamon G.H., Hars V., Wagner B.A., Hohl R.J., Lester E., Kirshner J.J., Vinciguerra V., and Paskett E. (1999). Phase I clinical study of fish oil fatty acid capsules for patients with cancer cachexia: cancer and leukemia group B study 9473. *Clin. Cancer Res.* 5, 3942-3947.

Burns T., Maciejewski S.R., Hamilton W.R., Zheng M., Mooss A.N., and Hilleman D.E. (2007). Effect of omega-3 fatty acid supplementation on the arachidonic acid:eicosapentaenoic acid ratio. *Pharmacotherapy* 27, 633-638.

Busstra M.C., Siezen C.L., Grubben M.J., van Kranen H.J., Nagengast F.M., and van't Veer P. (2003). Tissue levels of fish fatty acids and risk of colorectal adenomas: a case-control study (Netherlands). *Cancer Causes Control* 14, 269-276.

Chan T.A. (2003). Cyclooxygenase inhibition and mechanisms of colorectal cancer prevention. *Curr Cancer Drug Targets* 3, 455-463.

Chang W.C., Chapkin R.S., and Lupton J.R. (1997). Predictive value of proliferation, differentiation and apoptosis as intermediate markers for colon tumorigenesis. *Carcinogenesis* 18, 721-730.

Chang W.L., Chapkin R.S., and Lupton J.R. (1998). Fish oil blocks azoxymethane-induced rat colon tumorigenesis by increasing cell differentiation and apoptosis rather than decreasing cell proliferation. *J. Nutr.* 128, 491-497.

Cordain L., Eaton S.B., Sebastian A., Mann N., Lindeberg S., Watkins B.A., O'Keefe J.H., and Brand-Miller J. (2005). Origins and evolution of the Western diet: health implications for the 21st century. *Am. J. Clin. Nutr.* 81, 341-354.

Courtney E.D., Matthews S., Finlayson C., Di Pierro D., Belluzzi A., Roda E., Kang J.Y., and Leicester R.J. (2007). Eicosapentaenoic acid (EPA) reduces crypt cell proliferation and increases apoptosis in normal colonic mucosa in subjects with a history of colorectal adenomas. *International journal of colorectal disease* 22, 765-776.

De Robertis M., Massi E., Poeta M.L., Carotti S., Morini S., Cecchetelli L., Signori E., and Fazio V.M. (2011). The AOM/DSS murine model for the study of colon carcinogenesis: From pathways to diagnosis and therapy studies. *Journal of carcinogenesis* 10, 9.

Deschner E.E., Lytle J.S., Wong G., Ruperto J.F., and Newmark H.L. (1990). The effect of dietary omega-3 fatty acids (fish oil) on azoxymethanol-induced focal areas of dysplasia and colon tumor incidence. *Cancer* 66, 2350-2356.

Dmitry V. K., Joseph A. H., Charis L. U., Yu H. H., Robert C. M., and Smith W.L. (2013). Major Urinary Metabolites of 6-Keto-Prostaglandin F2alpha; in Mice. *Journal of lipid research* Accepted.

Dyerberg J., Bang H.O., Stoffersen E., Moncada S., and Vane J.R. (1978). Eicosapentaenoic acid and prevention of thrombosis and atherosclerosis? *Lancet* 2, 117-119.

Eberhart C.E., Coffey R.J., Radhika A., Giardiello F.M., Ferrenbach S., and DuBois R.N. (1994). Up-regulation of cyclooxygenase 2 gene expression in human colorectal adenomas and adenocarcinomas. *Gastroenterology* 107, 1183-1188.

Eritsland J., Arnesen H., Gronseth K., Fjeld N.B., and Abdelnoor M. (1996). Effect of dietary supplementation with n-3 fatty acids on coronary artery bypass graft patency. *Am. J. Cardiol.* 77, 31-36.

Faber J., Berkhout M., Fiedler U., Avlar M., Witteman B.J., Vos A.P., Henke M., Garssen J., van Helvoort A., Otten M.H. et al. (2013). Rapid EPA and DHA incorporation and reduced PGE2 levels after one week intervention with a medical food in cancer patients receiving radiotherapy, a randomized trial. *Clin. Nutr.* 32, 338-345.

Ferretti A., Flanagan V.P., and Reeves V.B. (1988). Occurrence of prostaglandin E3 in human urine as a result of marine oil ingestion: gas chromatographic-mass spectrometric evidence. *Biochim Biophys Acta* 959, 262-268.

Ferretti A., Judd J.T., Ballard-Barbash R., Nair P.P., Taylor P.R., and Clevidence B.A. (1991). Effect of fish oil supplementation on the excretion of the major metabolite of prostaglandin E in healthy male subjects. *Lipids* 26, 500-503.

Ferretti A., Judd J.T., Taylor P.R., Schatzkin A., and Brown C. (1989). Modulating influence of dietary lipid intake on the prostaglandin system in adult men. *Lipids* 24, 419-422.

Fini L., Piazzi G., Ceccarelli C., Daoud Y., Belluzzi A., Munarini A., Graziani G., Fogliano V., Selgrad M., Garcia M. et al. (2010). Highly purified eicosapentaenoic acid as free fatty acids strongly suppresses polyps in Apc(Min/+) mice. *Clin. Cancer Res.* 16, 5703-5711.

Fischer S., von Schacky C., and Schweer H. (1988). Prostaglandins E3 and F3 alpha are excreted in human urine after ingestion of n - 3 polyunsaturated fatty acids. *Biochim Biophys Acta* 963, 501-508.

Gallai V., Sarchielli P., Trequattrini A., Franceschini M., Floridi A., Firenze C., Alberti A., Di Benedetto D., and Stragliotto E. (1995). Cytokine secretion and eicosanoid production in the peripheral blood mononuclear cells of MS patients undergoing dietary supplementation with n-3 polyunsaturated fatty acids. *Journal of neuroimmunology* 56, 143-153.

Gee J.M., Watson M., Matthew J.A., Rhodes M., Speakman C.J., Stebbings W.S., and Johnson I.T. (1999). Consumption of fish oil leads to prompt incorporation of eicosapentaenoic acid into colonic mucosa of patients prior to surgery for colorectal cancer, but has no detectable effect on epithelial cytokinetics. *J. Nutr.* 129, 1862-1865.

Geelen A., Schouten J.M., Kamphuis C., Stam B.E., Burema J., Renkema J.M., Bakker E.J., van't Veer P., and Kampman E. (2007). Fish consumption, n-3 fatty acids, and colorectal cancer: a meta-analysis of prospective cohort studies. *Am. J. Epidemiol.* 166, 1116-1125.

GISSI. (1999). Dietary supplementation with n-3 polyunsaturated fatty acids and vitamin E after myocardial infarction: results of the GISSI-Prevenzione trial. Gruppo Italiano per lo Studio della Sopravvivenza nell'Infarto miocardico. *Lancet* 354, 447-455.

Good C.K., Lasko C.M., Adam J., and Bird R.P. (1998). Diverse effect of fish oil on the growth of aberrant crypt foci and tumor multiplicity in F344 rats. *Nutr. Cancer* 31, 204-211.

Gravaghi C., La Perle K.M., Ogrodwski P., Kang J.X., Quimby F., Lipkin M., and Lamprecht S.A. (2011). Cox-2 expression, PGE(2) and cytokines production are inhibited by endogenously synthesized n-3 PUFAs in inflamed colon of fat-1 mice. *J. Nutr. Biochem.* 22, 360-365.

Griffin M.D., Sanders T.A., Davies I.G., Morgan L.M., Millward D.J., Lewis F., Slaughter S., Cooper J.A., Miller G.J., and Griffin B.A. (2006). Effects of altering the ratio of dietary n-6 to n-3 fatty acids on insulin sensitivity, lipoprotein size, and postprandial lipemia in men and postmenopausal women aged 45-70 y: the OPTILIP Study. *Am. J. Clin. Nutr.* 84, 1290-1298.

Harris W.S., Assaad B., and Poston W.C. (2006). Tissue omega-6/omega-3 fatty acid ratio and risk for coronary artery disease. *Am. J. Cardiol.* 98, 19i-26i.

Hendrickse C.W., Keighley M.R., and Neoptolemos J.P. (1995). Dietary omega-3 fats reduce proliferation and tumor yields at colorectal anastomosis in rats. *Gastroenterology* 109, 431-439.

Heyer J., Yang K., Lipkin M., Edelmann W., and Kucherlapati R. (1999). Mouse models for colorectal cancer. *Oncogene* 18, 5325-5333.

Hibbeln J.R., Nieminen L.R., Blasbalg T.L., Riggs J.A., and Lands W.E. (2006). Healthy intakes of n-3 and n-6 fatty acids: estimations considering worldwide diversity. *Am. J. Clin. Nutr.* 83(6 Suppl), 1483S-1493S.

Hillier K., Jewell R., Dorrell L., and Smith C.L.. (1991). Incorporation of fatty acids from fish oil and olive oil into colonic mucosal lipids and effects upon eicosanoid synthesis in inflammatory bowel disease. *Gut* 32, 1151-1155.

Huang Y.C., Jessup J.M., Forse R.A., Flickner S., Pleskow D., Anastopoulos H.T., Ritter V., and Blackburn G.L. (1996). n-3 fatty acids decrease colonic epithelial cell proliferation in high-risk bowel mucosa. *Lipids* 31 Suppl, S313-317.

Hwang D.H., Chanmugam P.S., Ryan D.H., Boudreau M.D., Windhauser M.M., Tulley R.T., Brooks E.R., and Bray G.A. (1997). Does vegetable oil attenuate the beneficial effects of fish oil in reducing risk factors for cardiovascular disease? *Am. J. Clin. Nutr.* 66, 89-96.

Jacoby R.F., Marshall D.J., Newton M.A., Novakovic K., Tutsch K., Cole C.E., Lubet R.A., Kelloff G.J., Verma A., Moser A.R. et al. (1996). Chemoprevention of spontaneous intestinal adenomas in the Apc Min mouse model by the nonsteroidal anti-inflammatory drug piroxicam. *Cancer Res.* 56, 710-714.

James M.J., Gibson R.A., and Cleland L.G. (2000). Dietary polyunsaturated fatty acids and inflammatory mediator production. *Am. J. Clin. Nutr.* 71(1 Suppl), 343S-348S.

Kargman S.L., O'Neil G.P., Vickers P.J., Evans J.F., Mancini J.A., and Jothy S. (1995). Expression of prostaglandin G/H synthase-1 and -2 protein in human colon cancer. *Cancer Res.* 55, 2556-2559.

King I.B., Lemaitre R.N., and Kestin M. (2006). Effect of a low-fat diet on fatty acid composition in red cells, plasma phospholipids, and cholesterol esters: investigation of a biomarker of total fat intake. *Am. J. Clin. Nutr.* 83, 227-236.

Kitamura T., Itoh M., Noda T., Matsuura M., and Wakabayashi K. (2004). Combined effects of cyclooxygenase-1 and cyclooxygenase-2 selective inhibitors on intestinal tumorigenesis in adenomatous polyposis coli gene knockout mice. *Int. J. Cancer* 109, 576-580.

Kobayashi N., Barnard R.J., Henning S.M., Elashoff D., Reddy S.T., Cohen P., Leung P., Hong-Gonzalez J., Freedland S.J., Said J. et al. (2006). Effect of altering dietary omega-6/omega-3 fatty acid ratios on prostate cancer membrane composition, cyclooxygenase-2, and prostaglandin E2. *Clin. Cancer. Res.* 12, 4662-4670.

Kojima M., Wakai K., Tokudome S., Suzuki K., Tamakoshi K., Watanabe Y., Kawado M., Hashimoto S., Hayakawa N., Ozasa K. et al. (2005). Serum levels of polyunsaturated fatty acids and risk of colorectal cancer: a prospective study. *Am. J. Epidemiol.* 161, 462-471.

Kramer R.M., and Sharp J.D. (1997). Structure, function and regulation of Ca2+-sensitive cytosolic phospholipase A2 (cPLA2). *FEBS Lett* 410, 49-53.

Kris-Etherton P.M., Harris W.S., and Appel L.J. (2002). Fish consumption, fish oil, omega-3 fatty acids, and cardiovascular disease. *Circulation* 106, 2747-2757.

Kris-Etherton P.M., Taylor D.S., Yu-Poth S., Huth P., Moriarty K., Fishell V., Hargrove R.L., Zhao G., and Etherton T.D. (2000). Polyunsaturated fatty acids in the food chain in the United States. *Am. J. Clin. Nutr.* 71(1 Suppl), 179S-188S.

La Guardia M., Giammanco S., Di Majo D., Tabacchi G., Tripoli E., and Giammanco M. (2005). Omega 3 fatty acids: biological activity and effects on human health. *Panminerva Med.* 47, 245-257.

Lands W.E. (1995). Long-term fat intake and biomarkers. *Am. J. Clin. Nutr.* 61(3 Suppl), 721S-725S.

Lands W.E. (2007). Essential Fatty Acid Education: Eicosanoids/Diet Balance Interactive Chart.

http://efaeducation.nih.gov/sig/eicosa1.html.

Lands W.E., Libelt B., Morris A., Kramer N.C., Prewitt T.E., Bowen P., Schmeisser D., Davidson M.H., and Burns J.H. (1992). Maintenance of lower proportions of (n - 6) eicosanoid precursors in phospholipids of human plasma in response to added dietary (n - 3) fatty acids. *Biochim Biophys Acta* 1180, 147-162.

Latham P., Lund E.K., and Johnson I.T. (1999). Dietary n-3 PUFA increases the apoptotic response to 1,2-dimethylhydrazine, reduces mitosis and suppresses the induction of carcinogenesis in the rat colon. *Carcinogenesis* 20, 645-650.

Lopez-Ceron M., and Pellise M. (2012). Biology and diagnosis of aberrant crypt foci. *Colorectal Dis.* 14, e157-164.

Marszalek J.R., and Lodish H.F. (2005). Docosahexaenoic acid, fatty acid-interacting proteins, and neuronal function: breastmilk and fish are good for you. *Annu. Rev. Cell Dev. Biol.* 21, 633-657.

Minoura T., Takata T., Sakaguchi M., Takada H., Yamamura M., Hioki K., and Yamamoto M. (1988). Effect of dietary eicosapentaenoic acid on azoxymethane-induced colon carcinogenesis in rats. *Cancer Res.* 48, 4790-4794.

Monk J.M., Kim W., Callaway E., Turk H.F., Foreman J.E., Peters J.M., He W., Weeks B., Alaniz R.C., McMurray D.N. et al. (2012). Immunomodulatory action of dietary fish oil and targeted deletion of intestinal epithelial cell PPARdelta in inflammation-induced colon carcinogenesis. *Am. J. Physiol.* 302, G153-167.

Moore C.S., Bryant S.P., Mishra G.D., Krebs J.D., Browning L.M., Miller G.J., and Jebb S.A. (2006). Oily fish reduces plasma triacylglycerols: a primary prevention study in overweight men and women. *Nutrition* 22, 1012-1024.

Moreira A.P., Sabarense C.M., Dias C.M., Lunz W., Natali A.J., Gloria M.B., and Peluzio M.C. (2009). Fish oil ingestion reduces the number of aberrant crypt foci and adenoma in 1,2-dimethylhydrazine-induced colon cancer in rats. *Brazilian journal of medical and biological research* 42, 1167-1172.

Moser A.R., Pitot H.C., and Dove W.F. (1990). A dominant mutation that predisposes to multiple intestinal neoplasia in the mouse. *Science* 247, 322-324.

Murff H.J., Shrubsole M.J., Cai Q., Smalley W.E., Dai Q., Milne G.L., Ness R.M., and Zheng W. (2012). Dietary intake of PUFAs and colorectal polyp risk. *Am. J. Clin. Nutr.* 95, 703-712.

Murphey L.J., Williams M.K., Sanchez S.C., Byrne L.M., Csiki I., Oates J.A., Johnson D.H., and Morrow J.D. (2004). Quantification of the major urinary metabolite of PGE2 by a liquid chromatographic/mass spectrometric assay: determination of cyclooxygenase-specific PGE2 synthesis in healthy humans and those with lung cancer. *Anal Biochem.* 334, 266-275.

Nelson R.L., Tanure J.C., Andrianopoulos G., Souza G., and Lands W.E. (1988). A comparison of dietary fish oil and corn oil in experimental colorectal carcinogenesis. *Nutr. Cancer.* 11, 215-220.

Nkondjock A., Shatenstein B., Maisonneuve P., and Ghadirian P. (2003). Specific fatty acids and human colorectal cancer: an overview. *Cancer Detect Prev.* 27, 55-66.

Nussmeier N.A., Whelton A.A., Brown M.T., Langford R.M., Hoeft A., Parlow J.L., Boyce S.W., and Verburg K.M. (2005). Complications of the COX-2 inhibitors parecoxib and valdecoxib after cardiac surgery. *N. Engl. J. Med.* 352, 1081-1091.

Okita M., Hayashi M., Sasagawa T., Takagi K., Suzuki K., Kinoyama S., Ito T., and Yamada G. (2001). Effect of a moderately energy-restricted diet on obese patients with fatty liver. *Nutrition* 17, 542-547.

Oshima H., Oshima M., Kobayashi M., Tsutsumi M., and Taketo M.M. (1997). Morphological and molecular processes of polyp formation in Apc(delta716) knockout mice. *Cancer Res.* 57, 1644-1649.

Oshima M., Oshima H., Kitagawa K., Kobayashi M., Itakura C., and Taketo M. (1995a). Loss of Apc heterozygosity and abnormal tissue building in nascent intestinal polyps in mice carrying a truncated Apc gene. *Proc Natl Acad Sci U S A* 92, 4482-4486.

Oshima M., Takahashi M., Oshima H., Tsutsumi M., Yazawa K., Sugimura T., Nishimura S., Wakabayashi K., and Taketo M.M. (1995b). Effects of docosahexaenoic acid (DHA) on intestinal polyp development in Apc delta 716 knockout mice. *Carcinogenesis* 16, 2605-2607.

Pan D.A., Hulbert A.J., and Storlien L.H. (1994). Dietary fats, membrane phospholipids and obesity. *J. Nutr.* 124, 1555-1565.

Paulsen J.E., Elvsaas I.K., Steffensen I.L., and Alexander J. (1997). A fish oil derived concentrate enriched in eicosapentaenoic and docosahexaenoic acid as ethyl ester suppresses the formation and growth of intestinal polyps in the Min mouse. *Carcinogenesis* 18, 1905-1910.

Paulsen J.E., Namork E., Steffensen I.L., Eide T.J., and Alexander J. (2000). Identification and quantification of aberrant crypt foci in the colon of Min mice--a murine model of familial adenomatous polyposis. *Scandinavian journal of gastroenterology* 35, 534-539.

Petrik M.B., McEntee M.F., Chiu C.H., and Whelan J. (2000a). Antagonism of arachidonic acid is linked to the antitumorigenic effect of dietary eicosapentaenoic acid in Apc(Min/+) mice. *J. Nutr.* 130, 1153-1158.

Petrik M.B., McEntee M.F., Johnson B.T., Obukowicz M.G., and Whelan J. (2000b). Highly unsaturated (n-3) fatty acids, but not alpha-linolenic, conjugated linoleic or gamma-linolenic acids, reduce tumorigenesis in Apc(Min/+) mice. *J. Nutr.* 130, 2434-2443.

Raatz S.K., Bibus D., Thomas W., and Kris-Etherton P. (2001). Total fat intake modifies plasma fatty acid composition in humans. *J. Nutr.* 131, 231-234.

Rao C.V., Hirose Y., Indranie C., and Reddy B.S. (2001). Modulation of experimental colon tumorigenesis by types and amounts of dietary fatty acids. *Cancer Res.* 61,1927-1933.

Reddy B., and Maeura Y. (1984). Tumor promotion by dietary fat in azoxymethane-induced colon cancer. Carcinogenesis in female F-344 rats: Influence of amount and source of dietary fat. *JNCI* 72, 745-750.

Reddy B.S., Burill C., and Rigotty J. (1991). Effect of diets high in omega-3 and omega-6 fatty acids on initiation and postinitiation stages of colon carcinogenesis. *Cancer Res.* 51, 487-491.

Reddy B.S., and Sugie S. (1988). Effect of different levels of omega-3 and omega-6 fatty acids on azoxymethane-induced colon carcinogenesis in F344 rats. *Cancer Res.* 48, 6642-6647.

Rees D., Miles E.A., Banerjee T., Wells S.J., Roynette C.E., Wahle K.W., and Calder P.C. (2006). Dose-related effects of eicosapentaenoic acid on innate immune function in healthy humans: a comparison of young and older men. *Am. J. Clin. Nutr.* 83, 331-342.

Rose D.P., and Connolly J.M. (1999). Omega-3 fatty acids as cancer chemopreventive agents. *Pharmacol Ther.* 83, 217-244.

Sanderson P., Johnson I.T., Mathers J.C., Powers H.J., Downes C.S., McGlynn A.P., Dare R., Kampman E., Pool-Zobel B.L., Bingham S.A. et al. (2004). Emerging diet-related surrogate end points for colorectal cancer: UK Food Standards Agency diet and colonic health workshop report. *Br. J. Nutr.* 91, 315-323.

Sano H., Kawahito Y., and Wilder R.L. (1995). Expression of cyclooxygenase-1 and -2 in human colorectal cancer. *Cancer Res.* 55, 3785-3789.

Shikano M., Masuzawa Y., Yazawa K., Takayama K., Kudo I., and Inoue K. (1994). Complete discrimination of docosahexaenoate from arachidonate by 85 kDa cytosolic phospholipase A2 during the hydrolysis of diacyl- and alkenylacyl-glycerophosphoethanolamine. *Biochim Biophys Acta* 1212, 211-216.

Simopoulos A.P. (1999). Evolutionary aspects of omega-3 fatty acids in the food supply. *Prostaglandins Leukotrienes & Essential Fatty Acids* 60, 421-429.

Singh J., Hamid R., and Reddy B.S. (1998). Dietary fish oil inhibits the expression of farnesyl protein transferase and colon tumor development in rodents. *Carcinogenesis* 19, 985-989.

Slattery M.L., Potter J.D., Duncan D.M., and Berry T.D. (1997). Dietary fats and colon cancer: assessment of risk associated with specific fatty acids. *Int. J. Cancer.* 73, 670-677.

Smith W.L., and Murphy R. (2002). The eicosanoids: cyclooxygenase, lipoxygenase, and epoxygenase pathways. In: Vance D, and Vance J, editors. *Biochemistry of Lipids, Lipoproteins and Membranes*. Amsterdam: Elsevier. p 341-372.

Smith W.L. (2005). Cyclooxygenases, peroxide tone and the allure of fish oil. *Curr. Opin. Cell Biol.* 17, 174-182.

Solomon S.D., McMurray J.J., Pfeffer M.A., Wittes J., Fowler R., Finn P., Anderson W.F., Zauber A., Hawk E., and Bertagnolli M. (2005). Cardiovascular risk associated with celecoxib in a clinical trial for colorectal adenoma prevention. *N. Engl. J. Med.* 352, 1071-1080.

Sonoshita M, Takaku K, Oshima M, Sugihara K-i, and Taketo M, M. (2002). Cycloogenaze-2 Expression in Fibroblasts and Endothelial Cells of Intetinal Polyps. *Cancer Res.* 62, 6846-6849.

Swails W.S., Kenler A.S., Driscoll D.F., DeMichele S.J., Babineau T.J., Utsunamiya T., Chavali S., Forse R.A., and Bistrian B.R. (1997). Effect of a fish oil structured lipid-based diet on prostaglandin release from mononuclear cells in cancer patients after surgery. *J. Parenter Enteral. Nutr.* 21, 266-274.

Takahashi M., Fukutake M., Isoi T., Fukuda K., Sato H., Yazawa K., Sugimura T., and Wakabayashi K. (1997). Suppression of azoxymethane-induced rat colon carcinoma development by a fish oil component, docosahexaenoic acid (DHA). *Carcinogenesis* 18, 1337-1342.

Takahashi M., Minamoto T., Yamashita N., Yazawa K., Sugimura T., and Esumi H. (1993). Reduction in formation and growth of 1,2-dimethylhydrazine-induced aberrant crypt foci in rat colon by docosahexaenoic acid. *Cancer Res* 53, 2786-2789.

Takeda H., Sonoshita M., Oshima H., Sugihara K-i., Chulada P.C., Langenbach R., Oshima M., and Taketo M.M. (2003). Cooperation of Cyclooxygenase 1 and Cyclooxygenase 2 in Intestinal Polyposis. *Cancer Res.* 63, 4872-4877.

Tappel A. (2007). Heme of consumed red meat can act as a catalyst of oxidative damage and could initiate colon, breast and prostate cancers, heart disease and other diseases. *Med. Hypotheses* 68, 562-564.

Trebble T.M., Wootton S.A., Miles E.A., Mullee M., Arden N.K., Ballinger A.B., Stroud M.A., Burdge G.C., and Calder P.C. (2003). Prostaglandin E2 production and T cell function after fish-oil supplementation: response to antioxidant cosupplementation. *Am. J. Clin. Nutr.* 78, 376-382.

Tsuzuki T., Igarashi M., and Miyazawa T. (2004). Conjugated eicosapentaenoic acid (EPA) inhibits transplanted tumor growth via membrane lipid peroxidation in nude mice. *J. Nutr.* 134, 1162-1166.

Vanamala J., Glagolenko A., Yang P., Carroll R.J., Murphy M.E., Newman R.A., Ford J.R., Braby L.A., Chapkin R.S., Turner N.D. et al. (2008). Dietary fish oil and pectin enhance colonocyte apoptosis in part through suppression of PPARdelta/PGE2 and elevation of PGE3. *Carcinogenesis* 29, 790-796.

Vidgren H.M., Agren J.J., Schwab U., Rissanen T., Hanninen O., and Uusitupa M.I. (1997). Incorporation of n-3 fatty acids into plasma lipid fractions, and erythrocyte membranes and platelets during dietary supplementation with fish, fish oil, and docosahexaenoic acid-rich oil among healthy young men. *Lipids* 32, 697-705.

Wada M., DeLong C.J., Hong Y.H., Rieke C.J., Song I., Sidhu R.S., Yuan C., Warnock M., Schmaier A.H., Yokoyama C. et al. (2007). Enzymes and receptors of prostaglandin pathways with arachidonic acid-derived versus eicosapentaenoic acid-derived substrates and products. *J. Biol. Chem.* 282, 22254-22266.

Ward P.S., Fuller R.W., Ritter J.M., Cashman S.J., Rees A.J., and Dollery C.T. (1991). Excretion of metabolites of prostacyclin and thromboxane by rats with nephrotoxic nephritis: effects of interleukin-1. *Br. J. Pharm.* 103, 1663-1668.

Wargovich M.J., and Felkner I.C. (1982). Metabolic activation of DMH by colonic microsomes: a process influenced by type of dietary fat. *Nutr. Cancer.* 4, 146-153.

Warstedt K., Furuhjelm C., Duchen K., Falth-Magnusson K., and Fageras M. (2009). The effects of omega-3 fatty acid supplementation in pregnancy on maternal eicosanoid, cytokine, and chemokine secretion. *Pediatric Res.* 66, 212-217.

Wasan H.S., Novelli M., Bee J., and Bodmer W.F. (1997). Dietary fat influences on polyp phenotype in multiple intestinal neoplasia mice. *Proc Natl Acad Sci U S A* 94, 3308-3313.

West N.J., Clark S.K., Phillips R.K., Hutchinson J.M., Leicester R.J., Belluzzi A., and Hull M.A. (2010). Eicosapentaenoic acid reduces rectal polyp number and size in familial adenomatous polyposis. *Gut* 59, 918-925.

Willett W.C. (2007). The role of dietary n-6 fatty acids in the prevention of cardiovascular disease. *J. Cardiovasc. Med.* (Hagerstown) 8 Suppl 1, S42-45.

Yang P., Chan D., Felix E., Cartwright C., Menter D.G., Madden T., Klein R.D., Fischer S.M., and Newman R.A. (2004). Formation and antiproliferative effect of prostaglandin E(3) from eicosapentaenoic acid in human lung cancer cells. *J. Lipid Res.* 45, 1030-1039.

Yin H., Liu W., Goleniewska K., Porter N.A., Morrow J.D., and Peebles R.S., Jr. (2009). Dietary supplementation of omega-3 fatty acid-containing fish oil suppresses F2-isoprostanes but enhances inflammatory cytokine response in a mouse model of ovalbumin-induced allergic lung inflammation. *Free Radical Biology & Medicine* 47, 622-628.

Young L.R., Kurzer M.S., Thomas W., Redmon J.B., and Raatz S.K. (2011). Effect of dietary fat and omega-3 fatty acids on urinary eicosanoids and sex hormone concentrations in postmenopausal women: a randomized controlled feeding trial. *Nutr. Cancer* 63, 930-939.

In: Omega-3 Fatty Acids
Editor: Md. Wasim Khan
ISBN: 978-1-62948-516-4

Chapter 5

Epidermal Barrier Function and Omega-3 Fatty Acid Supplementation

Audrey Lajoinie[1,2]***, Patrick Gelas***[3]***, Marek Haftek***[1] ***and Fabrice Pirot***[1,2]

[1]University Lyon 1 – Claude Bernard, " Physiological, Clinical and therapeutic aspects of the skin barrier function", Lyon, France
[2]Hospices Civils de Lyon –Edouard Herriot Hospital, Pharmacy, Plateform FRIPHARM, Lyon, France
[3]Hospices Civils de Lyon, Croix-Rousse Hospital, Clinical Nutrition Intensive Care Unit, Lyon, France

Abstract

The first historical *in vivo* study which has revealed the necessity of dietary fats for skin barrier function was conducted in 1929 by Burr and Burr: rats deprived of dietary fats developed typical skin symptoms, such as scaly skin and increased transepidermal water loss. These symptoms reversed after diet supplementation with polyunsatured fatty acids (PUFAs), showing that these essential fatty acids (EFAs) play a major role in skin barrier function and integrity. Numerous *in vivo* studies showed that PUFA intakes improve the skin barrier function in humans, rats or dogs with EFA deficiency or skin diseases. The principal limitation of these studies was to evaluate PUFA supplementation effects on the skin without taking into consideration the differences between the two existing PUFA families, *i.e.*: the omega-6 and the omega-3. However, omega-6 and omega-3 PUFAs play totally distinct roles in many organs, including the skin. The omega-6 family is the most commonly encountered PUFA family in human epidermis. In particular, linoleic acid (LA) is a structural precursor for specific skin elements, known as ceramides, and thus plays the most important role in establishing structure and permeability of the epidermis. On the opposite, alpha-linolenic acid (ALA) and its derivatives of the omega-3 family are not essential compounds of the epidermis but are best known for their immunomodulative properties. Consequently, it is particularly important to distinguish between these two lipid family effects of on the skin barrier function.

A recent study conducted by our research team has demonstrated that the skin barrier function was highly improved in patients receiving intravenous lipid emulsions, which contained only omega-6 PUFAs, compared to patients without intravenous lipid intake. Interestingly, the skin barrier function was not significantly improved in patients receiving intravenous lipid emulsions, which contained both omega-6 and oemga-3 PUFAs. Our results underline the importance of exploring the influence of omega-6 and omega-3 PUFAs on the skin barrier function. Numerous hypotheses can be proposed to explain the observed differences but further investigations still need to be conducted in this field. Establishing of the epidermal barrier and improvement of its function is a priority during development and in individuals requiring skin repair, *e.g.*, premature neonates or severely burn patients. Investigations on PUFA supplementation should further improve our knowledge on mechanisms of omega-6 and omega-3 PUFAs incorporation into skin, leading to the improvement of the epidermal barrier function. New hypotheses could then be proposed for comprehension and treatment of widespread dermatological diseases.

Keywords: transepidermal water loss; lipid emulsions; parenteral nutrition; cutaneous barrier function; Stratum corneum; essential fatty acids

Introduction

Polyunsatured fatty acids (PUFAs), also known as essential fatty acids (EFAs), include two distinct families: the omega-6 and the omega-3 fatty acids. These two PUFA families have important effects on the structural organization and chemical properties of lipid biological membranes. They are also involved in numerous biological functions through eicosanoid synthesis, *via* cyclooxygenase (COX) and lipooxygenase (LOX) pathways, as well as in many intercellular signal transduction processes (Spector 1999). Linoleic acid (LA, 18:2 n-6) – the parent fatty acid of the omega-6 PUFA family - and alpha-linolenic acid (ALA, 18:3 n-3) – the parent of the omega-3 PUFA family - were historically considered as the only two EFAs, since, (i) they cannot be neosynthetized in mammals, making necessary external supply sources, and (ii) they are required for normal body development and biological functions. LA and ALA can be converted via a common desaturase-elongase metabolism pathway to long-chain PUFAs, considered as "conditionally EFAs" as they can also be supplied by external sources, and, interestingly, could have negative or positive effects on mammalian bodies depending on organ and pathological conditions (Cunnane 2003).

The lack of external supply of PUFAs in mammals is responsible for a very serious chronic illness, called "Essential Fatty Acid Deficiency". Characteristic cutaneous abnormalities are the major symptom of EFA deficiency, including mainly dermatitis, skin hyperproliferation and decreased skin barrier function against transepidermal water loss (Elias and Friend 1975; Cunnane 2003; Feingold 2007). These clinical observations provide evidence that PUFAs are required for the normal structure and permeability of human skin. Advances in understanding the skin barrier function have lead researchers to investigate the efficacy of the different omega-6 and omega-3 PUFAs, alone or in combination, in various chronic skin diseases, such as atopic dermatitis, contact dermatitis or psoriasis (Tanaka, Zhen et al. 1997; Das 2010; Foster, Hardy et al. 2010; Kawamura, Ooyama et al. 2011; Tretter and Mueller 2011; Weise, Heunemann et al. 2011; Popa, Remoue et al. 2012). These studies have confirmed that topical or oral PUFA intakes can improve epidermal barrier function in humans, rats or dogs with EFA deficiency. However, efficacy of PUFAs in cutaneous diseases is highly controversial and their structural, chemical and biological roles are still incompletely understood (Bamford, Gibson et al. 1985; Morse, Horrobin et al. 1989; Van Gool, Zeegers et al. 2004). Therefore, a better understanding of physiological roles of PUFAs on the epidermis is needed. Because epidermis is a highly active site for lipid metabolism, the major challenge in the interpretation of PUFA's effects on the epidermal barrier is to determine if provided PUFAs act directly or from their active metabolites, including long-chain PUFA derivatives and eicosanoid metabolites, well known for their numerous biological properties. Most of PUFAs having multiple roles, biophysical and biological effects of the different PUFAs and their metabolites in the epidermis are still unclear. Omega-6 and omega-3 PUFAs play totally distinct roles in many organs, including the skin. Skin lipid composition and function analysis revealed that LA, the most commonly encountered fatty acids in mammalian skin, play the most important role in maintaining structure and permeability of the epidermis (Chapkin and Ziboh 1984; Hansen and Jensen 1985; Chapkin, Ziboh et al. 1986; Feingold 2009). In the opposite, omega-3 PUFAs are considered as not essential structural compounds of the epidermis from chemical and physical investigations of the skin barrier function. Knowing the wide range of omega-3 PUFA effects on various physiological and pathological phenomena (*e.g.* cardiovascular diseases, tumour growth, insulin resistance and immunomodulation), it is of major interest to further investigate omega-3 PUFA's influence on epidermal barrier function. The increasing use of omega-3 PUFA in diet products and clinical nutrition emphasizes the importance of such investigations, especially for patients with impaired skin barrier function, such as the premature neonates and burn patients. This chapter will focus mainly on the chemical and structural role of both omega-6 and omega-3 PUFAs in the epidermal barrier function, attempting to identify main differences between the two PUFA's effects. PUFA's effects will be described on healthy skin and on affected skin, taking the atopic dermatitis as a model because of high evidence of impaired barrier function in the chronic skin disease physiopathology.

The Epidermal Barrier

The Stratum Corneum Barrier Function

The skin permeability barrier of mammals is primarily localized to the outermost 10-13 μm thick layer of the epidermis, the *Stratum corneum (SC)* (Elias and Friend 1975; Elias, Goerke et al. 1977; Feingold 2009). The major function of the *SC* is to form a barrier preventing water and electrolyte loss and protect the body against external aggressions, including pathogenic microorganism, xenobiotic, ultraviolet rays or mechanical insults. Risks of infection and desiccation in premature newborns and burn patients confirm the need to keep effective *SC* for body survival. However, the *SC* water loss barrier is not absolute.

A passive flux of water persists toward the more superficial *SC* layers, which have relatively low water content (Kalia, Pirot et al. 1996). This phenomena is the so-called «transepidermal water loss» (TEWL). The TEWL is constitutive in mammals and is independent of sweating. TEWL variation is the more extensively used criteria to monitor changes in the *SC* barrier function *in vitro* and *in vivo* and is a very useful tool to evaluate the effects of a number of products and conditions on this skin barrier function efficacy (Pirot, Berardesca et al. 1998). The measurement of the TEWL is performed with non-invasive bioengineering methods, measuring the water flux across the *SC*, expressed in $g.m^{-2}.h^{-1}$ (grams of water loss per square meter of skin surface and per hour), **by** applying Fick's law of diffusion. In healthy subjects, the TEWL value ranges between 10 $g.m^{-2}.h^{-1}$ and 12 ± 10 $g.m^{-2}.h^{-1}$. In atopic subjects, the TEWL reaches extreme values of 38 ± 10 $g.m^{-2}.h^{-1}$ on healthy skin surfaces and 70 ± 7 $g.m^{-2}.h^{-1}$ on damaged skin surfaces (Tanaka, Zhen et al. 1997). Any qualitative or quantitative abnormality of the *SC* structure is responsible of the TEWL increase. Elias et al. conjectured that water permeability differences may be attributable to fatty acid composition variations of the *SC* lipid composition (Elias, Cooper et al. 1981).

The Stratum Corneum Barrier Physiology and Synthesis

The *SC* can be considered as an elegant two compartment model, where "bricks" are embedded to an impermeable mortar (Simonetti, Hoogstraate et al. 1995). The bricks consist in corneocytes, which are dead keratinocytes containing water and keratin. The mortar is a multilamellar envelope of a heterogeneous mixture of different lipids, chemically bound to the corneocytes (Gray and Yardley 1975). This lipid envelope is known as the "*Stratum corneum* multilamellar lipid matrix".

The *SC* continuously regenerates its composition from the innermost layers – nearest to the dermis - to the outermost layers – nearest to the external environment - similarly to a treadmill. The time required for complete renewal of healthy *SC* is about two weeks. The underlying layer of the SC - the *Stratum granulosum (SG)* – is the point of origin of the *SC* renewal. Lipids of the *SC* multilamellar lipid matrix are actively biosynthesized from precursors localized in the keratinocyte lamellar bodies of the SG (Wertz 1992; Feingold 2009). These lamellar bodies primarily contain (i) polar lipids (phospholipids, ceramides and cholesterol), synthetized from circulating fatty acids and cholesterol and (ii) the enzymes needed to convert these precursors into *SC* intercellular lipids. At the interface *SC*/SG, the lamellar bodies fuse with the keratinocyte plasma membrane and polar lipids are released in the *SC* intercellular space. Enzymes then transform those polar lipids into non polar lipids assembled in hydrophobic intercellular lamellar structure, the *SC* multilamellar lipid matrix. It is particularly noticeable that the composition of lipids dramatically changes during this component apical migration.

The Multilamellar Lipid Matrix of the Stratum Corneum

The lipids of the mammalian *SC* have been extensively studied and analysed over the last decades. The permeability barrier of the epidermis is only located in the *SC* multilamellar lipid matrix (Haftek 2003). It has become widely understood that both architecture and hydrophobic properties of the extracellular lipid matrix give rise to the *SC*'s dense, impermeable structure acting as barrier to the movement of water and electrolytes (Wertz 2000; Feingold 2007). The lipids that constitute this extracellular lipid matrix have a unique and specific composition, differing strongly from the lipids constituting most of biological membranes (phospholipids, sphingolipids and cholesterol). Indeed, human *SC* consists of roughly equimolar proportions of ceramides, cholesterol and free fatty acids, with a respective amount of 50%, 25% and 15% of the total weight of *SC* lipids. In contrast to most biological membranes, very little phospholipid is present in the *SC* lipid matrix. Deficiency of one of the three indispensable lipid classes is responsible for epidermal barrier alteration, reflected by a TEWL increase (Holleran, Takagi et al. 2006). Most of skin diseases (*e.g.* atopic dermatitis, psoriasis, contact dermatitis) are related to altered lipid composition and structural organization of the *SC* multilamellar lipid matrix compared to healthy skin (Imokawa, Abe et al. 1991; Pilgram, Vissers et al. 2001; Macheleidt, Kaiser et al. 2002).

The composition of free fatty acids in mammalian *SC* is unique and totally contrasts with the composition of fatty acids in serum and sebum (Coderch, Lopez et al. 2003). Among extracellular matrix lipids, various fatty acids are indispensable for a normal epidermal barrier function, as well in a free form that bound to the ceramides (Mao-Qiang, Elias et al. 1993). The free fatty acids in human *SC* are predominantly long chain saturated, range from 14 to 28 carbons, and are characterized by a low inter-individual variability. The only unsatured fatty acid found in the *SC* are oleic acid and LA, respectively representing 6% and 2% of lipid matrix free fatty acids (Menon, Cleary et al. 2012). The other lipid class of the *SC* lipid matrix is the ceramides, which represent almost half of total lipid weight of *SC* lipids. Ceramide structure consists of a fatty acid bound by an amine to a sphingoid base. The *SC* contains numerous species of ceramides, differing in their sphingoid base and hydrocarbon chain length. At least eight different subclasses of ceramides have been identified in the human *SC* (Coderch, Lopez et al. 2003; Choi and Maibach 2005). Ceramides play the most important role in establishing structure and permeability of the epidermis (Hansen and Jensen 1985; Grubauer, Feingold et al. 1989). In addition to their structural role, they induce cell proliferation and apoptosis (*i.e.* programmed cell death) through complex extracellular *signalling* pathway (Coderch, Lopez et al. 2003).

Role of Pufas in Epidermal Barrier Function

The small amount of PUFAs present in normal mammalian epidermis suggests that they are not essential components for the epidermal barrier function. However, the severe skin disorders observed in EFAD strongly contradict this assumption. The first historical *in vivo* study that has revealed the necessity of dietary fats to maintain effective skin barrier function was conducted in 1929 by Burr and Burr (Burr and Burr 1973). In this study, rats deprived of dietary fats for a period of several months developed severe cutaneous and tail symptoms, such as erythematous eruption, scaly skin and increased water permeability of the epidermis. Extra-cutaneous symptoms were also associated, the most notables being growth retardation and reproductive failure. Cutaneous and tail symptoms rapidly improved before to completely reverse after diet supplementation with corn and linseed oil, containing both LA and ALA, the two parents of respectively omega-6 and omega-3 PUFA families. In the opposite, diet supplementation with coconut oils and butter, containing only saturated fatty acids, was inefficient on EFA deficiency cutaneous symptoms. This *in vivo* study demonstrated for the first time with a high level of evidence that PUFAs are of crucial importance for mammalian skin normal barrier function. Numerous studies have subsequently confirmed this finding (Gray and Yardley 1975; Friedman, Shochat et al. 1976; Chapkin and Ziboh 1984; Chapkin, Ziboh et al. 1986; Tanaka, Zhen et al. 1997; Feingold 2009; Das 2010; Foster, Hardy et al. 2010; Kawamura, Ooyama et al. 2011; Tretter and Mueller 2011; Weise, Heunemann et al. 2011; Popa, Remoue et al. 2012). In addition, advances in scientific knowledge on ceramide function in maintaining epidermal barrier efficacy and viability and the identification of specific linoleate-containing ceramides suggest that the role of PUFAs in epidermal barrier function is actually related to these specific compounds.

Omega-6 PUFA Family

Linoleic Acid and Its Derivatives

LA is the most common encountered unsatured fatty acid in the mammalian epidermis. The presence of LA in the *SC* is directly correlated with the permeability of the epidermal barrier (Hansen and Jensen 1985). The main long-chain PUFAs of the omega-6 family, synthesized by LA desaturation-elongation pathway, and the most studied, are the gamma-linolenic acid (GLA, 18:3 n-6), the dihomo-gamma-linolenic acid (DGLA, 20:3 n-6) and the arachidonic acid (AA, 20:4 n-6). Long-chain PUFAs of the omega-6 family can be precursors of prostaglandins metabolites from COX pathway and LOX pathway.

The first advanced hypothesis to explain the complete reverse of EFA deficiency's cutaneous symptoms after LA and ALA diet supplementation was the increasing endogenous synthesis of AA and its immunomodulative prostaglandin metabolites (Ziboh and Hsia 1972). Indeed, AA is the second most abundant PUFA in the epidermis after the LA and represents almost 10% of the total amount of fatty acids in human skin (Vroman, Nemecek et al. 1969; Yardley and Summerly 1981; Chapkin, Ziboh et al. 1986). In the epidermis, AA is primarily a component of phosphatidylinositol and phosphatidylserine of keratinocyte's membranes and is a major source of epidermal prostaglandins of 2-series (PGE_2, $PGF_{2\alpha}$, and PGD_2) and leukotriene $(LT)B_4$, *via* COX pathway. These eicosanoids are responsible for skin hyperproliferation and inflammation (Wertz 1992; Ziboh, Miller et al. 2000). The production of AA eicosanoid metabolites via the LOX pathway is minor. The functional role of AA depends largely on its generation of prostaglandins and their biological activities. At low concentrations, AA prostaglandins modulate physiological skin function, while, at high concentrations, they induce inflammatory processes implicated in many dermatoses, such as atopic dermatitis or psoriasis. The first interest for AA in clinical dermatology had been prompted by the identification of important omega-6 PUFA abnormalities atopic dermatitis patient epidermis. These abnormalities included a decreased plasma level of AA associated to elevated concentrations of eicosanoids in cutaneous lesions, responsible for the overproduction of IgE and the release of histamine observed in atopic dermatitis (Wright 1985; Lindskov and Holmer 1992; Van Gool, Zeegers et al. 2004). These findings highly incriminated the AA cascade in atopic dermatose pathogenesis.

One of the characteristic of the skin is its deficiency in desaturase, the first enzyme involved in the desaturase-elongation metabolism pathway of both omega-3 and omega-6 PUFAs. Therefore, LA is not metabolized in significant amount of AA in mammalian epidermis (Chapkin, Ziboh et al. 1986). Looking back to EFA deficiency's symptoms, this discovery has deeply challenges the first assumption of AA role in the improvement of EFA deficiency's cutaneous symptòm after diet supplementation with corn and linseed oils, AA free. In fact, the major identified metabolite of LA in the epidermis is the 13-hydroxyoctadecadienoic acid (13-HODE). This LA metabolism pathway is specific to the skin. Hypothesis has been advanced that 13-HODE would modulate the activity of the epidermal protein kinase C, leading to a decrease of the epidermal hyperproliferation (Cho and Ziboh 1994; Ziboh, Miller et al. 2000). This activity could have participated to the reverse of EFA's deficiency's cutaneous symptoms observed after diet supplementation with LA. To our knowledge, no study has investigated *in vivo* the impact of 13-HODE on the epidermal barrier function.

Other long-chain PUFA derivatives of LA are small constituent of the human epidermis. However, the diet supplementation and topical application of GLA has shown its efficacy in chronic skin disorders, and especially in atopic dermatitis. One of the pathological mechanisms under investigation in atopic dermatitis was the insufficiency of LA conversion into GLA, due to the lack of desaturase in the epidermis. The first proposed option was to switch LA desaturation stage in directly supplying GLA, through evening primrose or borage oil intakes (Horrobin 1992). Expected benefit of GLA supplementation was its conversion into DGLA *via* the epidermal elongation pathway. Indeed, DGLA is a precursor of prostaglandins PGE_1 and 15-hydroxyeicosatrienoic acid (15-HETrE), both having marked anti-inflammatory properties that may reduce the symptoms of inflammatory skin diseases (Wright and Burton 1982; Schalin-Karrila, Mattila et al. 1987; Ziboh, Miller et al. 2000). In addition, because of epidermis deficiency in desaturase enzyme, DGLA cannot be converted in AA, limiting the production pro-inflammatory prostaglandins metabolites. However, the efficacy of evening primrose oil on inflammatory dermatoses is still controverted (Kerscher and Korting 1992; Bamford, Ray et al. 2013).

Linoleic Acid as a Structural Component of the Ceramides

The second advanced hypothesis, and the most currently widespread, to explain PUFAs supplementation efficacy on EFA deficiency's cutaneous symptoms is the structural role of LA as a component of SC lipid matrix's ceramides. In the epidermis, the majority of LA is bound to ceramides (Grubauer, Feingold et al. 1989). The two major linoleate-containing ceramide subclasses (Cer1 and Cer4), both bounded to the corneocytes, are known to have a critical function in *SC* multilamellar lipid matrix synthesis and structural maintenance. They are involved in the epidermis chemical structure and macromolecular organization, in SC cell adhesion and in epidermal differentiation (Meguro, Arai et al. 2000). The two ceramide synthesis pathways are (i) the degradation of glucosylceramides and sphingomyelinases and (ii) the neosynthesis from serine and palmitic acid. Ceramide synthesis activity is indispensable for a normal epidermal barrier function (Holleran, Feingold et al. 1991).

In EFA deficiency, the replacement of linoleate-containing ceramide by oleic-containing ceramides results in marked *SC* desquamation and permeability barrier function abnormalities, associated with an anarchic organization of the multilamellar lipid matrix (Hou, Mitra et al. 1991; Feingold 2007). To date, numerous studies report abnormal composition or amount of SC lipid matrix's ceramides in altered skin, mainly in atopic dermatitis (Melnik, Hollmann et al. 1988; Imokawa, Abe et al. 1991; Yamamoto, Serizawa et al. 1991; Di Nardo, Wertz et al. 1998; Matsumoto, Umemoto et al. 1999; Macheleidt, Kaiser et al. 2002). Ceramide amount in the atopic dermatitis skin is decreased compared to healthy skin, leading to SC lipid matrix disorganization and TEWL increase. The linoleate-containing Cer1 is the most significantly reduced in atopic skin. Abnormalities of SC lipid composition in omega-6 PUFAs observed in atopic dermatitis are likely to be the source of ceramide amount decrease. Additionally, enzymatic synthesis pathway of ceramides could be altered in atopic dermatitis, explaining abnormalities observed in ceramide metabolism (Murata, Ogata et al. 1996).

Omega-3 PUFA Family

The main PUFAs of the omega-3 family are the ALA and its long-chain PUFA derivatives, the eicosapentaenoic acid (EPA, 20:5 n-3) and the docosahexaenoic acid (DHA, 22:6 n-3). Both of them are prostaglandin and leukotriene precursors. Interestingly, the metabolism process of ALA is not efficient in the body, especially for DHA production (Chan, McDonald et al. 1993; Emken, Adlof et al. 1994). Indeed, less than 10% and 5% of ALA are respectively metabolized to EPA and DHA in healthy adults (Wijendran and Hayes 2004; Brenna, Salem et al. 2009). EPA and DHA have therefore to be supplied by other sources than ALA intakes, such as fish oil intakes.

Contrary to omega-6 PUFAs, ALA and its two major derivatives are not found in substantial amount in the epidermis. They represent less than 2% of total epidermal fatty acids and are roughly absent from the *SC* normal lipid matrix (Ziboh, Miller et al. 2000). The low content of EPA and DHA highly reflect the *SC* lipid matrix composition. The distribution of omega-3 PUFAs in mammalian organs is dependent on their lipid composition: whereas ALA is found associated to triglycerides, EPA is found in cholesterol, triglycerides and phospholipids and DHA is mainly found in phospholipids (Simopoulos 1991). The very low rate of phospholipids in the *SC* extracellular matrix explains the lack of EPA and DHA in this structure. The more extensively studied omega-3 PUFAs in dermatology are EPA and DHA, mainly because of their immunomodulative properties. Inversely, the role of ALA in the skin barrier function and in skin diseases has been poorly investigated and is still unclear.

A study conducted on rat pups has compared the influence of maternal dietary with high ALA content to maternal dietary with low ALA content on PUFA composition of whole body, including the skin (Bowen and Clandinin 2000). Rat pups that received maternal dietary with low ALA content had significantly higher LA content in the skin, associated to remarkably higher ALA and EPA content. This first observation substantiates the hypothesis that omega-3 PUFAs intake results in a decrease of omega-6 incorporation into the skin, leading to epidermal barrier function alteration (Lajoinie, Gelas et al. 2013). Indeed, omega-6 and omega-3 PUFAs are in competition for the whole PUFA metabolism pathway, omega-3 being preferentially metabolized (Simopoulos 1991). However, no change in skin DHA content was found. The specific analysis of the epidermis showed no statistically significant difference in LA and DHA epidermal content between high and low ALA content maternal dietary. Interestingly, rat pups receiving high ALA content maternal dietary had significantly higher rate of ALA and EPA in the epidermis. Finally, it appears that the skin is one of the important organs for ALA accumulation after diet supplementation. From these results, DHA doesn't play a major role in the *SC* lipid matrix structure. Finally, oral omega-3 intake seems to substantially modify the *SC* lipid matrix PUFA composition. The low content of western diet in ALA and EPA could explain that those two long-chain PUFAs are not present in marked proportions in healthy skin.

In this study, omega-3 PUFAs were provided only through ALA containing maternal dietary. Knowing inefficient metabolism of ALA to produce EPA and DHA, fish oil intake would likely have different impact on skin PUFA composition. Oral fish oil effects in the skin have been more studied from a biological perspective. Fish oil intake **is** responsible for an increase of prostaglandin metabolites production in the epidermis. Eicosanoids produced from EPA have anti-inflammatory, anti-aggregating, vasodilatory and antiarythmic and immunomodulative properties. Mainly anti-inflammatory properties possibly explain beneficial effects of fish oil on cutaneous inflammation of atopic dermatitis (Bjorneboe, Soyland et al. 1989; Miller, Tang et al. 1991).

Omega-3 in Clinical Nutrition and Epidermal Barrier Function

The identification of numerous health benefits of omega-3 PUFAs provided by fish oil over the past few years have conducted pharmaceuticals industry and physician to consider as necessary the introduction of omega-3 PUFAs in clinical nutrition, especially for their liver protection and immunomodulative properties. As a consequence, most of infant formulas as well as enteral and parenteral nutrition (PN) products have been enriched with omega-3 PUFAs. Patients receiving lipids through a parenteral nutrition are interesting subjects to evaluate PUFA's effects on epidermal barrier function. Firstly, the absence of intestinal or cutaneous absorption greatly facilitates the interpretation of observations. Then, the only source of lipids for patients receiving total PN being the intravenous lipid emulsions (ILE), numerous of confounding factors are eliminated, such as oral intakes and pharmacokinetic parameters. Principal differences in main generations of commercialized ILE related to PUFAs composition, brought by long chain triglycerides. First generations of ILE only contained omega-6 PUFAs from olive and/or soybean oils. More recently, the enthusiasms in public health and nutrition care for omega-3 PUFAs has conducted to the development of the latest generation, containing omega-3 PUFAs provided by fish oil.

For patients receiving PN, the lipids, including PUFAs, have to be provided by ILE. Indeed, patients receiving lipid-free PN, containing only amino acids and carbohydrates as macronutrient, develop EFA deficiency, leading in a long-term to dermatitis, fatty liver, hematologic disturbances and immune system deregulations (Adolph, Hailer et al. 1995; Mascioli, Lopes et al. 1996). LA deficiency was long considered as the only responsible for major EFA deficiency symptoms. Indeed, LA supplementation was usually sufficient to prevent EFA deficiency, as demonstrated with the use of the first ILE generation, containing only omega-6 PUFAs. Contradicting this theory, infants receiving intravenous fish oil as the sole source of lipids did not developed any EFA deficiency's cutaneous symptoms (De Meijer, Le et al. 2010). Finally, Findings of studies conducted on ILE containing only fish oil as lipid source suggest that fish oil LA content, corresponding to approximately half of the recommended intake, is sufficient to prevent EFA deficiency (Strijbosch, Lee et al. 2008). The use of fish oil as only lipid source in clinical nutrition seems to be safe and could provide to patients the omega-3 health benefits.

In recent study conducted by our research team, the skin barrier function of patients receiving various ILE was compared to those of patients receiving lipid-free PN, served as control group (Lajoinie, Gelas et al. 2013). The objective of this study was to evaluate the influence of lipids provided by ILE on the *SC* barrier function, evaluated through TEWL measurement. The results showed that patients receiving ILE had a more effective *SC* barrier function than patients receiving lipid-free PN, due to PUFA supply by ILE. In addition, in 5 patients hospitalized for ILE introduction, the TEWL was systemically decreased one month after ILE introduction, reflecting the *SC* barrier function improvement through ILE lipids. A TEWL analysis was conducted depending on administered ILE. Surprisingly, patients receiving fish oil containing ILE had a less effective epidermal barrier function that patients receiving fish oil free ILE. Finally, patients receiving fish oil free ILE had a more effective *SC* barrier function compared to control patients, receiving lipid-free PN. Inversely, no significant improvement of the *SC* barrier function was found between patients receiving fish oil containing ILE and the same control group. These results suggest that only intravenous fish oil intakes hinder the *SC* barrier function improvement observed with fish oil free ILE. These results contradict arguments in favour of systematic use of omega-3 PUFAs.

The first hypothesis to explain lower skin barrier efficacy in patients receiving fish oil containing ILE is that DHA and EPA alter structural organization of the SC lipid matrix. EPA influence on cell membrane properties is primarily linked to its immunomodulative properties, through prostaglandin production, and its competition with AA metabolism, producing inflammatory derivatives (Obata, Nagakura et al. 1999). Main effects of EPA on membrane cells have been observed on immune cell, further as a result of eicosanoid production than due to cell membrane structure alteration. Inversely, DHA play a key role in membrane structure and fluidity of numerous cell membranes, such as eye rod cells and brain axon (Hulbert, Rana et al. 2002; Boesze-Battaglia, Damek-Poprawa et al. 2008). This PUFA is the longest and the most highly unsatured fatty acids of all encountered PUFAs in biological systems. These structural specificities are responsible for the two main effects of DHA on cell membranes: a decrease in membrane stability, due to membrane lipid packing distortion, and an increase in membrane permeability, probably linked to membrane stability perturbations (Stillwell and Wassall 2003). DHA effects on lipid membrane permeability could be at the origin of the TEWL increase in patients receiving fish oil containing ILE (Lajoinie, Gelas et al. 2013). To our knowledge, none of research on DHA effects on lipid layer properties has been conducted *in vivo* or *ex vivo* on the SC lipid matrix structure. In cell bilayer lipid membranes, DHA is rapidly and primarily linked to phospholipids of the plasma membrane (Zerouga, Stillwell et al. 2002). Given the negligible proportion of phospholipids in the *SC* lipid matrix, the results on various bilayer lipid membranes cannot be easily extrapolated to the *SC* lipid matrix. Finally, despite the large number of studies that have related DHA effects on biological membranes, numerous studies have failed to demonstrate DHA impact on fluidity (*e.g.* erythrocytes, platelets, hepatocyte) (Stillwell and Wassall 2003). Further investigations have to be conducted to verify the hypothesis of DHA influence on the epidermal barrier permeability.

The second hypothesis that could explain the negative impact of fish oil on the epidermal barrier function is the competition between omega-6 and omega-3 PUFAs for desaturase-elongase pathway, which is responsible for their respective long-chain PUFAs metabolites, as well as for COX and LOX pathway, responsible for eicosanoids production from long-chain PUFAs (Mantzioris, James et al. 1995). Epidermis being deficient in desaturase enzyme, the competition between omega-6 and omega-3 PUFAs concerns elongation and COX and LOX pathway enzymes. It should be noted that enzymes have higher affinity for omega-3 PUFAs than omega-6 PUFAs (Simopoulos 1991). Consequently, omega-3 intakes would be responsible for a decrease of the enzymatic metabolism and incorporation of omega-6 PUFAs the epidermis. Excess intakes of omega-3 PUFAs could deprive the *SC* lipid matrix of essential omega-6 PUFAs. This hypothesis is supported by Bowen *et al.* findings, showing that an oral ALA intake is responsible for a decrease of epidermal LA content (Bowen and Clandinin 2000).

Conclusion

Before systemizing the use omega-3 PUFAs in clinical nutrition, there is a need to clarify how omega-3 intakes affect the skin barrier function. Indeed, skin lipids and their metabolism pathway are very different from those found in most of body organs. In addition, the high metabolic activity and the continuous renewal of the epidermis explain the major influence of the quality of dietary fat intakes on the SC lipid matrix composition. The example of negative impact on fish oil intakes on the epidermal barrier function clearly shows that omega-3 PUFAs, considered as "good" for human health, could also have deleterious effects on certain organs in certain conditions. In skin physiology, the omega-6 PUFAs, usually considered as "bad" for global health, are beneficial and indispensable for epidermal barrier function. Since ALA transformation into DHA and EPA is known to be extremely low and considering the significant food intakes of ALA in western diet compared to EPA and DHA intakes, more work is required to identify ALA and its derivatives effects on healthy and affected skin.

References

Adolph, M., Hailer, S., Echart, J., (1995). Serum phospholipid fatty acids in severely injured patients on total parenteral nutrition with medium chain/long chain triglyceride emulsions. *Ann. Nutr. Metab.* 39, 251-260.

Bamford, J. T., Gibson, R. W., Renier, C. M., (1985). Atopic eczema unresponsive to evening primrose oil (linoleic and gamma-linolenic acids). *J. Am. Acad. Dermatol.* 13, 959-965.

Bamford, J. T., Ray, S., Musekiwa, A., van Gool, C., Humphreys, R., Ernst, E., (2013). Oral evening primrose oil and borage oil for eczema. *Cochrane Database Syst. Rev.* 4, CD004416.

Bjorneboe, A., Soyland, E., Bjorneboe, G. E., Rajka, G., Drevon, C. A., (1989). Effect of n-3 fatty acid supplement to patients with atopic dermatitis. *J. Intern. Med. Suppl.* 731, 233-236.

Boesze-Battaglia, K., Damek-Poprawa, M., Mitchell, D. C., Greeley, L., Brush, R. S., Anderson, R. E., Richards, M. J., Fliesler, S. J., (2008). Alteration of retinal rod outer segment membrane fluidity in a rat model of Smith-Lemli-Opitz syndrome. *J. Lipid Res.* 49, 1488-1499.

Bowen, R. A., Clandinin, M. T., (2000). High dietary 18:3n-3 increases the 18:3n-3 but not the 22:6n-3 content in the whole body, brain, skin, epididymal fat pads, and muscles of suckling rat pups. *Lipids*. 35, 389-394.

Brenna, J. T., Salem, N., Jr., Sinclair, A. J., Cunnane, S. C., (2009). alpha-Linolenic acid supplementation and conversion to n-3 long-chain polyunsaturated fatty acids in humans. *Prostaglandins Leukot. Essent. Fatty Acids*. 80, 85-91.

Burr, G. O., Burr, M. M., (1973). Nutrition classics from The Journal of Biological Chemistry 82:345-67, 1929. A new deficiency disease produced by the rigid exclusion of fat from the diet. *Nutr. Rev*. 31, 248-249.

Chan, J. K., McDonald, B. E., Gerrard, J. M., Bruce, V. M., Weaver, B. J., Holub, B. J., (1993). Effect of dietary alpha-linolenic acid and its ratio to linoleic acid on platelet and plasma fatty acids and thrombogenesis. *Lipids*. 28, 811-817.

Chapkin, R. S., Ziboh, V. A., (1984). Inability of skin enzyme preparations to biosynthesize arachidonic acid from linoleic acid. *Biochem. Biophys. Res. Commun*. 124, 784-792.

Chapkin, R. S., Ziboh, V. A., Marcelo, C. L., Voorhees, J. J., (1986). Metabolism of essential fatty acids by human epidermal enzyme preparations: evidence of chain elongation. *J. Lipid. Res*. 27, 945-954.

Cho, Y., Ziboh, V. A., (1994). 13-Hydroxyoctadecadienoic acid reverses epidermal hyperproliferation via selective inhibition of protein kinase C-beta activity. *Biochem Biophys Res Commun*. 201, 257-265.

Choi, M. J., Maibach, H. I., (2005). Role of ceramides in barrier function of healthy and diseased skin. *Am. J. Clin. Dermatol*. 6, 215-223.

Coderch, L., Lopez, O., de la Maza, A., Parra, J. L., (2003). Ceramides and skin function. *Am. J. Clin. Dermatol*. 4, 107-129.

Cunnane, S. C., (2003). Problems with essential fatty acids: time for a new paradigm? *Prog. Lipid. Res*. 42, 544-568.

Das, U. N., (2010). Polyunsaturated fatty acids and atopic dermatitis. *Nutrition*. 26, 719-720.

De Meijer, V. E., Le, H. D., Meisel, J. A., Gura, K. M., Puder, M., (2010). Parenteral fish oil as monotherapy prevents essential fatty acid deficiency in parenteral nutrition-dependent patients. *J. Pediatr. Gastroenterol. Nutr*. 50, 212-218.

Di Nardo, A., Wertz, P., Giannetti, A., Seidenari, S., (1998). Ceramide and cholesterol composition of the skin of patients with atopic dermatitis. *Acta Derm. Venereol*. 78, 27-30.

Elias, P. M., Cooper, E. R., Korc, A., Brown, B. E., (1981). Percutaneous transport in relation to stratum corneum structure and lipid composition. *J. Invest. Dermatol*. 76, 297-301.

Elias, P. M., Friend, D. S., (1975). The permeability barrier in mammalian epidermis. *J. Cell. Biol*. 65, 180-191.

Elias, P. M., Goerke, J., Friend, D. S., (1977). Mammalian epidermal barrier layer lipids: composition and influence on structure. *J. Invest. Dermatol*. 69, 535-546.

Emken, E. A., Adlof, R. O., Gulley, R. M., (1994). Dietary linoleic acid influences desaturation and acylation of deuterium-labelled linoleic and linolenic acids in young adult males. *Biochim. Biophys. Acta*. 1213, 277-288.

Feingold, K. R., (2007). Thematic review series: skin lipids. The role of epidermal lipids in cutaneous permeability barrier homeostasis. *J. Lipid Res.* 48, 2531-2546.

Feingold, K. R., (2009). The outer frontier: the importance of lipid metabolism in the skin. *J. Lipid Res.* 50 Suppl, S417-422.

Foster, R. H., Hardy, G., Alany, R. G., (2010). Borage oil in the treatment of atopic dermatitis. *Nutrition.* 26, 708-718.

Friedman, Z., Shochat, S. J., Maisels, M. J., Marks, K. H., Lamberth, E. L., Jr., (1976). Correction of essential fatty acid deficiency in newborn infants by cutaneous application of sunflower-seed oil. *Pediatrics.* 58, 650-654.

Gray, G. M., Yardley, H. J., (1975). Different populations of pig epidermal cells: isolation and lipid composition. *J. Lipid Res.* 16, 441-447.

Gray, G. M., Yardley, H. J., (1975). Lipid compositions of cells isolated from pig, human, and rat epidermis. *J. Lipid Res.* 16, 434-440.

Grubauer, G., Feingold, K. R., Harris, R. M., Elias, P. M., (1989). Lipid content and lipid type as determinants of the epidermal permeability barrier. *J. Lipid Res.* 30, 89-96.

Haftek, M., (2003). [Structural and ultrastructural data on human cutaneous lipids]. *Pathol. Biol. (Paris).* 51, 264-266.

Hansen, H. S., Jensen, B., (1985). Essential function of linoleic acid esterified in acylglucosylceramide and acylceramide in maintaining the epidermal water permeability barrier. Evidence from feeding studies with oleate, linoleate, arachidonate, columbinate and alpha-linolenate. *Biochim. Biophys. Acta.* 834, 357-363.

Holleran, W. M., Feingold, K. R., Man, M. Q., Gao, W. N., Lee, J. M., Elias, P. M., (1991). Regulation of epidermal sphingolipid synthesis by permeability barrier function. *J. Lipid Res.* 32, 1151-1158.

Holleran, W. M., Takagi, Y., Uchida, Y., (2006). Epidermal sphingolipids: metabolism, function, and roles in skin disorders. *FEBS Lett.* 580, 5456-5466.

Horrobin, D. F., (1992). Nutritional and medical importance of gamma-linolenic acid. *Prog. Lipid Res.* 31, 163-194.

Hou, S. Y., Mitra, A. K., White, S. H., Menon, G. K., Ghadially, R., Elias, P. M., (1991). Membrane structures in normal and essential fatty acid-deficient stratum corneum: characterization by ruthenium tetroxide staining and x-ray diffraction. *J. Invest. Dermatol.* 96, 215-223.

Hulbert, A. J., Rana, T., Couture, P., (2002). The acyl composition of mammalian phospholipids: an allometric analysis. *Comp. Biochem. Physiol. B. Biochem. Mol. Biol.* 132, 515-527.

Imokawa, G., Abe, A., Jin, K., Higaki, Y., Kawashima, M., Hidano, A., (1991). Decreased level of ceramides in stratum corneum of atopic dermatitis: an etiologic factor in atopic dry skin? *J. Invest. Dermatol.* 96, 523-526.

Kalia, Y. N., Pirot, F., Guy, R. H., (1996). Homogeneous transport in a heterogeneous membrane: water diffusion across human stratum corneum in vivo. *Biophys. J.* 71, 2692-2700.

Kawamura, A., Ooyama, K., Kojima, K., Kachi, H., Abe, T., Amano, K., Aoyama, T., (2011). Dietary supplementation of gamma-linolenic acid improves skin parameters in subjects with dry skin and mild atopic dermatitis. *J. Oleo. Sci.* 60, 597-607.

Kerscher, M. J., Korting, H. C., (1992). Treatment of atopic eczema with evening primrose oil: rationale and clinical results. *Clin. Investig.* 70, 167-171.

Lajoinie, A., Gelas, P., Salmon, D., Bergoin, C., Roussel, L., Falson, F., Chambrier, C., Souquet, J. C., Pivot, C., Haftek, M., Pirot, F., (2013). Impact of intravenous lipid emulsion infusion on the stratum corneum barrier function in patients receiving parenteral nutrition. *Eur. J. Dermatol.*

Lindskov, R., Holmer, G., (1992). Polyunsaturated fatty acids in plasma, red blood cells and mononuclear cell phospholipids of patients with atopic dermatitis. *Allergy*. 47, 517-521.

Macheleidt, O., Kaiser, H. W., Sandhoff, K., (2002). Deficiency of epidermal protein-bound omega-hydroxyceramides in atopic dermatitis. *J. Invest. Dermatol.* 119, 166-173.

Mantzioris, E., James, M. J., Gibson, R. A., Cleland, L. G., (1995). Differences exist in the relationships between dietary linoleic and alpha-linolenic acids and their respective long-chain metabolites. *Am. J. Clin. Nutr.* 61, 320-324.

Mao-Qiang, M., Elias, P. M., Feingold, K. R., (1993). Fatty acids are required for epidermal permeability barrier function. *J. Clin. Invest.* 92, 791-798.

Mascioli, E. A., Lopes, S. M., Champagne, C., Driscoll, D. F., (1996). Essential fatty acid deficiency and home total parenteral nutrition patients. *Nutrition*. 12, 245-249.

Matsumoto, M., Umemoto, N., Sugiura, H., Uehara, M., (1999). Difference in ceramide composition between "dry" and "normal" skin in patients with atopic dermatitis. *Acta Derm. Venereol.* 79, 246-247.

Meguro, S., Arai, Y., Masukawa, Y., Uie, K., Tokimitsu, I., (2000). Relationship between covalently bound ceramides and transepidermal water loss (TEWL). *Arch. Dermatol. Res.* 292, 463-468.

Melnik, B., Hollmann, J., Plewig, G., (1988). Decreased stratum corneum ceramides in atopic individuals--a pathobiochemical factor in xerosis? *Br. J. Dermatol.* 119, 547-549.

Menon, G. K., Cleary, G. W., Lane, M. E., (2012). The structure and function of the stratum corneum. *Int. J. Pharm.* 435, 3-9.

Miller, C. C., Tang, W., Ziboh, V. A., Fletcher, M. P., (1991). Dietary supplementation with ethyl ester concentrates of fish oil (n-3) and borage oil (n-6) polyunsaturated fatty acids induces epidermal generation of local putative anti-inflammatory metabolites. *J. Invest. Dermatol.* 96, 98-103.

Morse, P. F., Horrobin, D. F., Manku, M. S., Stewart, J. C., Allen, R., Littlewood, S., Wright, S., Burton, J., Gould, D. J., Holt, P. J., et al., (1989). Meta-analysis of placebo-controlled studies of the efficacy of Epogam in the treatment of atopic eczema. Relationship between plasma essential fatty acid changes and clinical response. *Br. J. Dermatol.* 121, 75-90.

Murata, Y., Ogata, J., Higaki, Y., Kawashima, M., Yada, Y., Higuchi, K., Tsuchiya, T., Kawainami, S., Imokawa, G., (1996). Abnormal expression of sphingomyelin acylase in atopic dermatitis: an etiologic factor for ceramide deficiency? *J. Invest. Dermatol.* 106, 1242-1249.

Obata, T., Nagakura, T., Masaki, T., Maekawa, K., Yamashita, K., (1999). Eicosapentaenoic acid inhibits prostaglandin D2 generation by inhibiting cyclo-oxygenase-2 in cultured human mast cells. *Clin. Exp. Allergy*. 29, 1129-1135.

Pilgram, G. S., Vissers, D. C., van der Meulen, H., Pavel, S., Lavrijsen, S. P., Bouwstra, J. A., Koerten, H. K., (2001). Aberrant lipid organization in stratum corneum of patients with atopic dermatitis and lamellar ichthyosis. *J. Invest. Dermatol.* 117, 710-717.

Pirot, F., Berardesca, E., Kalia, Y. N., Singh, M., Maibach, H. I., Guy, R. H., (1998). Stratum corneum thickness and apparent water diffusivity: facile and noninvasive quantitation in vivo. *Pharm. Res.* 15, 492-494.

Popa, I., Remoue, N., Osta, B., Pin, D., Gatto, H., Haftek, M., Portoukalian, J., (2012). The lipid alterations in the stratum corneum of dogs with atopic dermatitis are alleviated by topical application of a sphingolipid-containing emulsion. *Clin. Exp. Dermatol.*

Schalin-Karrila, M., Mattila, L., Jansen, C. T., Uotila, P., (1987). Evening primrose oil in the treatment of atopic eczema: effect on clinical status, plasma phospholipid fatty acids and circulating blood prostaglandins. *Br. J. Dermatol.* 117, 11-19.

Simonetti, O., Hoogstraate, A. J., Bialik, W., Kempenaar, J. A., Schrijvers, A. H., Bodde, H. E., Ponec, M., (1995). Visualization of diffusion pathways across the stratum corneum of native and in-vitro-reconstructed epidermis by confocal laser scanning microscopy. *Arch. Dermatol. Res.* 287, 465-473.

Simopoulos, A. P., (1991). Omega-3 fatty acids in health and disease and in growth and development. *Am. J. Clin. Nutr.* 54, 438-463.

Spector, A. A., (1999). Essentiality of fatty acids. *Lipids.* 34 Suppl, S1-3.

Stillwell, W., Wassall, S. R., (2003). Docosahexaenoic acid: membrane properties of a unique fatty acid. *Chem. Phys. Lipids.* 126, 1-27.

Strijbosch, R. A., Lee, S., Arsenault, D. A., Andersson, C., Gura, K. M., Bistrian, B. R., Puder, M., (2008). Fish oil prevents essential fatty acid deficiency and enhances growth: clinical and biochemical implications. *Metabolism.* 57, 698-707.

Tanaka, Zhen, Tagami, (1997). Normal recovery of the stratum corneum barrier function following damage induced by tape stripping in patients with atopic **dermatitis**. *Br. J. Dermatol.* 136, 966-967.

Tanaka, M., Zhen, Y. X., Tagami, H., (1997). Normal recovery of the stratum corneum barrier function following damage induced by tape stripping in patients with atopic dermatitis. *Br. J. Dermatol.* 136, 966-967.

Tretter, S., Mueller, R. S., (2011). The influence of topical unsaturated fatty acids and essential oils on normal and atopic dogs. *J. Am. Anim. Hosp. Assoc.* 47, 236-240.

Van Gool, C. J., Zeegers, M. P., Thijs, C., (2004). Oral essential fatty acid supplementation in atopic dermatitis-a meta-analysis of placebo-controlled trials. *Br. J. Dermatol.* 150, 728-740.

Vroman, H. E., Nemecek, R. A., Hsia, S. L., (1969). Synthesis of lipids from acetate by human preputial and abdominal skin in vitro. *J. Lipid Res.* 10, 507-514.

Weise, C., Heunemann, C., Loddenkemper, C., Herz, U., van Tol, E. A., Worm, M., (2011). Dietary docosahexaenoic acid in combination with arachidonic acid ameliorates allergen-induced dermatitis in mice. *Pediatr. Allergy Immunol.* 22, 497-504.

Wertz, P. W., (1992). Epidermal lipids. *Semin. Dermatol.* 11, 106-113.

Wertz, P. W., (2000). Lipids and barrier function of the skin. *Acta Derm. Venereol. Suppl. (Stockh).* 208, 7-11.

Wijendran, V., Hayes, K. C., (2004). Dietary n-6 and n-3 fatty acid balance and cardiovascular health. *Annu. Rev. Nutr.* 24, 597-615.

Wright, S., (1985). Atopic dermatitis and essential fatty acids: a biochemical basis for atopy? *Acta Derm. Venereol. Suppl. (Stockh).* 114, 143-145.

Wright, S., Burton, J. L., (1982). Oral evening-primrose-seed oil improves atopic eczema. *Lancet.* 2, 1120-1122.

Yamamoto, A., Serizawa, S., Ito, M., Sato, Y., (1991). Stratum corneum lipid abnormalities in atopic dermatitis. *Arch. Dermatol. Res.* 283, 219-223.

Yardley, H. J., Summerly, R., (1981). Lipid composition and metabolism in normal and diseased epidermis. *Pharmacol. Ther.* 13, 357-383.

Zerouga, M., Stillwell, W., Jenski, L. J., (2002). Synthesis of a novel phosphatidylcholine conjugated to docosahexaenoic acid and methotrexate that inhibits cell proliferation. *Anticancer. Drugs.* 13, 301-311.

Ziboh, V. A., Hsia, S. L., (1972). Effects of prostaglandin E 2 on rat skin: inhibition of sterol ester biosynthesis and clearing of scaly lesions in essential fatty acid deficiency. *J. Lipid Res.* 13, 458-467.

Ziboh, V. A., Miller, C. C., Cho, Y., (2000). Metabolism of polyunsaturated fatty acids by skin epidermal enzymes: generation of antiinflammatory and antiproliferative metabolites. *Am. J. Clin. Nutr.* 71, 361S-366S.

In: Omega-3 Fatty Acids
Editor: Md. Wasim Khan
ISBN: 978-1-62948-516-4

Chapter 6

Protective Effect of Omega-3 Polyunsaturated Fatty Acids against Antibiotic, Antihypertensive and Anti-Cancer Drug-Induced Nephrotoxicity and Oxidative Damage: A Focus on Cisplatin

Ashreeb Naqshbandi[1], *Md. Wasim Khan*[1,2], *Sana Rizwan*[1], *Sara Anees Khan*[1], *Shuhbha Priyamvada*[1] *and Ahad Noor Khan Yusufi*[1*]

[1]Department of Biochemistry, Faculty of Life Sciences, Aligarh Muslim University, Aligarh, UP, India

[2]DST INSPIRE Faculty, Cell Biology and Physiology Division, CSIR-Indian Institute of Chemical Biology, Kolkata, India

Abstract

Long-term clinical use of gentamicin (GM), sodium nitroprusside (SNP) and cisplatin (CP) is limited by the development of nephrotoxicity and oxidative damage. Dietary consumption of omega-3-polyunsaturated fatty acids (PUFA) and/or fish oil (FO) is known to have extensive health benefits. They have been shown to retard the progression of various forms of cancers, depression, diabetes, arthritis, and asthma and cardiovascular, neurological, and renal disorders and have been found to improve kidney functions. Here we discussed the protective effects of FO enriched in ω-3 PUFA consumption on GM-, SNP-, and CP-induced nephrotoxicity and oxidative damage.

* Corresponding Author: Professor A. N. K. Yusufi, (Retired). Department of Biochemistry, Faculty of Life Sciences. Aligarh Muslim University, Aligarh-202002, UP, India. E-mail: yusufi@lycos.com.

GM/SNP/CP administration resulted in severe renal damage as evident by an increase in serum creatinine, cholesterol, and blood urea nitrogen. Also, GM/SNP/CP administration led to a decrease in certain enzymes of carbohydrate metabolism, brush border membrane (BBM), and oxidative stress. BBM Pi transport in rat kidney was also reduced. However, dietary FO consumption markedly reversed GM/SNP/CP-induced adverse alterations by increasing energy metabolism, BBM integrity, and Pi transport. These drugs are known to generate ROS/RNOS that cause damage to mitochondria, microsomes, peroxisomes, lysosomes and BBM of renal proximal tubules thus alter their functions. Dietary ω-3 fatty acids affect membrane organization, fluidity, permeability and other properties by altering membrane fatty acids composition. FO-induced changes, at least in part, can also be attributed to improved antioxidant defense mechanisms that accelerate repair and regeneration of damaged cellular membranes. In conclusion, we propose that dietary FO consumption may provide a cushion for prolonged clinical use of these drugs by ameliorating drug-induced nephrotoxic and other deleterious effects.

Keywords: Fish oil; ω-polyunsaturated fatty acids; antioxidants; gentamicin; sodium nitroprusside; cisplatin; nephrotoxicity; oxidative damage; energy metabolism

Abbreviations

ACPase	Acid Phosphatase
ALP	Alkaline Phosphatase
ATP	Adenosine 5'-triphosphate
BBM	Brush border membrane
BBMV	Brush border membrane vesicles
BUN	Blood urea nitrogen
FBPase	Fructose,1,6,bisphosphatase
G6Pase	Glucose-6-phosphatase
G6PDH	Glucose-6-phosphate dehydrogenase
GGTase	γ-glutamyl transferase
GM	Gentamicin
GT	Green tea
HK	Hexokinase
HMP	Hexose monophosphate
LAP	Leucine aminopeptidase
LDH	Lactate dehydrogenase
LPO	Lipid peroxidation
MDH	Malate dehydrogenase
ME	Malic enzyme
Pi	Inorganic phosphate
ROS	Reactive oxygen species

SOD	Superoxide dismutase
TCA	Tricarboxylic acid cycle
w/v	weight/volume
SNP	Sodium nitroprusside
FO	Fish oil
EPA	Eicosapentaenoic acid
DHA	Docosahexaenoic acid
PUFA	Polyunsaturated fatty acids

1. Introduction

The last two decades have witnessed a major drift in the interest of the scientific community towards providing better means to containing the health risks of the human race. From ancient times, the physicians and scholars in Asia have understood that food have both preventive and therapeutic value and is an integral part of human health. Hippocrates (460-377 BC), the father of medicine, had said centuries ago that "Let thy food be thy medicine and thy medicine be thy food". The concept of disease prevention using naturally occurring substances that can be included in the diet consumed by human populations is gaining much interest. Of late, renewed interest and vigor has been showered on the role of omega fatty acids in health and disease. Fish oil enriched in ω-3 PUFA (EPA and DHA) provide one such dietary source of biologically active components that has been shown to be co-preventative and co-therapeutic in a wide variety of ailments (Doughman et al., 2007).

A number of environmental contaminants, chemicals and drugs including heavy metals, antibiotics, anti-hypertensive and anti-cancer drugs have been found to dramatically alter the structure and function of various tissues and produce multiple adverse effects in the kidney, liver and intestine (Parker, 1990; Ozturk et al., 1997; Kohn et al., 2005; Banday et al., 2008a, b). Histopathological and biochemical studies have revealed that exposure of certain chemicals/heavy metals such as trichloroethylene, chromium, uranium, and platinum, etc cause severe damage to kidney and alter the structure and functions of its proximal tubule leading to acute renal failure (ARF) (Sanchez et al., 2001; Fatima et al., 2004, 2005; Barbier et al., 2005; Banday et al., 2008a, b; Khan et al., 2009c). Gentamicin (GM) is an important aminoglycoside antibiotic effective against life-threatening bacterial infections (Mingeot-Leclercq and Tulkens, 1999a); cisplatin (CP) is an antineoplastic agent against diverse spectrum of malignancies (Cohen and Lippard, 2001) and sodium nitroprusside (SNP) is well-known anti-hypertensive drug (Chow and Hong, 2002). However, use of these drugs is limited by the development of nephrotoxicity and other adverse effects. Although both GM and CP alter the structure and function of kidney, they appear to be two dissimilar models of acute renal failure (ARF) (Annie et al., 2005). GM produces marked histopathological damage particularly to S1 and S2 segments, whereas CP primarily affects S3 segment of renal proximal tubules resulting in the loss of microvilli, alterations in the number and size of lysosomes and mitochondria, and tubular necrosis accompanied by functional disturbances of various cell organelles (Abdel-Gayoum et al., 1994; Kuhlmann et al., 1997; Townsend et al., 2003; Banday et al., 2008a). GM alters the structure and function of phospholipids (Cronin et al., 1986), CP causes renal damage by interacting with enzyme thiol groups in membranes and by targeting genomic DNA (Nakano and Gemba, 1989) whereas SNP induces adverse toxic effects by excessive production of NO (Khan et al., 2012a).

There is a continuous search for agents which can provide protection against renal impairment caused by GM, CP or other toxins. Many different agents and strategies have been used to ameliorate GM/CP nephrotoxicity in experimental animals (Ali, 2003, Nagai and Takano 2004; Ali and Al-Moundhri 2006). These strategies were mainly focused on the use of various antioxidants, including extracts from medicinal plants with antioxidant properties, agents interfering with metabolic pathways etc. However, a clinically viable approach has not yet been established.

It has been established that diet supplemented with fish oil (FO) enriched in ω-3 PUFA has profound beneficial health effects against various pathologies including cancers, cardiovascular disorders, diabetes, depression, arthritis, asthma and inflammatory and immune disorders of the kidney and intestine (De Caterina et al., 1994; Tsujikawa et al., 2000; Barbosa et al., 2003; Bhattacharya et al., 2006). Recently, ω-3 PUFA from some plants/seeds e.g. flaxseed oil (FXO) showed many similar health benefits as demonstrated by ω-3 PUFA from FO (Freese and Mutanen 1997; Ide et al., 2000; Kim and Choi 2001; Nannicini et al., 2006; Paschos et al., 2007). Although ω-3 PUFA have been shown as environmental deterrents (Watkins et al., 2007) for environmental contaminants, studies on their possible beneficial effects against drug/chemical induced nephrotoxicity are especially very limited. Preliminary reports showed that FO prevents GM and cyclosporine induced nephrotoxicity (Abdel-Gayoum et al., 1995; Thakkar et al., 2000). Recently it has been documented that ω-3 fatty acids protects against gentamicin and uranyl nitrate-induced nephrotoxicity (Priyamvada et al., 2008, 2010). Here we reviewed the protective effects of ω-3 PUFA from fish oil against GM/SNP/CP induced nephrotoxicity and oxidative damage with special focus on cisplatin.

1.1. Omega Fatty Acids - "Fats of Life"

"We are what we eat". The nature of the relation between diet and health/disease is the subject of great controversy. The dietary and body fat are essential for life. Dietary fat is one environmental variable that affects our health in different ways. The most dramatic change in what we eat has happened in the past century, with industrialization and development of the food industry, the intake of saturated fat, trans fat and especially ω-6 enriched refined oils have enormously increased whereas the consumption of ω-3 PUFA has considerably declined. These dietary imbalances in fat intake, in fact, are prime cause of modern sufferings like cancer, hypertension, diabetes, and depression, cardiovascular and renal disorders (Simopoulos, 2002a). Saturated fats mostly from animal based foods have been blamed for various pathologies whereas unsaturated fats in the diet were shown to have numerous health benefits (Deckelbaum et al., 2006). Early clues to the possible importance of unsaturated fatty acids especially polyunsaturated fatty acids (PUFA) came from studies showing that young humans and experimental animals experienced impaired growth when all fatty acids were removed from the diet. Small amounts of ω-3 (α-linolenic acid) or ω-6 (linoleic acid) fatty acids prevented that impaired growth and thus were termed as "essential fatty acids" (Burr and Burr, 1930). Polyunsaturated fatty acids (PUFA) were relatively found to be several-fold more effective (Turpeinen, 1937).

They are classified as follows:

(a) **Gifts from the sea**: omega-3 fatty acids (EPA and DHA)

The long chain ω-3 PUFA, eicosapentaenoic acid (EPA, C20:5 ω-3) and docosahexaenoic acid (DHA, C22:6 ω-3) are predominantly found in certain fish and sea mammals (tuna, trout, herring, salmon, sardine etc.). Fish oils are concentrated source of EPA and DHA (Simopoulos *et al.,* 1986).

(b) **Gifts from the land**: (i) omega-3 fatty acids (ii) omega-6 fatty acids

(i) *Omega-3 fatty acids: ALA*

α□-Linolenic acid (ALA, C18:3 ω-3) is found in chloroplast of green leafy vegetables (purslane, spinach etc.) and in oily seeds of canola, hempseed, walnuts and most notably in flaxseeds (richest source) etc. (Davis and Kris-Etherton, 2003) (ii) *Omega-6 fatty acids: LA*

Linoleic acid (LA, C18:2 ω-6) is plentiful in nature and is found in oily seeds of most plants except coconut, cocoa and palm. It is most predominant PUFA in western diet. Cooking oils with the greatest ω-6 fatty acid content include safflower oil; grape seed oil; sunflower oil; corn oil; cottonseed oil and soybean oil (Davis and Kris-Etherton, 2003).

1.1.1. Beneficial Health Effects of ω-3 PUFA from Marine Sources (EPA and DHA): "Medicinal fatty acids"

A number of studies have shown that both ω-3 and ω-6 PUFA have numerous health benefits. However, ω-3 PUFA especially from fish/marine foods were found to be more effective in lowering incidence of various pathologies compared to ω-6 PUFA rich diets (Simopoulos, 2001; Deckelbaum et al., 2006). This stems from the observations of the different prevalence of coronary heart disease and other chronic diseases (including psoriasis, bronchial asthma, diabetes mellitus and thyrotoxicosis) in the Greenland Inuit (Eskimo) population relative to Western populations (Kromann and Green, 1980). ω-3 PUFA (e.g. EPA and DHA) from marine fish and mammals were indicated as the main dietary factor responsible for such differences (Dyerberg et al., 1975; Bang et al., 1976).

The role of fish oil/ω-3 PUFA have been extensively investigated and received a great deal of attention in recent times as therapeutic option in variety of clinical situations ranging from CVD to hyperlipidemia, cancer, inflammatory and immune disorders , respiratory diseases, depression and diabetes and disorders of renal and gastrointestinal systems (Connor, 2000; Kakar et al., 2008). A number of investigations have demonstrated that ω-3 fatty acids present in fish oil have profound beneficial effects on murine function including both cell and humoral mediated immunity in autoimmune diseases such as lupus and rheumatoid arthritis (Grande and Donadio, 1998, 2001). FO/ω-3 fatty acids targeted cell cycle regulation/cell signaling may also underlie its protective effects (Yusufi et al., 2003). Recently it has been shown that FO may have a therapeutic potential as an adjunct therapy for human AIDS (Xi and Chen, 2000). DHA has been found to be essential in normal functional development of retina and brain (Simopoulos, 2000). The intake of ω-3 PUFA/fish oil or consumption of fish was shown to have inverse association with growth and development and rate of depression and other mental deficiencies (Simopoulos, 1991). Some of the specific beneficial effects of ω-3 PUFA are described as follows:

(A) Blood lipids and cardiovascular effects: The potential role of FO in CVD risk reduction first came from the observation involving Inuits in Greenland (Mouratoff et al., 1967). Epidemiological data also indicate a very low incidence of ischemic heart disease in Eskimos and Japanese population that consume a diet rich in seafood (Bang et al., 1976; Dyerberg and Bang, 1982; Schmidt, 1993). It has been reported that fish consumption reduce the incidence of deaths from coronary artery disease via effects on blood pressure, atherosclerosis and thrombogenesis (Kinsella et al., 1990; Connor, 1994; Leaf, 2008a, b). One of the most remarkable and consistent effects of ω-3 fatty acid supplementation in humans is their ability to decrease plasma concentration of triglycerides and VLDL (Harris, 1997; Foulon et al., 1999). PUFA have been shown to reduce total plasma cholesterol and improves LDL to HDL ratio (Dupont et al., 1990; Ruiz-Gutierrez et al., 1999). Dietary ω-3 FA retards the development of hypertension and related proteinuria (Rayner and Howe, 1995). Beneficial effects of ω-3 fatty acids from both plant (flaxseed oil) and marine sources have been demonstrated on all coronary artery disease, fatal and non fatal myocardial infarction, stroke, sudden cardiac death and all cause mortality (Schmidt, 1993; Lanzmann-Petithory, 2001; Nannicini et al., 2006; Posta et al., 2006;). Thus ω-3 fatty acids are "*Heart healthy*".

(B) Dietary lipids and cancer: During the past several decades, research surveys and animal studies have indicated that high-fat diets are associated with increased incidence and accelerated development of certain tumors (Fernandis and Venkatraman, 1993). However, unsaturated dietary fats are inversely related to the development and progression of various forms of cancers. Although dietary lipid itself is not usually considered a true carcinogen, it is postulated to enhance or promote development of tumor cells. In animal models ω-6 PUFA diets have stimulated development of four types of solid tumors (breast, colon, pancreas and prostrate), whereas ω-3 PUFA diets have diminished them (Cave, 1991). It has been found that ω-3 acids prevent the worsening of colon cancer while ω-6 fatty acids promote the growth of colon tumors (Anti et al., 1994). ω-3 fatty acids in combination with other nutrients may prove to be of particular value for preventing and treating breast cancer (Zhu et al., 1995). DHA and EPA can inhibit the growth of prostate cancer (Aronson et al., 2001). EPA and DHA have been shown to inhibit carcinogenesis by several molecular mechanisms including decreased/increased production of free radicals and ROS; suppression of arachidonic acid derived eicosanoid biosynthesis, gene expression (decreasing the expression of AP-1) and cell cycle regulation (Siddiqui et al., 2003), influences on transcription factor activity, signal transduction pathway and alteration of estrogen metabolism (Hardman, 2002; Larsson et al., 2004).

(C) Depression and other mental disorders: There are several lines of evidence relating diet with depression and other mental conditions. After all, *brain is all fat* and fats in the brain could easily be altered by diet. At present in the western societies the diets contain 15-20 times as much ω-6 as ω-3 PUFA and this imbalance is considered to be responsible for altered mental health (Hibbeln, 1998). The consumption of fish has now been linked with the rate of depression among various populations. In countries where *people eat the least fish the rate of depression is highest, and vice versa* (Hibbeln, 1998; Small, 2002). This correlation holds true across the world. The average New Zealander, for example, eats only 18 kg of fish a year and 6% of the population suffers from depression. In Japan, where they eat 64 kg of fish a year, depression accounts <1% of population. Low levels of ω-3 PUFA, especially DHA have been associated with several neurological and behavioral disorders, such as

depression, schizophrenia, Alzheimer's disease and attention deficit hyperactivity disorder (Conquer and Holub, 1997; Zamaria, 2004; Bourre, 2005), memory deficit (Gamoh et al., 1999), learning disability (Carrie et al., 1999) and visual acuity (Neuringer, 2000). Thus, it appears that ω- 3 PUFA in the diet ensures a "*happy and healthy brain*".

(D) Omega fatty acids and renal disorders: Recent studies have demonstrated that dietary supplementation with ω-3 PUFA retards renal disease progression in human and experimental animals (Donadio et al., 1994; Grande and Donadio, 2001; Donadio, 2001). In IgA nephropathy, the most common glomerulonephritis worldwide (D'Amico, 1987), the rate of renal disease progression was significantly reduced in patients given fish oil supplements (Donadio et al., 2001). Dietary ω-3 competes for arachidonic acid oxygenation pathway resulting in formation of biologically inactive end products thus exhibiting a favorable effect on glomerular injury, various forms of acute and chronic renal failure, allograft rejection, structural and functional renal damage in cyclosporine nephrotoxicity, hepatorenal syndrome, diabetic nephropathy, renal hypertension, vascular access thrombosis of end stage renal disease, idiopathic calcium urolithiasis, idiopathic immunoglobulin A nephropathy and lupus nephritis (Azar et al., 1989; De Caterina and Craven, 1993; De Caterina et al., 1994; Donadio, 2001; Baggio et al., 2002). Fish oil was also linked to the management of type 2 diabetes (Montori et al., 2000; Mostad et al., 2006).

(E) Omega fatty acids and gastrointestinal pathologies: In past few years there have been many advances in the area of small bowel physiology, pathology and therapy. Dietary PUFA influence the natural history of intestinal inflammatory diseases. Inflammatory bowel diseases (IBD) include: Crohn's disease (CD) and ulcerative colitis. ω-3 fatty acids found in fish oil have an anti-inflammatory activity and have given spectacular results in treatment of patients with CD (Belluzzi et al., 1996; Tsujikawa et al., 2000) and have shown marked clinical improvement of active, mild to moderate ulcerative colitis (Salomon et al., 1990; Aslan and Triadafilopoulos, 1992; Barbosa et al., 2003). Dietary FO has been shown to protect against ethanol-induced gastric mucosal injury (Leung, 1992).

1.1.2 Beneficial Health Effects of ω-3 PUFA from Plant Sources (ALA) in Health and Diseases -"Seeds of hope" for Vegetarians

The health benefits of dietary consumption of food rich in marine ω-3 PUFA are well established. ω-3 PUFA from certain plants/seeds e.g., walnut, canola and most notably from flax seeds (linseed) showed many similar health benefits as demonstrated by ω-3 PUFA from fish/fish oil. α-Linolenic acid (ALA), is a major ω-3 FA in plants (Harris, 2005). It is a precursor of EPA but not DHA and must be converted to EPA for its bioeffector role (Mantzioris et al., 1994). ALA is metabolized to EPA in humans to a significant extent (Valsta et al., 1996).

ALA from vegetable oils and EPA and DHA from fish oil have several parallel beneficial health effects (Freese and Mutanen, 1997; Ide et al., 2000; Kim and Choi, 2001). Recent epidemiologic studies have shown that dietary ALA is associated with a lower risk of CVD in men and women (Djousse et al., 2001). ALA has been associated with a lower rate of fatal and non-fatal coronary events (Nannicini et al., 2006). Hypotensive effect of flaxseed oil have been documented lately thus constituting another mechanism for cardioprotective effect of ALA (Paschos et al., 2007). The consumption of ALA rich flaxseed oil offers protective effects against cardiovascular disorders compared to LA-rich sunflower oil via their ability to decrease the tendency of platelet to aggregate (Allman, 1995) and inhibition of progression of arterial atherosclerosis (Lorgeril and Salen, 2004; Mozaffarian, 2005). ω-3 ALA has been shown to have greater hypolipidemic effect than ω-6 LA (Kim and Choi, 2001). Role of ALA in flaxseed oil in prevention of cancer has also been demonstrated (Kelley et al., 1991; Williams et al., 2007). Utility of flax and flaxseed oil in lupus nephritis and renal failure have also been accounted (Kelley et al., 1991; Basch et al., 2007).

Taken together, it appears that ω-3 PUFA both from marine and plant sources have many similar beneficial effects and that plant ω-3 PUFA are a great relief for those who are largely vegetarians.

1.2. Gentamicin

1.2.1. Pathophysiology

Aminoglycoside such as gentamicin (GM) are the most commonly used antibiotics worldwide in the treatment of gram-negative bacterial infections (Nagai and Takano, 2004). The high incidence of associated nephrotoxicity and ototoxicity (Lang and Liu, 1997) represents an important concern in the use of aminoglycoside antibiotics which have been implicated as one of the primary causes of drug induced acute-renal failure (Appel, 1990). The specificity of GM for renal toxicity is apparently related to its accumulation in the renal proximal convoluted tubule causing a number of morphological and biochemical alterations in humans and experimental animals (Kacew and Bergeron, 1990; Ali, 1995; Mingeot-Leclercq and Tulkens, 1999a). GM has been shown to cause marked histological damage in particular to renal proximal convoluted tubules (S_1 and S_2 subsegments) (Humes, 1988; Abdel-Gayoum et al., 1994) resulting in swelling, vacuolization and necrosis of epithelial cells and accumulation of myelin-like bodies (Ali and Bashir, 1994; Pedraza-Chaverri et al., 2000). The adverse interaction of the drug with critical intracellular processes leads to renal cortical phospholipidosis disrupting functions of membranes and organelles including BBM, BLM, mitochondria, lysosomes and microsomes (Cronin and Henrich, 1996; Laurent et al., 1990; Mingeot-Leclercq and Tulkens, 1999b; Sundin et al., 2001).

GM nephrotoxicity is often characterized by increase in serum creatinine and blood urea nitrogen (BUN) associated with massive proteinuria, enzymuria, glucosuria and phosphaturia (Banday et al., 2008a). Gentamicin has been reported to inhibit the oxidative phosphorylation in the isolated renal cortex and to decrease the total ATP content of rat kidney (Simmons *et* al., 1980). Recently GM has been shown to affect various enzymes of carbohydrate metabolism including glycolysis, TCA cycle, HMP shunt pathway and gluconeogenesis (Banday et al., 2008a).

The damage caused by GM to lysosomes, BBM and BLM was demonstrated by decrease in the activities of their biomarker enzymes in renal cortex associated with their increase in the urine (Banday et al., 2008a Khundmiri et al., 1997). GM has been shown to generate ROS and produce oxidative stress leading to cellular injury (Walker et al., 1999; Cuzzocrea et al., 2002; Ali, 2003; Banday et al., 2008a). Decreased expression of multiple genes by GM treatment has been elucidated by toxicogenomic studies (Amin et al., 2004; Kramer et al., 2004).)

1.2.2. Protection against Gentamicin Toxicity

Gentamicin, an effective and widely used aminoglycoside antibiotic is known to be potentially nephrotoxic despite close attention to the pharmacokinetics and dosing schedules of the drug (Ishikawa et al., 1985). Several approaches, utilizing different mechanisms, have been attempted to reduce nephrotoxicity of GM and related aminoglycoside antibiotics. These mechanisms include decreasing or preventing drug accumulation by the kidneys or decreasing binding to BBM (Mingeot-Leclercq and Tulkens 1999b). Many different agents and strategies have been reported to ameliorate GM nephrotoxicity in experimental animals (Ali, 2003; Nagai and Takano, 2004). Among them, protection against GM nephrotoxicity was mainly focused on the use of various antioxidant agents including the extracts from medicinal plants with antioxidant properties (Ali, 2003). Such medicinal plants include garlic (Pedraza-Chaverri et al., 2000), as well as diallyl sulfide, a compound isolated from garlic (Pedraza-Chaverri et al., 2003) and *Nigella sativa* oil (Ali, 2004). In experimental animals several strategies to ameliorate the toxicity have been attempted (Ali, 2003) that include modification of the diet (Paquette et al., 2002) and co-administration of antibiotics and certain agents such as thyroid hormone (Cronin et al., 1986) to mitigate renal toxicity.

In past few years, much interest has been centered on the role of naturally occurring dietary substances for the control and management of various chronic diseases such as cancer and cardiovascular disorders (Connor, 2000; Hardman, 2002; Larsson et al., 2004; Williams et al., 2007; Kakar et al., 2008; Leaf, 2008a, b). Fish oil at high doses has been found to exert protection against gentamicin nephrotoxicity (Ali and Bashir, 1994). Treatment with green tea modified the biochemical changes that occurred during gentamicin nephrotoxicity and thus was shown to have potential protective effect (Upaganlawar et al., 2006; Khan et al., 2009a).

Recently Priyamvada et al. (2008) have studied GM induced nephrotoxic alterations and oxidative damage and possible mechanism by which FO consumption exerts its protective effect in ameliorating GM induced nephrotoxicity. Multiple injections of GM (80 mg/kg body weight/day) were administered to control and FO-feeding rats daily for 10 days (Priyamvada et al., 2008). In confirmation of earlier reports (Banday et al, 2008a), GM increased serum creatinine and BUN, cholesterol but decreased serum glucose and Pi accompanied by massive proteinuria, enzymuria, glucosuria and phosphaturia indicating GM-induced neprotxicity. GM administration resulted in severe oxidative damage to the kidney as reflected by enhanced oxidative stress parameters (lipid peroxidation, LPO) and suppressed antioxidative parameters (total SH, SOD, catalse and GSH-peroxidase) in renal cortex and medulla. It was brush border membranes (BBM) of renal proximal tubules which was badly affected by GM administration as indicated by significant decrease in Na-dependant transport of Pi and the activities of alkaline phosphatase, γ-glutamyl transferase and leucine aminopeptidase, the maker enzymes of BBM. GM administration decreased the activity of MDH (TCA cycle), G6Pase and FBPase (gluconeogenesis) and malic enzyme but increased the activity of LDH (glycolysis) and G6PDH (HMP shunt pathway) showing disturbances in cellular energy metabolism. The decrease in MDH along with the increase in LDH activity demonstrates a shift in energy dependence from aerobic to anaerobic metabolism most likely due to GM induced mitochondrial damage. However, dietary FO given to GM-treated rats reduced nephrotoxicity parameters, enhanced antioxidant defense and improved cellular metabolism, BBM integrity by reversing GM-induced effects as reflected by specific alterations in various enzymes and other parameters studied (Priyamvada et al., 2008).

Recently dietary fish consumption was shown to reduce uranyl nitrate (UN) induced-nephrotoxicity and oxidative damage (Priyamvada et al., 2010) Dietary ω-3 PUFA are known to incorporate in cellular membranes and cause alterations in membrane organization, fluidity, and permeability that stabilize membrane bound components including enzymes and transport proteins and reduce lipid peroxidation (Kaur et al., 1996: Carrillo-Tripp et al., 2005).

1.3. Sodium Nitroprusside

Nitric oxide (NO), a pluripotent gaseous free radical is involved in a myriad of reactions *in vivo* that are both beneficial and harmful. It exerts potent actions in the regulation of cell function and tissue viability (Ignarro, 1989; Moncada et al., 1991; Culotta and Koshland, 1992). NO has also been shown to participate in a large number of pathophysiological conditions such as arthritis, atherosclerosis, cancer, diabetes, numerous degenerative neuronal diseases, stroke, and myocardial infarction to name a few (Culotta and Koshland, 1992, Gross and Wolin, 1995).

Various metabolites and donors of nitric oxide (NO) have been used as useful chemicals for the treatment of various diseases (Nickerson, 1970; Chow and Hong, 2002). Nitrate and nitrite are used as food preservatives and coloring agents in meat, fish products and cheese (Lundberg et al., 2004). In addition, potassium nitrate has been used as a diuretic agent and nitrite as a vasodilator, circulatory depressant or an antidote for cyanide poisoning (Bruijns, 1982; Chow and Hong, 2002). Sodium nitroprusside, an anti-hypertensive agent spontaneously generates NO. Nitric oxide (NO), which is synthesized in vivo from L-arginine (ARG), can be oxidized to nitrite or nitrate (Lundberg et al., 2008). In acidic stomach nitrite is spontaneously converted to NO that that can be metabolized to nitrite (Lundberg et al., 2008). Thus nitrate, nitrite and NO can be interconvert and recycled back and forth. NO which is normally useful has been shown to be toxic in excess.

1.3.1. Pathophysiology

Sodium nitroprusside (SNP) is an NO donor compound that has been used as an anti-hypertensive agent since the 1920s (Schmidt et al., 1992). SNP has been documented to induce genotoxicity as assessed by the measurement of micro-nucleated lymphocytes (Andreassi et al., 2001) and to cause DNA strand breaks (Asaka et al., 1997). SNP generates ROS during the redox cycling of nitroprusside (Ramakrishna et al., 1996). It is converted to nitrite, NO, cyanide and oxygen free radicals such as superoxide and hydroxyl radicals (Ramakrishna *et al.,* 1996). Since SNP is a NO donor, it can also be converted to peroxynitrite (ONOO) and thus can cause mutagenesis (Lin et al., 1998). The potential of SNP to induce apoptosis directly from NO liberation has been well established (Feldman et al., 1997). Thus NO and its derivatives (e.g. ONOO) can induce a variety of toxic effects including DNA damage, inhibition of DNA synthesis, cell cycle arrest (Wink et al., 1991; Nguyen et al., 1992) and inhibition of respiratory chain (Brown and Borutaite, 2002). On the other hand SNP has been shown to induce stimulation of glycolysis and glycolytic production of ATP (Maletic et al., 2000). However it may not be sufficient to compensate decrease in ATP production due to inhibition of oxidative phosphorylation. Hence total energy production was found to be decreased by SNP and SNP-induced NO formation. Thus, the mitochondria appear to be the primary site of SNP effects.

1.3.2. Protective Effect of Fish Oil ω-PUFA on Sodium Nitroprusside Induced Nephrotoxicity and Oxidative Damage in Rat Kidney

In past few years, much interest has been centered on the role of naturally occurring dietary substances for the control and management of various chronic diseases. A number of investigations have shown that ω-3 fatty acids retard the progression of various forms of cancers, depression, arthritis, and asthma, cardiovascular and renal disorders (De Caterina et al., 1994; Kakar et al., 2008). Recently it has been documented that ω-3 fatty acids protects against gentamicin and uranyl nitrate-induced nephrotoxicity (Priyamvada et al., 2008, 2010). Attempts were also made to reduce SNP-induced toxicity with the use of resveratrol and red wine and tea due to their antioxidant properties (Bastianetto et al, 2000; Paquay et al, 2000).

Recently Khan et al. (2012a) have studied SNP induced nephrotoxic alterations and oxidative damage and possible mechanism by which FO consumption exerts its protective effect in ameliorating SNP induced nephrotoxicity. SNP (1.5 mg/kg body weight/day) was administered to control and FO-feeding rats for 7 days (Khan et al., 2012a). SNP increased nephrotoxicity (serum creatinine and BUN) and oxidative stress parameters (lipid peroxidation, LPO) in renal tissues and suppressed antioxidative parameters (total SH, catalse and GSH-peroxidase). However, the activity of superoxide dismutase (SOD) increased in the cortex but decreased in the medulla indicating that severe damage has occurred to the kidney. It was brush border membranes (BBM) of renal proximal tubules which was badly affected by SNP administration as indicated by significant decrease in Na-dependant transport of Pi and the activities of alkaline phosphatase, γ-glutamyl transferase and leucine aminopeptidase, the maker enzymes of BBM. SNP administration resulted in down-regulation of malate dehydrogenase (MDH), glucose-6-phosphatase, fructose-1, 6-bisphosphatase and malic enzyme but up-regulated the activity of lactate dehydrogenase (LDH) and glucose-6-phosphate dehydrogenase indicating disturbances in cellular energy metabolism. The decrease in MDH might be due to SNP induced mitochondrial dysfunction whereas the increase in LDH activity demonstrates a shift in energy dependence from aerobic to anaerobic metabolism due to SNP induced mitochondrial damage. However, dietary FO given to SNP-treated rats reduced nephrotoxicity parameters and enhanced antioxidant defense, improved cellular metabolism and BBM integrity (Khan et al., 2012a). Similar protective effects were obtained when flaxseed oil was used as dietary source of ω-3 PUFA (Khan et al., 2012a).

Recently, Khan et al. (2012b; 2013) showed that dietary ω-3 PUFA enriched fish oil and flaxseed oil also prevented L-arginine and nitrite (NO donor/metabolite, respectively) induced nephrotoxicity and oxidative damage. The nephrotoxic and other adverse effects caused by NO donors/metabolites can be attributed, at least in part, to drug elicited ROS/RNOS species. Dietary ω-3 PUFA enriched fish oil and flaxseed oil to a larger extent prevented NO donors/metabolites induced nephrotoxicity parameters by enhancing nutrition/energy metabolism and ^{32}Pi transport capacity, by strengthening antioxidant defense mechanism and by altering membrane organization/fluidity/permeability.

1.4. Cisplatin

1.4.1. Pathophysiology

Cisplatin (cis-diamminedichloroplatinum II, CP) is a potent antitumor agent and is highly effective agent against a diverse spectrum of malignancies (Cohen and Lippard, 2001). However, the use of this agent in combating cancer is limited by the development of nephrotoxicity and ototoxicity (Ward and Fauvie, 1976; Ali and Al-Moundhri, 2006), along with various untoward side effects including nausea, vomiting, diarrhea and myelo-suppression (McKeage, 1995; Markman, 2003).

Cisplatin induced nephrotoxicity is associated with necrosis of the terminal portion of the proximal tubule and apoptosis, predominantly in cells in the distal nephron (Arany et al., 2004; Taguchi et al., 2005). Light and electron microscopy have shown that the CP induced injury and necrosis in the rat kidney are predominantly localized in the S3 segment of proximal tubule in the cortical-medullary region with or without accompanying distal changes (Townsend et al., 2003). A spectrum of changes was seen including focal loss of brush border, cellular swelling, and condensation of nuclear chromatin and focal areas of necrosis (Dobyan et al., 1980). The mitochondria, lysosomes and microsomes appeared to be critical targets of CP (Zhang and Lindup, 1993; Leibbrandt et al., 1995). Different studies have shown that CP-induced cytotoxicity is probably due to combination of insults including peroxidation of cell membrane, inhibition of protein synthesis and DNA damage in the kidney (Santos et al., 2007). It has been shown that CP induces renal damage by free radical generation, such as hydroxyl radical and superoxide anion, by altering arginine metabolism and by apoptosis (Fatima et al., 2004, Santos et al., 2007, Khan et al., 2009b). However, CP-induced oxidative stress was shown to be strongly involved in acute renal damage (Gonzales et al., 2005).

1.4.2. Protective Effects of ω-3 Fatty Acids Enriched Fish Oil (FO) against CP Induced Nephrotoxicity and Oxidative Damage

Several agents have been tested to ameliorate cisplatin-induced nephropathy and other adverse effects (Ali and Al-Moundhri, 2006). These include antioxidants, modulators of nitric oxide, agents interfering with metabolic pathways of CP, diuretics and cytoprotective and apoptotic agents (Ali and Al-Moundhri, 2006). However, none of these strategies were found to be safe for clinical use in protecting CP induced nephrotoxicity. Increasing number of studies has demonstrated that naturally occurring chemical moieties (neutraceuticals) hold great promise in combating several disorders. Green tea (GT) and ω-3 PUFA/fish oil are such biologically active dietary sources that have been shown to have extensive health benefits against several human pathologies (Simopoulos, 1991; Khan and Mukhtar, 2007) including cardiovascular diseases, respiratory diseases, diabetes, depression, cancers, inflammatory and immune renal disorders (De Caterina et al., 1994, Small, 2002). Preliminary reports showed that dietary fish oil prevents GM and cyclosporine induced nephrotoxicity (Abdel-Gayoum, 1995, Thakkar, 2000). Recently Khan et al. (2009a, b) have shown that GT consumption prevent GM- and CP-induced nephrotoxicity and oxidative damage whereas Priyamvada et al. (2008, 2010) have demonstrated protection against GM- and uranyl nitrated induced nephrotoxicity by dietary fish oil. The present work was undertaken to study detailed biochemical events/cellular response/mechanisms of cisplatin nephropathy and its protection by dietary fish oil. We now hypothesized that fish oil (FO) would prevent cisplatin-induced nephrotoxicity due to its intrinsic biochemical and antioxidant properties that would lead to improved metabolism and antioxidant defense mechanism of the kidney. The results of the present study demonstrate marked reversal of CP induced nephrotoxicity parameters in the serum and alterations in the enzymes of carbohydrate metabolism, BBM and antioxidant system by FO diet.

Animal experiments were conducted according to the guidelines of the Committee for Purpose of Control and Supervision of Experiments on Animals (CPCSEA), Ministry of Environment and Forests, Government of India. Adult male Wistar rats were used in the study. Multiple injections of CP (3 mg/kg body weight/day), were administered to control and fish oil (15%, w/v) consuming rats to induce nephrotoxicity every fifth day to control and FO-feeding rats over a period of 25 days as described earlier (Priyamvada et al., 2008; Khan et al., 2009b). The prolonged feeding of fish oil and multiple CP injections protocol was used that may be needed for cancer therapy. Various serum parameters, enzymes of carbohydrate metabolism, brush border membrane (BBM), mitochondria, lysosomes, and parameters of oxidative stress were analyzed to ascertain potential beneficial effects of FO against CP–induced nephrotoxicity and oxidative damage in the rat kidney. In general, the rats remained active and alert throughout the study. The daily food and fluid intake was similar in various experimental groups of rats (data not shown).

Effect of Dietary Fish Oil on CP Induced Nephrotoxicity in Serum Parameters

CP administration caused nephrotoxicity as characterized by increased serum creatinine, cholesterol, and BUN accompanied by a significant decrease in serum glucose and inorganic phosphate (Pi) (Table 1). Prolonged FO-diet consumption by CP-treated rats prevented CP induced increased concentrations of serum creatinine, cholesterol, and BUN indicating reduction in drug-induced nephropathy. Serum glucose and Pi, however, profoundly increased (Table 1).

Effect of Dietary Fish Oil on CP Induced Alterations in BBM and Lysosomal Enzymes in the Renal Homogenates and in Isolated BBMV

The renal proximal tubular BBM along with lysosomes and mitochondria have been shown as the major sites of CP-induced renal injury (Arany and Safirstein 2003; Fatima *et al.*, 2004 Khan et al, 2009b). Since BBM contains a number of hydrolytic enzymes and transport systems, the structural/functional integrity was assessed by the status of specific biomarker enzymes in the homogenates and isolated BBM vesicles under various experimental conditions. CP administration caused significant decrease in specific activities of ALP, GGTase, and LAP (BBM enzyme) in renal cortical as well as in medullary homogenates (Table 2). Feeding of FO diet alone significantly increased the activity of GGTase and LAP whereas ALP activity slightly decreased compared to control rats (Priyamvada et al., 2008). Moreover, CP treatment to FO- feeding rats markedly prevented CP induced decrease in various BBM enzyme activities both in the cortex and medulla (Table 2). The effect of CP and their combination on BBM marker enzymes was further analyzed in isolated BBMV preparations.

Table 1. Effect of dietary fish oil (FO) on serum parameters with / without CP treatment

Group / Groups	Creatinine (mg/dl)	BUN (mg/dl)	Cholesterol (mg/dl)	Phospholipids (mg/dl)	Phosphate (μmol/ml)	Glucose (mg/dl)
Control	1.00± 0.03	10.04 ± 0.90	89.53 ± 2.02	125.6 ± 3.46	2.56 ± 0.07	103.4 ± 6.71
CP	1.60 ± 0.06* (+60%)	19.71 ± 1.64* (+96%)	122.95 ± 5.42* (+37%)	188 ± 18.53* (+50%)	1.90 ± 0.06* (-26%)	92.69 ± 3.27* (-10%)
CPF	1.21 ± 0.03*† (+21%)	14.97 ± 1.36*† (+49%)	101.82 ± 2.98*† (+14%)	140.25 ± 3.91* (+12%)	2.21 ± 0.04*† (+14%)	125.7 ± 7.96*† (+22%)

Results are Mean ± SEM for five different preparations.

*Significantly different from control, †significantly different from CP at $p < 0.05$ by one way ANOVA. Values in parentheses represent percent change from control.

Table 2. Effect of dietary fish oil (FO) on biomarker enzymes of BBM and lysosomes in homogenates in a) Cortex and b) Medulla with/without CP treatment

Enzyme / Groups	ALP (μmol/mg protein/h)	GGTase (μmol/mg protein/h)	LAP (μmol/mg protein/h)	ACPase (μmol/mg protein/h)
a) Cortex Control	7.38 ± 0.14	22.36 ± 0.76	19.23 ± 0.29	4.36 ± 0.20
CP	4.63 ± 1.24* (-37%)	15.6 ± 2.04 (-30%)	8.4 ± 1.59* (-56%)	5.38 ± 0.12* (+23%)
CPF	5.4 ± 0.22*† (-27%)	20.64 ± 1.28† (-8%)	16.94 ± 3.42† (-12%)	4.63 ± 0.13† (+6%)
Medulla Control	52.53 ± 2.28	30.95 ± 2.09	19.16 ± 0.92	1.41 ± 0.08
CP	23.88 ± 2.22* (-55%)	20.51 ± 0.86* (-34%)	6.78 ± 0.33* (-65%)	2.3 ± 0.12* (+63%)
CPF	38.28 ± 0.81*† (-27%)	24.9 ± 0.07*† (-20%)	17.54 ± 0.94† (-8%)	1.89 ± 0.08 (+34%)

Results (specific activity expressed as μmoles / mg protein/h) are Mean ± SEM for five different preparations.

*Significantly different from control, †significantly different from CP at $p < 0.05$ by one way ANOVA. Values in parentheses represent percent change from control.

Table 3. Effect of dietary fish oil (FO) on activities of biomarker enzymes of BBM in cortical BBMV with / without CP treatment

Groups	ALP (μmol/mg protein/h)	GGTase (μmol/mg protein/h)	LAP (μmol/mg protein/h)
Control	67.36 ± 4.64	77.07 ± 4.65	43.21 ± 3.83
CP	28.89 ± 2.29* (-57%)	50.98 ± 3.76* (-34%)	25.2 ± 2.28* (-42%)
CPF	52.47 ± 1.87*† (-22%)	63.55 ± 3.37† (-17%)	37.87 ± 1.71 (-12%)

Results (specific activity expressed as μmoles / mg protein/h) are Mean ± SEM for five different preparations.

*Significantly different from control, †significantly different from CP at $p < 0.05$ by one way ANOVA. Values in parentheses represent percent change from control.

A similar activity pattern of the enzymes was observed as in cortical homogenates, however, the magnitude of the effects of CP and CP+FO were much more pronounced in BBMV than in the respective homogenate values (Table 3). The data suggests that CP caused severe damage to the structural architecture of BBM as reported earlier (Fatima et al., 2004; Khan et al., 2009b). FO feeding was able to maintain BBM integrity by lessening the damage caused or by increasing the regeneration processes (Priyamvada et al., 2008).

The effect of CP and CP+FO in was also determined on the activities of ACPase in renal tissues. The activity of ACPase (lysosomal enzyme) significantly increased by CP treatment by altering the number and size of lysosomes which results in modified lysosomal functions (Kuhlmann et al., 1997) which were significantly reduced by FO diet both in cortical and medullary homogenates albeit to different extent (Table 2).

Effect of Dietary Fish Oil on CP Induced Alterations on Metabolic Enzymes in Renal Cortex and Medulla

Kidneys play an essential role in the maintenance of total body fluid volume, its composition, and acid-base balance by selective reabsorption of various solutes, including Na and Pi (Coux et al., 2001).

Since these transports depend on structural integrity of BBM and available energy as ATP, which is generated by various metabolic pathways including glycolysis and oxidative metabolism, it is imperative that any alterations to these metabolic pathways caused by toxic insult or otherwise would modify the rate of these pathways thus affecting kidney functions (Khundmiri et al., 2004, 2005; Banday et al., 2008a, b).

The effect of CP and CP+ FO was determined on the activities of various enzymes of carbohydrate metabolism involved in glycolysis, TCA cycle, gluconeogenesis and HMP shunt pathway in renal cortex and medulla (Table 4). CP administration caused significant decrease in MDH, G6Pase, FBPase G6PDH but increase in LDH and to some extent in HK activity in both renal cortex and medulla. Although actual rates of glycolysis or TCA cycle were not determined, marked decrease in MDH activity appears to be due to CP-induced mitochondrial dysfunction (Kuhlmann et al., 1997; Khan et al., 2009b). At the same time, marked increase in LDH and to some extent in hexokinase activity appears to be an adaptive cellular effect for energy dependence from aerobic metabolism to mainly anaerobic glycolysis as shown by many toxicants (Banday et al., 2008a, b; Priyamvada et al., 2008, 2010, Khan et al., 2009a, b, c). When CP treatment was extended to FO-fed rats, FO diet consumption prevented CP-induced alterations in metabolic enzymes and improved over all energy situations (Table 4). FO alone has been shown to increase LDH, MDH, G6Pase, FBPase, and G6PDH activities (Priyamvada et al., 2008, 2010). Increased production of NADPH by G6PDH may support reductive anabolic reactions which might play a major role in strengthening antioxidant defense mechanism as reported earlier (Priyamvada et al., 2008; 2010).

Effect of Dietary Fish Oil on CP Induced Alterations in Antioxidant Defense Parameters in Renal Cortex and Medulla

A relationship between oxidative stress and nephrotoxicity has been well demonstrated in many experimental animal models (Devipriya and Shyamaladevi 1999). It has been reported that some drugs and toxicants, including GM. SNP and CP, exert their toxic effects by inducing generation of reaction oxygen species (ROS) and/or reactive nitrogen species RNOS (Fatima et al., 2004; Walker et al., 1999; Priyamvada et al., 2008; Khan et al., 2009a, b; Khan et al 2012a,).

Table 4. Effect of dietary fish oil (FO) on carbohydrate metabolic enzymes with / without CP treatment in homogenates of a) Cortex and b) Medulla

Groups	HK (µmol/mg protein/h)	LDH (µmol/mg protein/h)	MDH (µmol/mg protein/h)	G6Pase (µmol/mg protein/h)	FBPase (µmol/mg protein/h)	G6PDH (µmol/mg protein/h)	ME (µmol/mg protein/h)
a) Cortex							
Control	2.77±0.11	2.23 ± 0.2	17.17 ± 0.8	1.58± 0.09	3.72 ± 0.15	0.07± 0.01	0.51± 0.02
CP	3.37±0.09* (+22%)	3.8 ± 0.48* (+70%)	9.62 ± 0.73* (-46%)	1.21 ±0.09* (-23%)	3.01 ± 0.06 (-19%)	0.04± 0.01* (-43%)	0.69± 0.08* (+35%)
CPF	3.19 ±0.19 (+15%)	2.57 ± 0.11† (+13%)	15.84±0.32† (-8%)	1.27±0.1 (+5%)	3.05 ± 0.14* (+1%)	0.07±0.002*† (0%)	0.44±0.03*† (-14%)
b) Medulla							
Control	3.05±0.46	3.07±0.33	23.7 ± 0.76	2.73 ±0.16	3.75 ±0.22	0.96 ±0.24	0.932±0.011
CP	3.83±0.22* (+26%)	6.21±0.28* (+102%)	13.64±1.14* (-42%)	2.06±0.05* (-25%)	3.29 ± 0.1* (-38%)	1.12 ±0.03* (+17%)	0.687±0.017* (-27%)
CPF	3.12 ± 0.5† (+2%)	3.87±0.25† (+33%)	19.03±0.98*† (-20%)	2.4±0.29† (-12%)	3.49 ± 0.12 (+7%)	0.87±0.04*† (-9%)	0.796±0.013*† (-15%)

Results are Mean ± SEM for five different preparations. *Significantly different from control, †significantly different from CP at $p < 0.05$ by one way ANOVA. Values in parentheses represent percent change from control.

Table 5. Effect of dietary fish oil (FO) on enzymatic and non enzymatic antioxidant parameters in homogenates of a) Cortex and b) Medulla with / without CP treatment

Parameters / Groups	Lipid Peroxidation (nmol/gm tissue)	Total SH (μmol/gm tissue)	SOD (μmol/mg protein)	Catalase (μmol/mg protein/ min)	GSH-Px (μmol/mg protein/ min)
a) Cortex					
Control	47.111 ± 1.67	3.49 ± 0.55	60 ± 4.24	237 ± 7.95	0.15 ± 0.03
CP	148.48 ± 4.09* (+215%)	2.45. ± 0.77* (-30%)	38.14 ± 2.0* (-36%)	178.05 ± 7.85* (-25%)	0.1 ± 0.005* (-33%)
CPF	73.07 ± 5.62*† (+55%)	2.88 ± 0.28*† (-17%)	53.05 ± 0.87† (-12%)	215.12 ± 07.68† (-9%)	0.107 ± 0.01 (-29%)
b) Medulla					
Control	26.53 ± 0.81	1.44 ± 0.07	109.5 ± 6.43	198.95 ± 7.19	0.17 ± 0.01
CP	48.21 ± 1.07* (+82%)	1.13 ± 0.16* (-22%)	88.34 ± 4.31 (-19%)	133.15 ± 11.04* (-33%)	0.09 ± 0.01* (-47%)
CPF	38.38 ± 0.86*† (+45%)	1.82 ± 0.3*† (+26%)	118.07 ± 2.04† (+8%)	193.58 ± 13.71† (-3%)	0.12 ± 0.01*† (-29%)

Results are Mean ± SEM for five different preparations. *Significantly different from control, †significantly different from CP at $p < 0.05$ by one way ANOVA. Values in parentheses represent percent change from control.

However, oxidative stress can occur as a result of either increased ROS generation and/or decreased activity of antioxidant enzymes comprising SOD, catalase, and GSH-peroxidase. These antioxidant enzymes protect the cell against cytotoxic ROS. GM/SNP/CP have been shown to generate ROS and inhibit antioxidant enzymes in renal tissues (Fatima et al., 2004; Banday et al., 2008a; Priyamvada et al., 2008; Khan et al., 2009a, b; Khan et al., 2012a).

In agreement with previous studies, CP significantly decreased the activities of SOD, catalase and GSH-Px and increased LPO (measured in terms of malondialdehyde) both in the cortex and medulla (Table 5). CP caused significant decline in total-SH in these tissues. However, CP induced cellular damage was much more apparent in renal cortex than in medulla as reflected by greater increase in LPO and decrease in total-SH. In contrast, FO feeding along with CP treatment prevented CP elicited decline in SOD, catalase and GSH-Px activities. Dietary FO consumption also lowered drug induced LPO. The results indicate marked protection by FO diet against CP induced oxidative damage to renal tissues.

These biochemical studies clearly demonstrate that long-term CP administration (multiple injections) produces severe nephrotoxicity and causes profound damage to plasma membranes, mitochondria, peroxisomes, etc. of renal proximal tubules as reflected by the status of specific biomarkers whereas FO diet, major source of ω-3 fatty acids appeared to affect membrane organization and functions by altering membrane fatty acid composition as evident by increased activities of BBM enzymes and the enzymes of carbohydrate metabolism (Priyamvada et al., 2008, 2010).Most importantly, FO by activating endogenous antioxidant defense mechanism provided protection against CP-induced free radical attack. These results to larger extent are in agreement with recently published short-term (using single CP injection to induce nephrotoxicity) study on the protection of CP-induced (using single injection) nephrotoxicity by dietary fish oil (Naqshbandi et al., 2012).

Summary and Conclusion

As summarized in Figure 1, the present biochemical studies clearly demonstrate that long-term GM/SNP/CP administration elicited deleterious nephrotoxic effects by causing major damage to mitochondria, lysosomes, and in particular to BBM of renal proximal tubules as reflected by significant decrease in the activities of their specific biomarker enzymes, confirming morphologic and/or toxicogenomic observations that showed these organelles as prime targets.

These drugs, in general, seem to enhance the glycolytic enzyme LDH in the kidney in order to increase energy dependence on anaerobic glycolysis due to drug-induced mitochondrial dysfunction and depressed TCA cycle enzymes. A significant decrease in G6Pase and FBPase activities (gluconeogenic enzymes) also appeared to be due to mitochondrial damage. The decrease/increase in the activities of various enzymes by GM/SNP/CP might be due to suppression/induction of their synthesis as microsomes are reported to be possible targets. The drug induced decrease of serum phosphate and renal Pi transport suggests decrease in energy metabolism. The nephrotoxic and other adverse effects appeared to be mediated at least in part due to GM/SNP/CP elicited oxidative damage as evident by enhanced LPO and suppressed antioxidant enzymes. Although GM, SNP, and CP are different structures and exert different physiological effects but appear to have similar nephrotoxic effects. However, they seem to be different models of ARF as GM in known to affects early whereas CP affects late proximal tubule in the kidney. SNP appears to effect pars recta similar to CP.

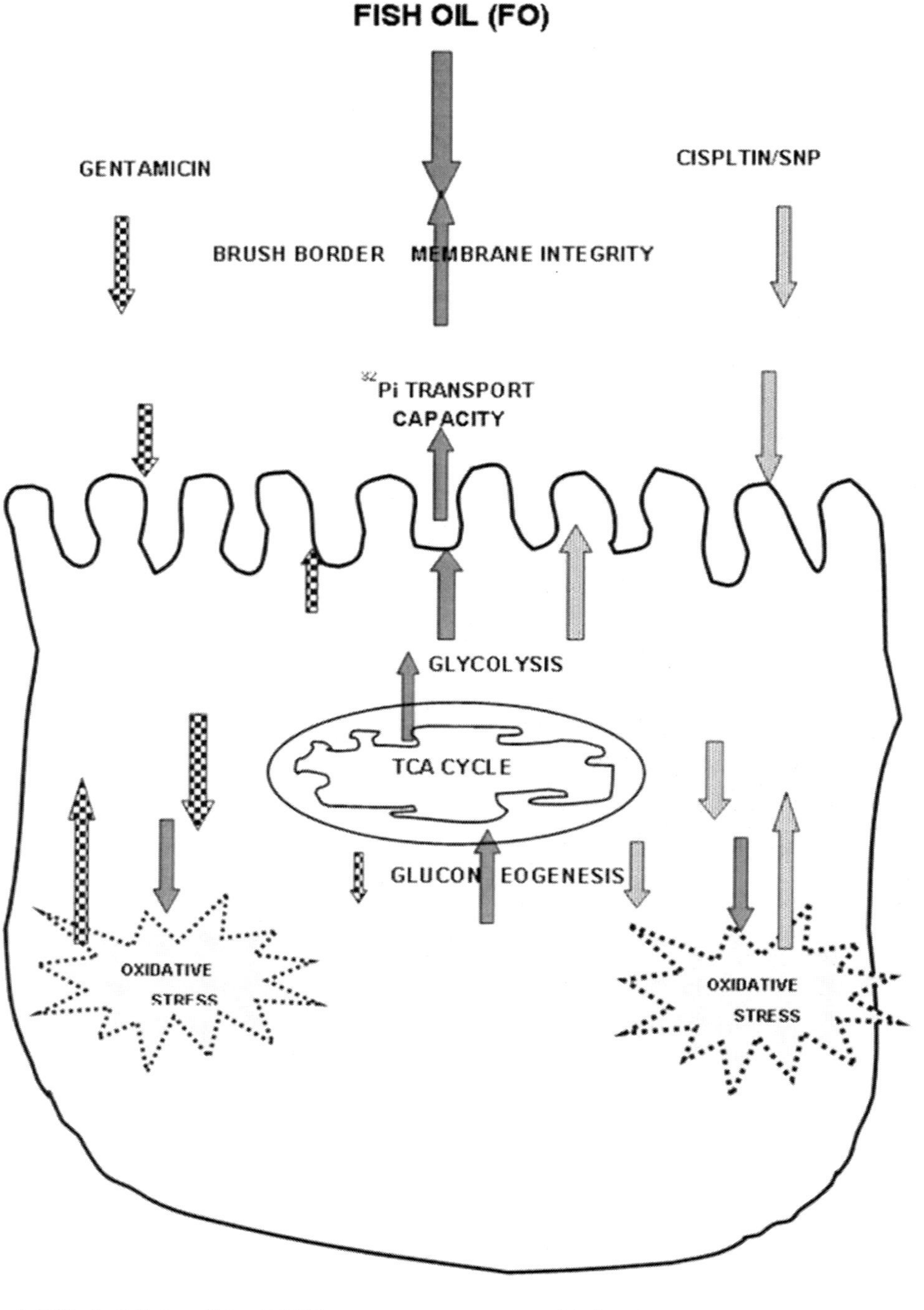

Figure 1. Effect of dietary fish oil (FO) on serum parameters with / without CP treatment.

In contrast, FO-diet feeding enriched in ω-3 fatty acids with CP treatment prevented CP-induced alterations in the activities of various enzymes of carbohydrate metabolism, BBM and oxidative stress. This is due to the fact that dietary ω-3 PUFA is incorporated in cellular membranes and changes their fatty acid compositions that affect membrane organization, fluidity, permeability and functions. FO appears to accelerate repair and/or regeneration of injured BBM and other cellular organelles as evident by increased activities of certain BBM and metabolic enzymes. Most importantly, by enhancing endogenous antioxidant defense mechanism FO provided protection from GM/SNP/CP-induced free radical attack and reduced lipid peroxidation whereas enhanced the activity of antioxidant enzymes. FO/ω-3 fatty acids targeted cell cycle regulation/cell signaling may also underlie its protective effects. Since FO has been known to possess antibacterial, antihypertensive and anticarcinogenic properties along with other known health benefits, it would be able to maximize the clinical use of GM, SNP and CP in exerting their physiological effects due to its intrinsic biochemical properties.

Taken together our present and past studies clearly demonstrate that diet has pivotal role in maintaining health and expression/prevention of disease. Dietary supplementation with omega-3 polyunsaturated fatty acids enriched fish oil/green tea, source of biologically active components provide an important option in the management of various diseases and drug induced toxicities with additive benefits and without harmful side effects. We suggest that to remain healthy and be protected from various diseases and environmental contaminants one should have frequent fish meal or FO supplement, a few kernels of walnuts, a snack of flaxseeds, and a few cups of green tea.

References

Abdel-Gayoum, A.A., Ali, B.H., Abdel-Razig, A.A., Bashir, A.A., Ghywarsha, K., 1994. Effect of gentamicin-induced nephrotoxicity on some carbohydrate metabolic pathways in the rat renal cortex. *Arch. Toxicol.* 68: 643-647.

Abdel-Gayoum, A.A., Bashir, A.A., El-Fakhri, M.M., 1995. Effect of fish oil and sunflower oil supplementations on gentamicin induced nephrotoxicity in rat. *Hum. Exp. Toxicol.* 14: 884-888.

Ali, B.H., Bashir, A.A., 1994. Effect of fish oil treatment on gentamicin nephrotoxicity in rats. *Ann. Nutr. Metab.* 38: 336-339.

Ali, B.H., Al-Moundhri, S.M., 2006. Agents Ameliorating or Augmenting The Nephrotoxicity Of Cisplatin And Other Platinum Compounds: A Review Of Some Recent Research. *Food Chem. Toxicol.* 44: 1173-1183.

Ali, B.H., 2003. Agents ameliorating or augmenting experimental gentamicin nephrotoxicity: some recent research. *Food Chem. Toxicol.* 41: 1447-1452.

Ali, B.H., 2004. The effect of Nigella sativa oil on gentamicin nephrotoxicity in rats. *Am. J. Clin. Med.* 32: 49-55.

Allman, M.A., Pena, M.M., Pang, D., 1995. Supplementation with flaxseed oil versus sunflower seed oil in healthy young men consuming a low fat diet: effects on platelet composition and function. *Eur. J. Clin. Nutr.* 49: 169-178.

Amin, R.P., Vickers, A.E., Sistare Frank, et al. 2004. Identification of putative gene-based markers of renal toxicity. *Environ. Health Perspect.* 112: 465-479.

Andreassi, M.G., Picano, E. Del Ry, S. et al., 2001. Chronic long-term nitrate therapy: possible cytogenetic effect in humans? *Mutagenesis* 16: 517–521.

Annie, S., Rajgopal, P.L., Malini, S., 2005. Effect of *Cassia auriculata* Linn. Root extract on cisplatin and gentamicin-induced renal injury. *Phytomedicine* 12: 555-560.

Anti, M., Arm Clau, F., Marra, G., et al., 1994. Effects of different doses of fish oil on rectal cell proliferation in patients with sporadic colonic adenomas. *Gastroenterol.* 107: 1892-1894.

Appel, G.B., 1990. Aminoglycoside nephrotoxicity. *Am. J. Med.* 88: 16S-20S.

Arany, I., Safirstein, R., 2003. Cisplatin nephrotoxicity. *Semin. Nephrol*. 23, 460–464.

Arany, I., Megyesi, J.K., Kaneto, H., Price, P.M., Safirstein, R.L., 2004. Cisplatin-induced cell death is EGFR/src/ERK signaling dependent in mouse proximal tubule cells. *Am. J. Physiol. Renal Physiol.* 287: F543-F549.

Aronson, W.J., Glaspy, J.A., Reddy, S.T., Reese, D., Heber, D., Bagga, D., 2001. Modulation of n-3 and n-6 polyunsaturated fatty acid ratios with dietary fish oils in men with prostrate cancer. *Urology* 58: 283-288.

Asaka, M., Takeda, H., Sugiyama, T., Kato, M., 1997. What role does Helicobacter pylori play in gastric cancer? *Gastroenterol.* 113: S56–S60.

Aslan, A., Triadafilopoulos, G., 1992. Fish oil fatty acid supplementation in active ulcerative colitis: A double-blind, placebo-controlled, crossover study. *Am J Gastroenterol.* 87: 432-437.

Azar, R., Dequeidt, F., Awada, J., Dequiedt, P., Tacquet, A., 1989. Effects of fish oil rich in polyunsaturated fatty acids on hyperlipidemia of hemodialysis patients. *Kid. Int.* 27: S239-S242.

Baggio, B., Budakovic, A., Priante, G., Gambaro, G., Manzato, E., Khan, S., 2002. Dietary fatty acid supplementation modulates the urinary excretion of calcium and oxalate in the rat. Insight into calcium lithogenesis. *Nephron.* 91: 486-491.

Banday, A.A., Farooq, N., Priyamvada, S., Yusufi, A.N.K., Khan, F., 2008a. Time dependent effects of gentamicin on the enzymes of carbohydrate metabolism, brush border membrane and oxidative stress in rat kidney tissues. *Life Sci.* 82: 450-459.

Banday, A.A., Priyamvada, S., Farooq, N., Yusufi, A.N.K., Khan, F., 2008b. Effect of uranyl nitrate on enzymes of carbohydrate metabolism and brush border membrane in different kidney tissues. *Food Chem. Toxicol.* 46: 2080-2088.

Bang, H.O., Dyerberg, J., Hjoorne, N., 1976. The composition of food consumed by Greenland Eskimos. *Acta. Med. Scand.* 200: 69-73.

Barbier, O., Jacquillet, G., Tauc, M., Cougnan, M., Poujeol, P., 2005. Effect of heavy metals on, and handling by kidney. *Nephron Physiol.* 99: 105-110.

Barbosa, D.S., Cecchini, R., El Kadri, M.Z., Rodriguez, M.A., Burini, R.C., Dichi, I., 2003. Decreased oxidative stress in patients with ulcerative colitis supplemented with fish oil omega-3 fatty acids. *Nutrition* 19: 837-842.

Basch, E., Bent, S., Collins, J., Dacev, C., Hammerness, P., Harrison, M., Smith, M., Szapary, P., Ulbricht. C., Vora, M., Weissner, W., 2007. Flax and Flaxseed Oil (Linum usitatissimum): A Review by the Natural Standard Research Collaboration. *J. Soc. Integr. Oncol.* 5: 92-105.

Bastianetto, S., Zheng, W.H., Quirion, R., 2000. Neuroprotective abilities of resveratrol and other red wine constituents against nitric oxide-related toxicity in cultured hippocampal neurons. *British J. Pharm.* 131: 711-720.

Belluzzi, A., Brignola, C., Campieri, M., Pera, A., Boschi, S., Miglioli, M., 1996. Effect of an enteric-caoted fish oil preparation on relapses in Crohn's disease. *The N. Eng. J. Med.* 334: 1557-1560.

Bhattacharya A, Ghosal S, Bhattacharya SK. Effect of fish oil on offensive and defensive factors in gastric ulceration in rats. *Prostaglandins Leukot Essent Fatty Acids* 2006; 74: 109-116.

Bourre, J.M., 2005. Where to find omega-3 fatty acids and how feeding animals with diet enriched in omega-3 fatty acids to increase nutritional value of derived products for human: what is actually useful? *J. Nutr. Health Aging* 9: 232-242.

Brown, G.C., Borutaite, V., 2002 Nitric oxide inhibition of mitochondrial respiration and its role in cell death. *Free Radic. Biol. Med.* 33: 1440-1450.

Bruijns E. Treatment of kidney stones with ammonium nitrate. *PhD Thesis*, University of Nijmegen 1982.

Burr, G.O., Burr, M.M., 1930. The nature and role of the fatty acids essential in nutrition. *J. Biol. Chem.* 86: 587-621.

Carrie. I., Clement, M., De Javel, D., Frances, H., Bourre, J.M., 1999. Learning deficits in first generation OF1 mice deficient in (n-3) polyunsaturated fatty acids do not result from visual alteration. *Neurosci. Lett.* 266: 69-72.

Carrillo-Tripp, M., and Feller, S.E., 2005. Evidence for a mechanism by which omega-3 polyunsaturated lipids may affect membrane protein function. *Biochemistry* 44: 10164-10169.

Cave, W.T., 1991. Dietary n-3 (ω-3) Polyunstaurated fatty acid, effects on animal tumorigenesis. *FASEB J* 5: 2160-2166.

Chow, C.K., Hong, C.B., 2002. Dietary vitamin E and selenium and toxicity of nitrite and nitrate. *Toxicol.* 180: 195-207.

Cohen, S.M., Lippard, S.J., 2001. Cisplatin: from DNA damage to cancer chemotherapy. *Prog. Nucleic Acid Res. Mol. Biol.* 67: 93-130.

Connor, W.E., 2000. Importance of n-3 fatty acids in health and disease. *Am. J. Clin. Nutr.* 71: 171S-175S.

Connor, W.E., 1994. n-3 fatty acids and heart disease. In: *Nutrition and disease update: Heart disease*, Kritchevsky D and Carroll KK (editors), Champaign, IL: American Oil Chemist's society, 7-42.

Conquer, J., Holub, B., 1997. Docosahexaenoic acid (n-3) and vegetarian nutrition. *Veg. Nutr. Int. J.* 1: 42-49.

Coux, G., Trurumper, L., Elias, M.M., 2001. Cortical Na^{+}-K^{+} ATPase activity, abundance and distribution after in vivo renal ischemia without reperfusion in rats. *Nephron* 89: 82-89.

Cronin, R.E., Henrich, W.L. Toxic nephropathy. In: *The kidney*, Brenner BM and Rector FC (editors), WB. Saunders Co. 1996; 2: 1680-1711

Cronin, R.E., Brown, D.M., Simonsen, R., 1986. Protection by thyroxine in nephrotoxic acute renal failure. *Am. J. Physiol.* 25: F408-F416.

Culotta, E., Koshland, D.E., 1992. NO news is good news. *Science* 258: 1862–1865.

Cuzzocrea, S., Mazzon, E., Dugo, L., Serraino, I., Paola, D.R., Britti, D., De Sarro, A., et al. 2002. A role for superoxide in gentamicin mediated nephropathy in rats. *Eur. J. Pharmacol.* 450: 67-76.

D'Amico G. The commonest glomerulonephritis in the world: IgA nephropathy. *Q J Med* 1987; 245: 709-727.

Davis BC and Kris-Etherton PM. Achieving optimal essential fatty acid status in vegetarians: current knowledge and practical implications. *Am J Clin Nutr* 2003; 78: 640S-646S.

De Caterina R and Craven PA. Eicosanoids in the pathogenesis of the functional and structural alterations of the kidney in diabetes. *Am J Kid Dis* 1993; 22: 727-735.

De Caterina R, Endres S, Kristensen SD, Schmidt EB. n-3 fatty acids and renal diseases. *Am J Kid Dis* 1994; 24: 397-415.

Deckelbaum RJ, Worgall TS, Seo T. n-3 fatty acids and gene expression. *Am J Clin Nutr* 2006; 83: 1520S-1524S.

Devipriya S, and Shyamaladevi CS. Protective effect of quercetin in cisplatin-induced cell injury in the rat kidney. *Indian J. Pharmaco.*1999; 31: 422-426.

Djoussé L, Pankow JS, Eckfeldt JH, Folsom AR, Hopkins PN, Province MA, Hong Y, Ellison RC. Relation between dietary linolenic acid and coronary artery disease in the National Heart, Lung, and Blood Institute Family Heart Study. *Am J Clin Nutr* 2001; 74: 612-619.

Dobyan DC, Levi J, Jacobs C, Kosek J, Weiner MW. Mechanism of cisplatinum nephrotoxicity: II. Morphologic observations. *J Pharmacol Exp Ther* 1980; 213: 551–6.

Donadio JV Jr, Larson TS, Bergstralh EJ, Grande JP. A randomized trial of high –dose compared with low-dose omega-3 fatty acids in severe IgA nephropathy. *J Am Soc Nephrol* 2001; 12: 791-799.

Donadio JV, Bergstralh EJ, Offord KP, Spencer DC, Holley KE. A controlled trial of fish oil in IgA nephropathy. *N Engl J Med* 1994; 331: 1194-1199.

Donadio JV. The emerging role of omega-3 polyunsaturated fatty acids in the management of patients with IgA nephropathy. *J Renal Nutr* 2001; 11: 122-128.

Doughman SD, Krupanidhi S, Sanjeevi CB. Omega-3 fatty acids for nutrition and medicine: considering microalgae oil as a vegetarian source of EPA and DHA. *Curr Diabetes Rev* 2007; 3: 198-203.

Dupont J, White PJ, Carpenter MP, Schaefer EJ, Meydani SN, Elson CF, Woods M, Gorbach SL. Food uses and health effects of corn oil. *J Am Coll Nutr* 1990; 9: 438-470.

Dyerberg J and Bang HO. A hypothesis on the development of acute myocardial infarction in Greenlanders. *Scand J Clin Lab Invest* 1982; 161: 7-13.

Dyerberg J, Bang HO, Hjorne N. Fatty acid composition of the plasma lipids in Greenland Eskimos. *Am J Clin Nutr* 1975; 28: 958-966.

Endres S and von Schacky C. n-3 polyunsaturated fatty acids and human cytokine synthesis. *Curr Opin Lipidol* 1996; 7: 48-52.

Fatima S, Arivarasu NA, Banday AA, Yusufi ANK, Mahmood R. Effect of potassium dichromate on renal brush border membrane enzymes and phosphate transport in rats. *Hum Exp Toxicol* 2005; 24: 631-638.

Fatima S, Yusufi ANK, Mahmood R. Effect of cisplatin on renal brush border membrane enzymes and phosphate transport. *Hum Exp Toxicol* 2004; 23: 547-554.

Feldmann G. Liver apoptosis. *J Hepatol* 1997; 26: 1-11.

Fernandis G and Venkatraman JT. Role of omega-3 fatty acids in health and disease. *Nutr Res* 1993; 13: S19-S45.

Foulon T, Richard MJ, Paven N, Bourrain JL, Beani JC, Laporte F, Hadjian A. Effects of fish oil fatty acids on plasma lipids and lipoproteins and oxidant-antioxidant imbalance in healthy subjects. *Scand J Clin Lab Invest* 1999; 59: 239-248.

Freese R and Mutanen M. Alpha-linolenic acid and marine long-chainn-3 fatty acids differ only slightly in their effects on haemostatic factors in healthy subjects. *Am J Clin Nutr* 1997; 66: 591-598.

Gamoh S, Hashimoto M, Sugioka K, Shahdat Hossain M, Hatta N, Misawa Y, Masumura S. Chronic administration of docosahexaenoic acid improves reference memory –related learning ability in young rats. *Neuroscience* 1999; 93: 237-241.

Gonzales R, Romay C, Borrego A, Hernande F, Zamor Z, and Rojas E. Lipid peroxides and antioxidant enzymes in cispaltin chronic nephrotoxicity in rats. *Mediators Inflamm.* 2005; 3: 139-143.

Grande JP and Donadio JV. n-3 Polyunsaturated fatty acid in the treatment of patients with IgA nephropathy, In: *Medicinal fatty acids in inflammation*, Kremer JM (editor), Birkhauser Veklag Basel/Switzerland, 1998; 125-140.

Grande JP and Donadio JV. Role of dietary fish oil supplementation in IgA nephropathy: mechanistic implications. *Minerva Urol Nephrol* 2001; 53: 201-209.

Gross SS and Wolin MS. Nitric oxide: Pathophysiological mechanisms. *Annu Rev Physiol*1995; 57: 737–769.

Hardman WE. Omega-3 fatty acids to augment cancer therapy. *J Nutr* 2002; 132: 3508S-3512S.

Harris WS. n-3 fatty acids and serum lipoproteins: human studies. *Am J Clin Nutr* 1997; 65: 1645S-1654S.

Harris WS. Alpha-linolenic acid: A gift from land? *Circulation* 2005; 111: 2872-2874.

Hibbeln J. Fish consumption and major depression. *The Lancet* 1998; 351: 1213.

Humes HD. Aminoglycoside nephrotoxicity. *Kid Int* 1988; 33: 900-911.

Ide T, Kobayashi H, Ashakumary L, Rouyer IA, Takahashi Y, Aoyama T, Hashimoto T, Mizugaki M. Comparative effects of perilla and fish oils on the activity and gene expression of fatty acid oxidation enzymes in rat liver. *Biochim Biophys Acta* 2000; 1485: 23-25.

Ignarro LJ. Endothelium-derived nitric oxide: Pharmacology and relationship to the actions of organic esters. *Pharm Res* 1989; 6: 651–659.

Ishikawa Y, Inui K, Hori R. Gentamicin binding to brush border and basolateral membranes isolated from rat kidney cortex. *J Pharm Dynam* 1985; 8: 931-941.

Kacew S and Bergeron MG. Pathogenic factors in aminoglycoside- induced nephrotoxicity. *Toxicol Lett* 1990; 51: 241-259.

Kakar P, Watson T, Gregory YH. New approaches to therapy with omega-3 fatty acids. *Curr Atheroscler Rep* 2008; 10: 79-87.

Kaur M, Kaur J, Ojha S, Mahmood A. Dietary fat effects on brush border membrane composition and enzyme activites in rat intestine. *Ann Nutr Metab* 1996; 40: 269-276.

Kelley DS, Branch LB, Love JE, Taylor PC, Rivera YM and Iacono JM. Dietary α-linolenic acid and immunocompetence in humans. *Am J Clin Nutr* 1991; 53: 40-46.

Khan MW, Priyamvada S, Khan SA, Khan S, Naqshbandi A, Yusufi ANK. Protective effect of ω-3 polyunsaturated fatty acids (PUFA) on Sodium nitroprusside induced nephrotoxicity and oxidative damage in rat kidney. *Human and Experimental Toxicology* 2012a; 31: 1035-1049.

Khan MW, Priyamvada S, Khan SA, Khan S, Naqshbandi A, Yusufi ANK. Protective effect of ω-3 polyunsaturated fatty acids (PUFA) on L-arginine induced nephrotoxicity and oxidative damage in rat kidney. *Human and Experimental Toxicology* 2012b; 10: 1022-1034.

Khan MW, Arivarasu NA, Priyamvada S, Khan SA, Khan S, Yusufi ANK. Studies on the protective effect of polyunsaturated fatty acids on sodium nitrite induced nephrotoxicity in rats. *Journal of Functional Foods* 2013; 2: 956-967.

Khan, N, and Mukhtar H. Tea polyphenols for health promotion. *Life Sci.* 2007; 81: 519-533.

Khan SA, Priyamvada S, Khan S, Khan MW, and Yusufi ANK. Protective effect of green tea extract on gentamicin-induced nephrotoxicity and oxidative stress in kidney and other rat tissues. *Pharmacol Res* 2009a; 59(4): 254-262.

Khan SA, Priyamvada S, Khan MW, Khan S, Farooq N, and Yusufi ANK. Studies on the protective effect of green tea against cisplatin induced nephrotoxicity. *Pharmacol Res* 2009b; 60: 382-391.

Khan S, Priyamvada S, Khan SA, Khan MW, Farooq N, Khan F and Yusufi ANK. Effect of trichloroethylene (TCE) toxicity on the enzymes of carbohydrate metabolism, BBM, and oxidative stress in kidney and other rat tissues. *Food Chem Toxicol* 2009c; 47:1562-1568.

Khundmiri, S.J., Asghar, M., Banday, A.A., Khan, F., Salim, S., Levi, M., Yusufi, A.N.K., 2004. Effect of ischemia and reperfusion on enzymes of carbohydrate metabolism in rat kidney. *J. Nephrol.* 17: 377-383.

Khundmiri, S.J., Asghar, M., Banday, A.A., Khan, F., Salim, S., Levi, M., Yusufi, A.N.K., 2005. Effect of reperfusion on sodium dependent phosphate transport in renal brush border membranes. *Biochim. Biophys. Acta* 1716: 19-28.

Khundmiri, S.J., Asghar, M., Khan, F., Salim, S., Yusufi, A.N.K., 1997. Effect of reversible and irreversible ischemia on marker enzymes of BBM from renal cortical PT subpopulations. *Am. J. Physiol.* 273: F849-F856.

Kim HK and Choi H. Dietary alpha-linolenic acid lowers postprandial lipid levels with increase of eicosapentaenoic and docosahexaenoic acid contents in rat hepatic membrane. *Lipids* 2001; 36: 1331-1336.

Kinsella, J.E., Lokesh, B., Stone, R.A., 1990. Dietary n-3 polyunsaturated fatty acids and amelioration of cardiovascular disease: possible mechanisns. *Am. J. Clin. Nutr.* 52: 1-28.

Kohn, S., Fradis, M., Robinson, E., Iancu, T.C., 2005. Hepatotoxicity of combined treatment with cisplatin and gentamicin in the guinea pig. *Ultrastruct. Pathol.* 29: 129-137.

Kramer, J.A., Pettit, S.D., Amin, R.P., et al. 2004. Overview of the application of transcription profiling using selected nephrotoxicants for toxicology assessment. *Environ. Health Perspect.* 112: 460-464.

Kromann, N., Green, A., 1980. Epidemiological studies in the Upernavik District, Greenland. *Acta. Med. Scand.* 208: 401-406.

Kuhlmann, K., Burkhardt, G., Kohler, H., 1997. Insights into potential cellular mechanisms of cisplatin nephrotoxicity and their clinical application. *Nephrol. Dial. Transplant.* 12, 2478-2480.

Lang, H., Liu, C., 1997. Apoptosis and hair cell degenerationin the vestibular sensoryepithelia of the guinea pig following gentamicin insult. Hear. Res. 111: 177-184.

Lanzmann-Petithory, D., 2001. Alpha-linolenic acid and cardiovascular diseases. *J. Nutr. Health Aging* 5: 179-183.

Larsson, S., Kulin, M., Ingelman-Sundberg, M., Wolk, A., 2004. Dietary long chain n-3 fatty acids for the prevention of cancer. *Am. Soc. Clin. Nutr.* 79: 935-945.

Laurent, G., Kishore, B.K., Tulkens, P.M., 1990. Aminoglycoside induced renal phospholipidosis and nephrotoxicity. *Biochem. Pharmacol.* 40: 2383-2392.

Leaf, A., Kang, J.X., Xiao, Y.F., 2008. Fish oil fatty acids as cardiovascular drugs. *Curr. Vasc. Pharmacol.* 6: 1-12.

Leaf, A., 2008. Historical overview of n-3 fatty acids and coronary heart disease. *Am. J. Clin. Nutr.* 87: 1978S-1980S.

Leibbrandt, M.E.I., Grushenka, H.I.W., Metz, A.L., Ozobia, A.A., Haskins, J.R., 1995. Critical subcellular targets of cisplatin and related platinum analogs in rat renal proximal tubule cells. *Kidney Int.* 48:761–70.

Leung, F.W., 1992. Fish oil protects against ethanol-induced gastric mucosal injury in rats. *Digest. Dis. Sci.* 37:636-637.

Lin, W., Xue, H., Liu, S., He, Y., Fu, J., Zhou, Z., 1998. Genotoxicity of nitric oxide produced from sodium nitroprusside. *Mutat. Res.* 413: 121–127.

Lorgeril de M., Salen, P., 2004. Alpha-linolenic acid and coronary heart disease. *Nutr. Metab. Cardiovasc. Dis.* 14: 162-169.

Lundberg, J.O., Weitzberg, E., Cole, J.A., Benjamin, N., 2004. Nitrate, bacteria and human health. *Nature Rev. Microbiol.* 2: 593–602.

Lundberg, J.O., Weitzberg, E., Gladwin, M.T., 2008. The nitrate-nitrite-nitric oxide pathway in physiology and therpeutics. *Nature* 7: 156-167.

Maletic, S.D., Dragicevic-Djokovic, L.J.M., Zikic, R.V., Stajn, A.S., Milenkovic, P., Kostic, M.M., 2000. Effects of nitric oxide donors on energy metabolism of rat erythrocytes. *J. Environ. Pathol. Toxicol. Oncol.* 19: 383-390.

Mantzioris, E., James, M.J., Gibson, R.A., Cheland, L.G., 1994. Dietary substitution with an ALA rich vegetable oil increases EPA concentrations in tissues. *Am. J. Clin. Nutr.* 59: 1304-1307.

Markman, M., 2003. Toxicities of the platinum antineoplastic agents. *Exp Opinion Drug Safety* 2: 597-607.

McKeage, M.J., 1995. Comparative adverse effect profiles of platinum drugs. *Drug Safety* 13: 228-224.

Mingeot-Leclercq, M., Tulkens, P.M., 1999a. Aminoglycoside nephrotoxicity. *Antimicrob. Agents Chemother.* 43: 1003-1012.

Mingeot-Leclercq, M., Tulkens, P.M., 1999b. Aminoglycosides: activity and resistance. *Antimicrob. Agents Chemother.* 43: 727-737.

Moncada, S., Palmer, R.M., Higgs, E.A., 1991. Nitric oxide: physiology, pathophysiology, and pharmacology. *Pharmacol. Rev.* 43: 109-42.

Montori, V.M., Wollan, P.C., Farmer, A., Dinneen, S.F., 2000. Fish oil supplementation in type 2 diabetes. *Diabetes Care* 23: 1407-1415.

Mostad, I.L., Bjerve, K.S., Bjorgass, M.R., Lydersen, S., Grill, V., 2006. Effects of n-3 fatty acids in subjects with type 2 diabetes: reduction of insulin sensitivity and time dependent alteration from carbohydrate to fat metabolism. *Am. J. Clin. Nutr.* 84: 540-550.

Mouratoff, G.J., Carroll, N.V., Scott, E.M., 1967. Diabetes mellitus in Eskimos. *JAMA* 199: 107-112.

Mozaffarian, D., 2005. Does alpha-linolenic acid intake reduce the risk of coronary heart disease? A review of the evidence. *Altern. Ther. Health Med.* 11: 24-30.

Nagai, J., Takano, M., 2004. Molecular aspects of renal handling of aminoglycosides and strategies for preventing nephrotoxicity. *Drug Metab. Pharmacokin.* 19: 159-170.

Nakano, S., Gemba, M., 1989. Potentiation of cisplatin-induced lipid peroxidation in kidney cortical slices by glutathione depletion. *Jpn. J. Pharmacol.* 50:67-92.

Nannicini, F., Sofi, F., Avanzi, G., Abbate, R., Gensini, G.F., 2006. Alpha-linolenic acid and cardiovascular diseases omega 3 fatty acids beyond eicosapentaenoic acid and docosahexaenoic acid. *Minerva Cardioangiol.* 54: 431-442.

Naqshbandi, A., Khan, M.W., Rizwan, S., Rehman, S.U., Khan, F., 2012. Studies on the protective effect of dietary fish oil on cisplatin induced nephrotoxicity in rats. *Food Chem. Toxicol.* 50: 265-273.

Neuringer, M., 2000. Infant vision and retinal function in studies of dietary long-chain polyunsaturated fatty acids: methods, results, and implications. *Am. J. Clin. Nutr.* 71: 256S-267S.

Nguyen, T., Brunson, D., Crespi, C.L., Penman, B.W., Wishnok, J.S., Tannenbaum, S.R., 1992. DNA damage and mutation in human cells exposed to nitric oxide in vitro. *Proc. Natl. Acad. Sci. USA* 89: 3030-3034.

Nickerson M. In: Goodman, L.S., Gilm1an, A. (Eds.), *The Pharmacological Basis of Therapeutics*, 4th edition. London, MacMillan, 1970; 745-763.

Ozturk, H.S., Kavutcu, M., Kacmaz, M., Canbolat, O., Durak, I., 1997. The effects of gentamicin on the activities of glutathione peroxidase and superoxide dismutase enzymes and malondialdehyde levels in heart tissues of guinea pigs. *Curr. Med. Res. Opin.* 14: 47-52.

Paquay, J.B.G., Haenen, G.R.M.M., Stender, G., Wiseman, S.A., Tijburg, L.B.M., Bast, A., 2000. Protection against Nitric Oxide Toxicity by Tea. *J. Agric. Food Chem.* 48(11): 5768–5772.

Paquette, M., Plante, I., Labrecque, G., Beauchamp, D., Thibault, L., 2002. Dietary composition alters gentamicin-induced nephrotoxicity in rats. *Physiol. Behav.* 77: 141-50.

Parker, D.S., 1990. Manipulation of the functional activity of the gut by dietary and other means (antibiotics/probiotics) in ruminants. *J. Nutr.* 120: 639-648.

Paschos, G.K., Magkos, F., Panagiotakos, D.B., Votteas, V., Zampelas, A., 2007. Dietary supplementation with flaxseed oil lowers blood pressure in dyslipidaemic patients. *Eur. J. Clin. Nutr.* 61: 1201-1206.

Pedraza-Chaverrí, J., González-Orozco, A.E., Maldonado, P.D., Barrera, D., Medina-Campos, O.N., Hernández-Pando, R., 2003. Diallyl disulfide ameliorates gentamicin-induced oxidative stress and nephropathy in rats. *Eur. J. Pharmacol.* 473: 71-78.

Pedraza-Chaverri, J., Maldonado, P.D., Medina-Campos, O.N., Olivares-Corichi, I.M., Grandos- Silvestre, M., Hernandez- Pando, R., Ibarra- Rubio, M.E., 2000. Garlic ameliorates gentamicin nephrotoxicity: relation to antioxidants. *Free Radic. Biol. Med.* 29: 602-611.

Posta, T.L., Gebauer, S.K., Kris-Etherton, P., 2006. Dietary omega-3 fatty acid intake and cardiovascular risk. *Am. J. Cardiol.* 98: 3i-18i.

Priyamvada, S., Khan, S.A., Khan, W.Md., Khan, S., Farooq, N., Yusufi, A.N.K., 2010. Studies on the protective effect of dietary fish oil on uranyl nitrate-induced nephrotoxicity and oxidative damage in rat kidney. *Prostaglandins Leukot. Essent. Fatty Acid* 82: 35-44.

Priyamvada, S., Priyadarshini, M., Arivarasu, N.A., Farooq, N., Khan, S., Khan, S.A., Khan, W.Md., Yusufi, A.N.K., 2008. Studies on the protective effect of dietary fish oil on gentamicin-induced nephrotoxicity and oxidative damage in rat kidney. *Prostaglandins Leukot. Essent. Fatty Acid* 78: 369-81.

Ramakrishna, R.D.N., Cederbaum, A.I., 1996. Generation of reactive oxygen species by the redox cycling of nitroprusside. *Biochim. Biophys. Acta.* 1289: 195–202.

Rayner, T.E., Howe, P.R.C., 1995. Purified ω-3 fatty acids retard the development of proteinuria in salt-loaded hypertensive rats. *J. Hypertens.* 13: 771-780.

Ruiz-Gutierrez, V., Perez-Espinosa, A., Vazquez, C.M., Santa –Maria, C., 1999. Effects of dietary fats (fish, olive and high-oleic acid sunflower oils) on lipid composition and antioxidant enzymes in rat liver. *Br. J. Nutr.* 82: 233-241.

Salomon, P., Kornbluth, A.A., Janowitz, H.D., 1990. Treatment of ulcerative colitis with fish oil n-3 omega fatty acid an open trial. *J. Clin. Gastroenterol.* 12: 157-161.

Sanchez, D.Z., Belles, M., Albina, M.L., Sirvent, J.J., Domingo, J.L., 2001. Nephrotoxicity of simultaneous exposure to mercury and uranium in comparison to individual effects of these metals in rats. *Biol. Trace Elem. Res.* 84: 139-154.

Santos, N.A., Catao, C.S., Martins, N.M., Curti, C., Bianchi, M.L., Santos, A.C., 2007.Cisplatin induced nephrotoxicity is associated with oxidative stress, redox state unbalance, impairment of energetic metabolism and apoptosis in rat kidney mitochondria. *Arch. Toxicol* 81: 495–504.

Schmidt, E.B., 1993. n-3 polyunsaturated fatty acids and ischaemic heart disease. *Curr. Opinion Lipidol.* 4: 27-33.

Schmidt, H.H., Warner, T.D., Murad, F., 1992. Double-edged role of endogenous nitric oxide. *Lancet* 339: 986.

Siddiqui, R.A., Jenski, L.J., Harvey, K.A., Wiesehan, J.D., Stillwell, W., Zaloga, G.P., 2003. Cell-cycle arrest in Jurkat Leukaemic cells: a possible role for docosahexaenoic acid. *Biochem. J.* 371: 621-629.

Simmons, C.F., Jr, Boqusky, R.T., Humes, H.D., 1980. Inhibitory effect of GM on renal mitochondrial oxidative phosphorylation. *J. Pharmacol. Exp. Ther.* 214: 709-715.

Simopoulos, A.P., Kifer, R.R., Martin, R.E. (editors): "*Health effects of polyunsaturated fatty acids in seafoods*". Orlando, FL: Academic Press, 1986.

Simopoulos, A.P., 2002b. Omega-3 fatty acids in wild plants nuts and seeds. *Asia Pac. J. Clin. Nutr.* 11: S163-S173.

Simopoulos, A.P., 2000. Human requirement for n-3 polyuinsaturated fatty acids. *Poult. Sci.* 79: 961-970.

Simopoulos, A.P., 1991. Omega-3 fatty acids in health and disease and in growth and development. *Am. J. Clin. Nutr.* 54: 438-463.

Simopoulos, A.P., 2001. The Mediterranean diets: what is so special about the diet of Greece? The scientific evidence. *J. Nutr.* 131: 3065S-3073S.

Small, M.F., 2002. The happy fat. *New Scientist* 34-37.

Sundin, D.P., Sandoval, R., Molitoris, B.A., 2001. Gentamicin inhibits renal protein and phospholipid metabolism in rats: implications involving intracellular trafficking. *J. Am. Soc. Nephrol.* 12: 114-123.

Taguchi, T., Nazneen, A., Abid, M.R., Razzaque, M.S., 2005. Cisplatin-associated nephrotoxicity and pathological events. *Contrib. Nephrol.* 148: 107-121.

Thakkar, R.R., Wang, O.L., Zerouga, M., Stillwell, W., Haq, A., Kissling, R, *et al.,* 2000. Docosahexaenoic acid reverses cyclosporin-A induced changes in membrane tructure and function. *Biochim. Biophys. Acta* 1474: 183-195.

Townsend, D.M., Deng, M., Zhang, L., Lapus, M.G., Hanigan, M.H., 2003. Metabolism of cisplatin to a nephrotoxin in proximal tubule cells. *J. Am. Soc. Nephrol.* 14:1–10.

Tsujikawa, T., Satoh, J., Uda, K., Ihara, T., Okamoto, T., Araki, Y., Sasaki, M., Fujiyama, Y., Bamba, T., 2000. Clinical importance of n-3 fatty acid rich diet and nutritional education for the maintenance of remission in Crohn's disease. *J. Gastroenterol.* 35: 99-104.

Turpeinen, O., 1937. Further studies on the unsaturated fatty acids essential in nutrition. *J. Nutr.* 15: 351-366.

Upaganlawar, A., Farswan, M., Rathod, S., Balaraman, R., 2006. Modification of biochemical parameters of gentamicin nephrotoxicity by coenzyme Q10 and green tea in rats. *Indian J. Exp. Biol.* 44: 416-418.

Valsta, L.M., Salminen, I., Aro, A., Mutanen, M., 1996. Alpha-linolenic acid in rapeseed oil partly compensates for the effect of fish restriction on plasma long chain n-3 fatty acids. *Eur. J. Clin. Nutr.* 50: 229-235.

Walker, P.D., Barri, Y., Shah, S.V., 1999. Oxidant mechanisms on gentamicin nephrotoxicity. *Renal Fail.* 21: 433-442.

Ward, J.M., Fauvie, K.A., 1976. The Nephrotoxic Effects of Cis-Diamminedichloroplatinum II in Male F344 Rats. *Toxicol. Appl. Pharmacol.* 38:535-547.

Watkins, B.A., Hannon, K., Ferruzzi, M., Li, Y., 2007. Dietary PUFA and flavonoids as deterrents for environmental pollutants. *J. Nutr. Biochem.* 18: 196-205.

Williams, D., Verghese, M., Walker, L.T, Boateng, J., Shackelford, L., Chawan, C.B., 2007. Flax seed oil and flax seed meal reduce the formation of aberrant crypt foci (ACF) in azoxymethane-induced colon cancer in Fisher 344 male rats. *Food Chem. Toxicol.* 45: 153-159.

Wink, D.A., et al. 1991. DNA deaminating ability and genotoxicity of nitric oxide and its progenitors. *Science* 254: 1001-1003.

Xi, S., Chen, L.H., 2000. Effects of dietary fish oil on tissue glutathione and antioxidant defense enzymes in mice with murine AIDS. *Nutr. Res.* 20: 1287-1299.

Yusufi, A.N.K., Cheng, J., Thompson, M.A., Walker, H.J., Gray, C.E, Warner, G.M., and Grande, J.P., 2003. Differential effects of low-dose docosahexaenoic acid and eicosapentaenoic acid on the regulation of mitogenic signaling pathways in mesangial cells. *J. Lab. Clin. Med.* 141: 318-329.

Zamaria, N., 2004. Alteration of polyunsaturated fatty acid status and metabolism in health and disease. *Reprod. Nutr. Dev.* 44: 273-282.

Zhang, J.G., Lindup, W.E., 1993. Role of mitochondria in cisplatin-induced oxidative damage exhibited by rat renal cortical slices. *Biochem. Pharmacol.* 45: 2215–22.

Zhu, Z.R., Mannisto, J.A.S., Pietinene, P. 1995. Fatty acid composition of breast adipose tissue in breast cancer patients and patients with benign breast disease. *Nutr. Cancer* 24: 151-160.

In: Omega-3 Fatty Acids
Editor: Md. Wasim Khan

ISBN: 978-1-62948-516-4

Chapter 7

Role of Supplementation with Omega-3 Fatty Acids in Preventing Cardiac Arrhythmias and Sudden Cardiac Death

Bernhard Rauch*[*1] and Jochen Senges[2]

[1]ZAR- Zentrum für Ambulante Rehabilitation am Klinikum der Stadt Ludwigshafen

[2]Institut für Herzinfarktforschung Ludwigshafen, Germany

Abstract

For several decades supplementation with omega-3 polyunsaturated fatty acids (Ω-3) has been believed to reduce cardiovascular risk in primary and secondary prevention, and potential risk reduction especially was attributed to prevention of tachyarrhythmias and sudden cardiac death (SCD). However, recent randomized controlled trials and meta-analyses challenge a general efficacy of Ω-3 supplementation in reducing cardiovascular risk. Especially, an anti-arrhythmic effect as proposed in earlier studies cannot be expected in general, and under certain conditions even pro-arrhythmic effects may occur. Based on molecular interactions of Ω-3 at the cellular level and results of earlier and recent experimental animal studies the following chapter will critically discuss recent clinical studies, emphasizing the individual background conditions that may determine the clinical outcome of Ω-3 supplementation. By this way the hypothesis will be delineated, that the efficacy of Ω-3 supplementation to prevent cardiac arrhythmias strongly depends on the underlying clinical and pharmacological conditions and – at the present stage of knowledge - may be difficult to be predicted in some cases.

Keywords: Omega-3 polyunsaturated fatty acids; cardiac arrhythmia; prevention; cardiovascular disease; myocardial infarction; death, sudden

---

* Corresponding author address: Email: rauch@zar-kardio-ludwigshafen.de.

Introduction

Omega-3 polyunsaturated fatty acids (Ω-3), especially eicosapentaenoic acid (EPA) and docosahexaenoic acid (DHA), if supplemented to daily nutrition, have been believed for decades to safely reduce cardiovascular risk. Indeed, this hypothesis was supported by observational and clinical studies as well as by experimental animal studies, and there was evidence that prevention of tachyarrhythmias could be a major mechanism of the Ω-3 related cardiovascular risk reduction.

In the meantime, a huge amount of data has been accumulated on this topic and an impressive number of position papers, reviews and meta-analyses have been published (Bucher HC et al. 2002; Whelton SP et al. 2004; Dhein S et al. 2005; Reiffel JA & McDonald A 2006; Wang C et al. 2006; Siddiqui RA et al. 2008; Marik PR & Varon J 2009; Mozaffarian D et al. 2011).

The primary enthusiasm, however, declined in recent years, as more and more clinical and experimental data were published, that challenged a general beneficial effect of Ω-3, and systematic reviews and meta-analyses conducted more recently were not able anymore to demonstrate a clear beneficial effect of Ω-3 supplementation (Yzebe D & Lievre M 2004;. Hooper L et al. 2006; Léon H et al. 2008; Jenkins DJA et al. 2008a; Cheng JWM & Santoni F 2008; Zhao YT et al. 2009; Filion KB et al. 2010; Kotwal S et al. 2012; Rizos EC et al. 2012). This partly may be due to the rapid development in prevention and treatment of cardiovascular disease including effective medical treatment (i.e. beta-blockers, statins, ACE-inhibitors, platelet aggregation inhibitors), early revascularization in acute myocardial infarction and finally increasing implementation of protective life style changes (i.e. exercise training, mediterranean diet including fish consumption, smoking cessation).

Even more important may be to reflect the interactions of Ω-3 at the molecular and cellular level. These interactions are complex and the biological effects strongly depend on the specific conditions at the cellular level, which makes it difficult to safely predict clinical effects (Lombardi F & Terranova P 2007; Den Ruijter HM et al. 2007; Rauch B & Senges J 2012, Billman GE et al. 2012).

Actual Knowledge from Experimental and Clinical Studies

Molecular and Cellular Interactions

The molecular and cellular interaction has been delineated extensively in several reviews (Leaf A et al. 2005b; Dhein S et al. 2005; McLennan PL & Abeywardena MY 2005; Lombardi F 2007; Den Ruijter HM et al. 2007; Siddiqui RA et al. 2008). In the following some important aspects potentially explaining heterogeneous clinical effects of Ω-3 on cardiac rhythm are summarized:

There are three major ways in which Ω-3 may interfere with cellular and membrane function, thereby potentially moderating cardiac rhythm (see also Figure 1):

a) *Direct interactions of Ω-3 with membrane bound proteins* like the fast sodium channel, the voltage-gated L-type Ca^{2+} channel, specific potassium channels, and the Na^{+}/Ca^{++}-exchanger (Hallaq H et al. 1990 & 1992; Honore E et al. 1994; Xiao YF et al. 1995 & 1997; Kang JX & Leaf A 1996, Leifert WR et al. 1999; Leaf A 2003; Den Ruijter HM et al. 2007; Wang RX et al. 2010). Such interactions may occur predominantly with circulating Ω-3, when it is delivered by acute administration and infusion.
b) *Incorporation into the phosholipid bilayer*, thereby potentially changing membrane fluidity, and/or forming Ω-3 rich micro-domains, and/or interacting with internal binding sites. This may result in a change of the function of membrane bound proteins like ion channels, receptors and signal transduction systems (McMurchie EJ et al. 1988; Croset M et al. 1989; Kinoshita I et al. 1994; Grynberg A et al. 1996; Leifert WR et al. 1999 &2000; McLennan PL 2001; Den Ruijter HM et al. 2007). Incorporation into the cellular membranes predominantly is achieved by dietary long-term administration of Ω-3.
c) *Interaction with intracellular pathways* including gene expression and metabolism of phospho-inositides (Judé S et al. 2006).

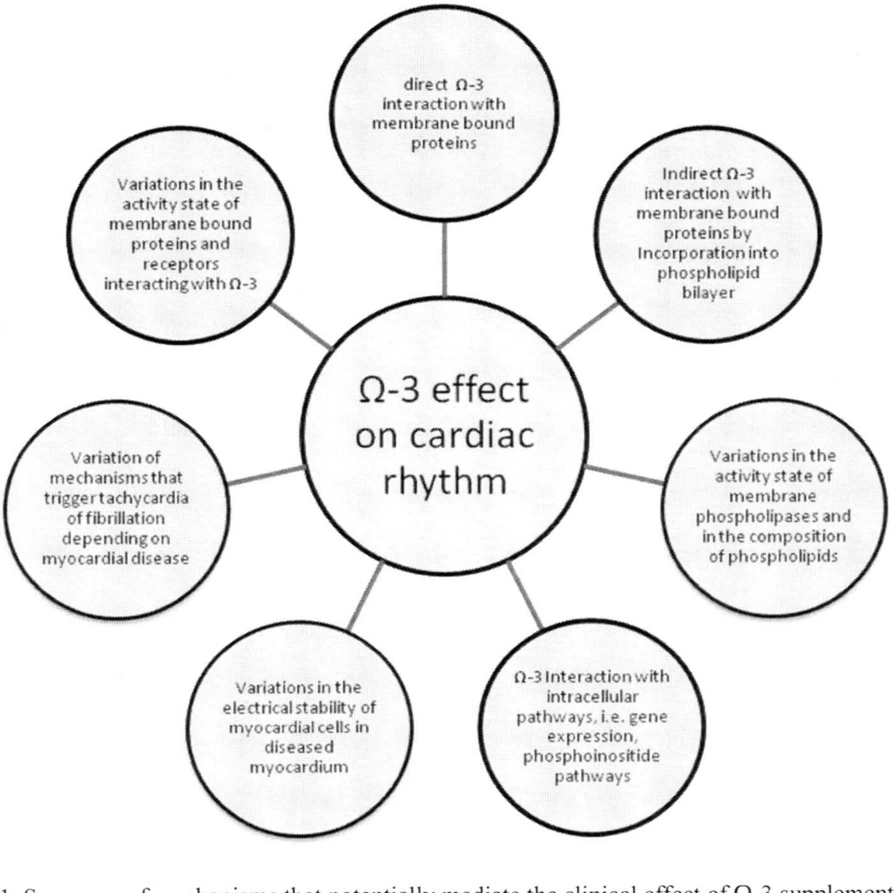

Figure 1. Summary of mechanisms that potentially mediate the clinical effect of Ω-3 supplementation of cardiac rhythm.

Circulating Ω-3 compounds are likely to have different electrophysiological effects, if compared to Ω-3 incorporated into the membranes (Den Ruijter HM et al. 2007 for review). For example, peak cardiac sodium current was reduced by 51% after acute administration of EPA and DHA in neonatal rat cardiomyocytes (Xiao YF et al. 1995), but remained unaffected by Ω-3 incorporated in pig and rat cardiomyocytes (Verkerk AO et al. 2006; Leifert WR et al. 2000). Differential effects of circulating versus incorporated Ω-3 have also been demonstrated with respect to various potassium channels and the regulation of calcium homeostasis (Den Ruijter HM et al. 2007 for review). Incorporated Ω-3, however, also may prevent further AP shortening induced by circulating Ω-3. Patients with high levels of incorporated Ω-3 therefore may not have a further benefit from short term Ω-3 supplementation (Den Ruijter HM et al. 2010). This mechanism may have contributed to negative results in some clinical studies including the JELIS trial (Yokoyama M et al.2007)

Apart from these considerations the molecular interactions of Ω-3 and their effects on cardiac rhythm may be influenced by a large variety of additional conditions (Figure 1):

a) *The various kinds of Ω-3 formulations* being used (re-esterified tri-acyl-glycerides, ethyl-esters or phospholipids; Neubronner J et al. 2011; Schuchardt JP et al. 2011)
b) The *activity state of membrane bound proteins and ligand occupation* of specific receptors involved in signal transduction (Den Ruijter HM et al. 2007; Xiao YF et al 1998; Rauch B et al. 1989), or the increased responsiveness of inhibitory G-proteins after ischemic preconditioning (Niroomand F et al. 1995).
c) The *activity of cellular phospholipases* and the presence of lysophosphatides that change phospholipid environment and function of membrane bound proteins (Chien KR et al. 1981; Corr PB et al. 1984; Rauch B et al. 1994). The activity of phospholipases may vary between different myocardial regions depending on the degree of ischemia and/or inflammation.
d) The *heterogeneity of electrical stability of myocardial cells in the diseased heart* muscle due to regional differences with regard to various degrees of ischemia and tissue damage, ischemic preconditioning, etc. (Dhein S et al. 2005). In this respect it should also be remembered, that in patients with coronary artery disease, myocardium is not presenting as a homogeneous and healthy tissue experiencing acute ischemia in a well-defined area, but rather as a mixture of healthy myocardium, hypertrophied tissue, scar tissue and ischemic myocardium and also includes areas of tissue with ischemic preconditioning, inflammation, various degrees of membrane phospholipid degradation and with more or less acute or chronic stretch etc. (Janse MJ et al. 2003).
e) The *species (human, various animals) being studied.* The characteristics of action potentials (APs) vary significantly between human and various animal myocardial cells and with gender (Karagueuzian HS et al. 1982; Shattock MJ & Bers DM 1989; Cheng J 2006; Tanaka H et al. 2008).
f) The *various mechanisms that trigger ventricular tachycardia and fibrillation.* Under clinical conditions ventricular tachycardia or fibrillation are predominantly caused by triggered activity or by reentry mechanisms. Fish oil shortens cardiac AP and accentuates the AP notch, which may lead to depression or even loss of the AP dome (Verkerk AO et al. 2006, 2007). Under clinical conditions, where AP is prolonged triggered activity may be the predominant pro-arrhythmic mechanism, which could

be inhibited by superfusion with Ω-3 of isolated cardiomyocytes from rabbits and from patients with end stage heart failure (Den Ruijter HM et al. 2008). Triggered activity also could be inhibited in pig cardiomyocytes (Den Ruijter HM et al. 2006). In keeping with these experimental results Ω-3 were effective in reducing the arrhythmic risk in patients with idiopathic dilated cardiomyopathy (Nodari S et al. 2009).

Conversely, AP shortening may be pro-arrhythmic by reducing the refractory period and thereby promoting re-entry. Supplementation with Ω-3 may increase a preexisting heterogeneity in AP duration and repolarization (Verkerk AO et al. 2009), as can be seen in acute ischemia (Yan GX et al. 2004). By this way the occurrence of unidirectional block and re-entry may be facilitated (Janse MJ& Wit AL 1989). In the clinical situation therefore supplementation with Ω-3 may prevent or facilitate ventricular tachyarrhythmias depending on the predominant underlying arrhythmic mechanism (Den Ruijter HM et al. 2007).

Based on these considerations it becomes apparent that Ω-3 do not have one single specific way to act, but rather possess multiple sites of potential actions, that may be influenced by a number of external conditions at the cellular and molecular level. Multiple sites of interaction between Ω-3 and myocardial tissue in combination with various possible ways of interference with these biochemical interactions, are unlikely to result in an unequivocally predictable and homogeneous beneficial effect on clinical outcomes. This is reflected by the results of animal studies and finally by the results of clinical trials.

Animal Studies

Animal studies (mostly using the rat or canine model) in general support an effect of Ω-3 supplementation on cardiac rhythm. An anti-arrhythmic effect of Ω-3 especially could be shown with respect to ischemia induced ventricular tachycardia (VT) or fibrillation (VF) (Matthan NR et al. 2005; Billman GE 2006). The clearest effect in the prevention of VF by Ω-3 was demonstrated in infusion studies using a special experimental canine model. In this model acute myocardial ischemia was induced at a site distant from a previous myocardial infarction during submaximal exercise thereby activating the autonomic nervous system (Billman GE 2006) and inducing VF.

However, these clear effects of Ω-3 under well-defined experimental conditions cannot simply be translated into the clinical situation, and several aspects have to be considered (Billman GE 2006):

a) In this canine model not only superfusion with Ω-3 but also the application of β-receptor antagonists, calcium channel blockers, and endurance exercise training – all interventions that are routinely used in actual clinical practice – were effective in VF prevention (Billman GE 2009).
b) Not all dogs were susceptible to ischemia induced VF in this model. Animals resistant to VF were characterized by reduced β-receptor responsiveness and an intact parasympathetic regulation, indicating that these are first line mechanisms to prevent ischemia induced tachyarrhythmias.

c) Finally, incorporation of Ω-3 into the phospholipid bilayer can be expected to be significantly less in infusion studies as compared to feeding studies.

Feeding studies more closely imitate the clinical situation, and under these conditions Ω-3 can be expected to exert their effect primarily after being incorporated into the cellular membrane (see chapter above). Numerous animal feeding studies have been published between 1987 -1999, and the results showed a considerable heterogeneity. Still, a meta-analysis of these studies suggests fish oil to prevent ischemia and ischemia-reperfusion induced VT/VF (Matthan NR et al. 2005). This conclusion however remains questionable as a recent feeding study using the above mentioned canine model failed to show a reduction of life-threatening ventricular arrhythmias in post myocardial infarction dogs at risk for ventricular fibrillation (Billman GE et al. 2012). Moreover Ω-3 treatment even increased the susceptibility to malignant arrhythmias in low risk dogs with or without acute myocardial infarction (Billman GE et al. 2012). In isolated hearts of pigs fed with fish oil for 8 weeks, spontaneous ischemia-induced sustained VT/VF also was facilitated in the Ω-3 group (Coronel R et al. 2007), whereas other studies report increased resistance to ischemia-reperfusion injury after dietary Ω-3 application, which also could be a basis to protect against reperfusion arrhythmias (Abdukeyum GG et al. 2008; Zeghichi-Hamri S et al. 2010).

In summary, the heterogenous results of animal studies strengthen the hypothesis that Ω-3 supplementation does not protect against serious cardiac arrhythmias under any condition. Moreover, Ω-3 even may facilitate malignant arrhythmias under certain conditions that include animals with primarily low arrhythmic risk (Billman GE et al 2012).

Earlier Clinical Studies

An inverse relationship between consumption of fish oil and cardiovascular risk was shown in early observational, case-control, and cohort studies, with respect to the occurrence of cardiovascular disease (Whelton SP et al. 2004), sudden and non-sudden cardiac death from coronary heart disease (Daviglus M et al. 1997), and with regard to SCD in apparently healthy persons (Siscovick D et al. 1995; Albert CM et al. 2002; Hu F et al. 2002; Mozaffarian D et al. 2003). Ω-3 levels in erythrocyte membranes were directly associated with a reduced rate of primary cardiac arrest (Siscovick D et al. 1995). Similarly, elevated Ω-3 blood levels were associated with a reduced risk of sudden death among men without evidence of prior cardiovascular disease (Albert CM et al. 2002).

These data were supported by prospective and randomized nutritional intervention studies of secondary prevention after acute myocardial infarction. In the Diet and Reinfarction Trial (DART) a diet rich in fish and cereals was associated with a significant 29% reduction of all-cause mortality within 2 years after acute myocardial infarction (Burr ML et al. 1989). In the Lyon Diet Heart Study the Mediterranean diet group (enriched in alpha-linolenic acid [Ω-3] and olive oil, more cereals, fresh fruit, vegetables and fish, limited intake of saturated fat and linoleic acid [Ω-6]) had a significantly lower rate of the combined endpoint cardiac death and nonfatal myocardial infarction, if compared to the control group taking a prudent western-type diet($p = 0.0001$; follow-up 27 months; de Lorgeril M et al. 1994; de Lorgeril M et al. 1998; de Lorgeril M et al. 1999).

It has to be noted that these trials primarily evaluated healthy nutrition including elevated fish consumption but not supplementation with purified Ω-3 compounds.

A predefined supplementation of Ω-3 was used in the large placebo-controlled, open labeled GISSI Prevenzione Trial (EPA+DHA 1g/day, or placebo; GISSI Prevenzione Investigators 1999), focusing on secondary prevention after acute myocardial infarction. In this study the intervention arms using Ω-3 showed a significant reduction of sudden cardiac death (SCD) - though this was not the primary endpoint of this trial but calculated by secondary analysis (Table 1). In combination with the older studies focusing on nutrition with increased fish consumption the GISSI Prevenzione Trial for several years served as a basis to treat patients after AMI with highly purified Ω-3 compounds and this even was recommended in earlier clinical guidelines (Kris-Etherton PM et al. 2003). In a consequence it was estimated that 5 -10% of the adult US population used fish oil supplements spending more than seven billion dollars by the end of 2011 (www.marketresearch.com, cited in Billman GE et al 2012). However research continued and the results of more recent clinical trials challenge a global beneficial effect of Ω-3 supplementation especially with respect to the prevention of cardiac arrhythmias.

Clinical Studies Using Patients with Implanted Cardioverter Defibrillator (ICD)

Three randomized prospective studies evaluating the effect of high doses of Ω-3 in patients with ICD-devices failed to give homogeneous results (Table 1).

In a study predominantly including patients with coronary artery disease, Ω-3 supplementation was associated with a non-significant reduction of the primary endpoint defined as time to the first ICD-event or death from any cause (reduction by28%; p=0.057). However, the death rates did not significantly differ between the study groups. Remarkably, in this study no significant effect of Ω-3 could be shown in the subgroups of patients without coronary artery disease or with a left ventricular ejection fraction above 30% (Leaf A et al. 2005a).

In another study recurrent VT events not due to myocardial ischemia were even more common in patients treated with fish oil (Raitt MH et al. 2005).

Finally the SOFA-study did not show a significant effect of Ω-3 supplementation on the primary endpoint defined as appropriate ICD-interventions for recurrent VT/VF or death from any cause. The majority of the patients included in the SOFA-study had coronary artery disease, more than 60% with previous myocardial infarction. Almost 40% of the study participants had various forms of cardiomyopathy or valvular heart disease (Brouwer IA et al. 2006). In a meta-analysis of these three studies all-cause mortality did not significantly differ between the fish oil and the control groups (relative risk 0.70; 75% CI 0.42 – 1.15; Jenkins AJA et al. 2008).

In a sub-study of the GISSI-HF trial (566 heart failure patients with implanted ICD-devices) a statistically non-significant trend towards a lower risk of ICD-discharges in patients treated with Ω-3 was shown. However, total mortality was similar in both groups (Ω-3 26.6%, placebo 24.3%), and there even was a trend for increased arrhythmia induced mortality (Ω-3 3.6%, placebo 2.1%; Table 1; Finzi AA et al 2011).

Table 1. Selected studies evaluating the effect of Ω-3 supplementation on clinical outcomes, especially cardiac arrhythmias, in primary and secondary prevention of cardiovascular disease (also see text)

Study	- Study population, - number of participants, - Intervention, - duration of follow-up	- Primary endpoints (EP1) - Secondary endpoints (EP2)	Results	Comments
Earlier clinical studies				
GISSI Prevenzione 1999	Secondary prevention after AMI n = 11,324 4 intervention arms: Ω-3, Ω-3 + vitamin E, vitamin E, placebo; Ω-3: 850-882 mg EPA+DHA (ratio 1:2) daily 3,5 years	EP1a: cumulative rates of all cause death, nonfatal AMI, nonfatal stroke, EP1b: cumulative rates of cardiovasc. death, nonfatal AMI, nonfatal stroke; secondary analysis for all cause death, SCD and others	EP1a: Ω-3 group: reduced rate RR 95% CI: 0.85 (0.74-0.98) EP1b: Ω-3 group: reduced rate RR 95% CI: 0.80 (0.68-0.95) Secondary analysis: Ω-3 group: reduced rate for SCD, RR 95% CI, 0.55 (0.40-0.76)	a) No "up to date" treatment of AMI: i.e. low rates of acute PCI, low rates of statin application b) SCD was no primary endpoint, but evaluated by secondary data analysis
Clinical studies using patients with implanted cardioverter defibrillator (ICD)				
Leaf A et al. 2005a	ICD-patients predominantly with CHD n =402 2.6 g Ω-3 ethyl esters daily 12 months	EP1: Time to first ICD-event for VT or VF or all cause death	EP1: Ω-3 group: Primary analysis: trend for a decreased event rate ($p=0.057$) If therapies for probable VT or VF were included: risk reduction by 31% ($p=0.033$) No Ω-3 effect in subgroups without CHD or LVEF < 30%	randomized, double blind study

Study	- Study population, - number of participants, - Intervention, - duration of follow-up	- Primary endpoints (EP1) - Secondary endpoints (EP2)	Results	Comments
Raitt MH et al.2005	ICD-patients with a recent episode of sustained VT of VF n =200 fish oil 1.8 g daily, containing 42% EPA, 30% DHA median 718 days; range 20 – 828 days	EP1: Time to first episode of ICD-event for VT or VF not due to ischemia	EP1: Ω-3 group: Increased rate of recurrent VT/VF ($p = 0.007$)	a) Randomized, placebo-controlled, double-blind study b) Ω-3 concentrations measured in red blood cells
Brouwer IA et al. 2006 (SOFA)	ICD-patients with recent episodes of VT or VF n=546 4 capsules of fish oil daily, each containing 464 mg EPA, 335 mg DHA, 162 mg other Ω-3 median 356 days; range 14-379 days	EP1: Appropriate ICD-interventions for recurrent VT/VF or all-cause death	EP1: Ω-3 group: No difference compared to control group HR, 95% CI: 0.86, 0.64-1.16	Randomized, placebo-controlled, double-blind study
Finzi AA et al. 2002	ICD-patients with heart failure, 57% with previous AMI n=566 Ω-3: 850-882 mg EPA+DHA (ratio 1:2) daily mean 928 days	EP1: Appropriate ICD-intervention EP2: All cause death Mortality for arrhythmias	EP1: Ω-3 group: No significant reduction HR, 95% CI: 0.80, 0.59-1,09 EP2: Ω-3 group: Total mortality: no effect HR, 95%CI: 1.25, 0.89-1.75	Substudy of the GISSI-HF study

Table 1. (Continued)

Study	- Study population, - number of participants, - Intervention, - duration of follow-up	- Primary endpoints (EP1) - Secondary endpoints (EP2)	Results	Comments
			Ω-3 group: Trend for increased mortality for arrhythmias: HR, 95% CI 1.84; 0.67-5.05	
Clinical trials for primary and secondary prevention including SCD in the era of primary PCI and intensified cardiovascular prevention				
Yokoyama M et al. 2007 (JELIS)	Hyperlipidemic patients without previous AMI, no serious cardiac disease but treated with statins n=18,645 1.8 g EPA daily mean 4.6 years	EP1: Any major coronary event, combining SCD, fatal and nonfatal AMI and others (see text) EP2: SCD, cardiac death, all cause death	EP1: Ω-3 group: Significant 19% relative risk reduction EP2: Ω-3 group: neither reduction of SCD, nor reduction of coronary death or all cause death	a) Prospective randomized trial b) Low rate of SCD of 0.2%; c) assumed high fish intake at baseline in both study groups
Kromhout Det al. 2010 (Alpha-Omega)	Patients 60 – 80 years of age in the chronic stable phase after AMI Four intervention arms: 1. EPA/DHA 400mg daily, 2. ALA 2 g daily, 3. EPA/DHA+ALA, 4. placebo n=4,837	EP1: combined endpoint including death, non-fatal CV events, cardiac intervention	EP1: Ω-3 group: no reduction	a) Randomized, placebo-controlled, double-blind trial b) In a secondary analyses less VT-events in patients after AMI plus diabetes
Rauch B et al.2010 (OMEGA)	Patients directly after AMI 1 g EPA/DHA ethyl esters daily n =3,851 one year	EP1: SCD EP2: All cause death, major cardiovascular and cerebrovascular events and others	EP1: no difference between the Ω-3 group and the placebo group	a) randomized, placebo-controlled, double-blind trial

Study	- Study population, - number of participants, - Intervention, - duration of follow-up	- Primary endpoints (EP1) - Secondary endpoints (EP2)	Results	Comments
			EP2: no difference between the Ω-3 and the placebo group.	b) Study underpowered for EP1 due to unexpected low rates of SCD; c) a secondary analysis showed increased rate of non sudden cardiac death in Ω-3 group (Rauch G et al 2012)
Clinical trials for prevention of atrial fibrillation				
Mozzafarian D et al 2004	Adults > 65 years; assessment of dietary intake 1989 and 1990 n=4,815 tuna, other broiled or baked fish 12 years	EP1: incidence of AF during 12 years follow-up	EP1: Fish-group: lower incidence of AF	a) Prospective polulation-based cohort study b) control of plasma Ω-3 levels
Nodari S et al.2001	Patients after electrical conversion for persistent AF treated with amiodaron and ACE-inhibitors n =199 850-882 mg EPA/DHA (ratio 0.9-1.5) daily one year	EP1: probability of maintenance of SR 1 years after cardioversion	EP1: Ω-3 group: improved probability for SR maintenance	Randomized, double-blind, placebo-controlled study
Macchia A et al. 2008	Post AMI patients free of AF during the acute event Ω-3 population: n=215	EP1: Death from any cause; hospitalization for AF	EP1: Group with elevated Ω-3 intake: Reduced EP1 with respect to both, rehospitalization and all cause death	Prospective population study, using propensity score

Table 1. (Continued)

Study	- Study population, - number of participants, - Intervention, - duration of follow-up	- Primary endpoints (EP1) - Secondary endpoints (EP2)	Results	Comments
	Control: 3.027 At least one prescription of Ω-3 preceding or following index hospitalization 360 days			
Saravanan D et al. 2010	Patients undergoing bypass surgery n=103 2g EPA/DHA ethyl esters (ratio 1.2/1), start 5 days before surgery Follow-up 5 days after bypass surgery	EP1: occurrence of episodes of AF post surgery	EP1: No difference between study groups	Randomized, double-blind, placebo controlled study
Kowey PR et al. 2010	Outpatient participants with paroxysmal (n=542) or persistent AF (n=121), no structural heart disease, SR at baseline Loading: 7 days 8 capsules of 1 g Ω-3 (EPA/DHA 465/375 mg) Continuous treatment: 24 weeks 4 capsules per day) 6 months	EP1: Patients with paroxysmal AF: First symptomatic recurrence of AF.	EP1: No difference between study groups	Randomized, double-blind, placebo controlled study
Bianconi L et al. 2011	Patients after electrical conversion for persistent AF n =204	EP1: recurrence rate of AF	EP1: No difference between study groups	Randomized, double-blind, placebo controlled study

Study	- Study population, - number of participants, - Intervention, - duration of follow-up	- Primary endpoints (EP1) - Secondary endpoints (EP2)	Results	Comments
	850 mg EPA/DHA (ratio 0.9-1.5) 3-times daily, 7 days before cardioversion and 6 months thereafter			
Farquharson AL et al. 2011	Patients after cardiac surgery n = 194 4.6 g dietary Ω-3 daily, start 3 weeks before surgery 6 days post surgery	EP1: incidence of AF	EP1: No difference between study groups	Randomized, double-blind placebo-controlled study
Mozaffarian D et al 2012	Patients after cardiac surgery n =1,516 1 g capsule containing EPA 465 mg + DHA 375 mg daily; 10g as loading dose over 3-5 days pre surgery 10 days post- surgery or hospital discharge	EP1: occurrence of post-surgery AF,duration > 30 seconds EP2: a variety of clinical events including 30-day mortality	EP1: Ω-3 group: no reduction of post- surgery AF EP2: no effect of Ω-3 supplementation on any of the secondary endpoints	Randomized, double-blind placebo-controlled trial

AF, atrial fibrillation; ALA, alpha-linolenic acid; AMI, acute myocardial infarction; DHA, docosahexaenoic acid; EPA, eicosapentaenoic acid; ICD, implanted cardioverter defibrillator; SCD, sudden cardiac death; SR, sinus rhythm; VT, ventricular tachycardia; VF, ventricular fibrillation.

In conclusion, the apparent heterogeneity in the response of ICD-patients to Ω-3 supplementation may be regarded as a consequence of different study populations and different arrhythmic origins (ischemic versus non ischemic). However, heterogeneity also could be the result of different concomitant medications of the study populations including β-blockers, digoxin, and amiodaron.

Clinical Trials for Primary and Secondary Prevention Including Sudden Cardiac Death in the Era of Primary PCI and Intensified Cardiovascular Prevention (Table 1)

The multicenter Japan EPA Lipid Intervention Study (JELIS; Yokoyama M et al. 2007) investigated the consumption of 1.8 g EPA per day over a mean period of 4.6 years in hyperlipidemic patients without previous AMI and no apparent serious heart disease but treated with statins. The data showed a reduction of the combined endpoint of major cardiovascular events (including sudden cardiac death, fatal and nonfatal myocardial infarction, and other nonfatal events including unstable angina, angioplasty, stenting, or coronary bypass grafting) from 3.5% to 2.8% (p=0.011; hazard ratio [HR] 0.81, 95% CI 0.69-0.95). The limitation of using combined competing endpoints becomes apparent, as subgroup analyses neither showed a reduction of sudden cardiac death (0.2 % in both study arms; HR 1.06, 95% CI 0.55-20.07) nor of coronary death (0.3 % in both study arms; HR 0.94, 95% CI 0.57-1.56) or all-cause mortality (control 2.8%, EPA-group 3.1 %; HR 1.09, 95% CI 0.92-1.28).

Compared to the GISSI Prevenzione trial death rates and especially the rates of sudden cardiac death were very low in both groups and therefore may be difficult to be further reduced by any intervention (rates for sudden cardiac death: JELIS 0.2 % in both groups; GISSI 2.2%/2.9% Ω-3 versus control; Yokoyama M et al. 2007; GISSI-Prevenzione Investigators 1999). Furthermore, these low event rates may at least in part be the result of a high fish consumption of the Japanese population at baseline. As it has been reported that the major part of risk reduction already occurs at about 250 mg EPA/DHA per day (Mozaffarian D & Rimm EB 2006) a further increase of Ω-3 intake may not have a substantial additional effect on cardiac death reduction (Mozaffarian D 2007).

In the Alpha-Omega-Study 4,837 patients in the chronic stable phase after myocardial infarction were randomly assigned to one of four trial arms. Margarine was used in all trial arms, supplemented with either EPA/DHA, alpha-linolenic acid, EPA/DHA + ALA, or placebo, respectively (Kromhout D et al. 2010; Table 1). The rates of the primary combined endpoint (death, nonfatal cardiovascular events, or cardiac intervention) did not differ between the study groups. In addition, in all secondary endpoints, including ventricular-arrhythmia and total death, there was no significant difference between the study groups. Importantly, a high percentage of the patients received "state of the art medication", including statins.

In a post hoc analysis after unblinding of the data in the subgroup of patients with diabetes ventricular arrhythmia related events tended to be reduced in the EPA-DHA group (HR 0.51; 95% CI 0.24-1.11) and significantly were reduced in the ALA-group (HR 0.39; 95% CI 0.17-0.88). In a secondary analysis of the Alpha-Omega Trial taking high risk patients with previous myocardial infarction and diabetes the EPA-DHA plus ALA group experienced significantly less ventricular arrhythmia-related events (HR 0.16; 95% CI 0.04-0.69; Kromhout D et al. 2011). These differential results again underscore the necessity to exactly define the clinical conditions under which supplementation of Ω-3 may be beneficial.

The OMEGA trial tested the effect of supplementation with 1 g/day of esterified EPA/DHA on the rate of SCD and other clinical events within one year after acute myocardial infarction (AMI; 3,851 patients, 25.6% female, mean age 64.0 years; Rauch B et al.2006, 2010). A one year follow-up was chosen, as the risk of cardiac death after acute myocardial infarction including a presumed arrhythmic death is highest in the first three months after the event (Solomon SD et al. 2005; Pouleur AC et al. 2010). Furthermore, in the GISSI trial significance in lowering SCD by Ω-3 had already been reached within 120 days after AMI (Marchioli R et al. 2002). Following guidelines for the management of acute myocardial infarction and secondary prevention 77% of the patients in the OMEGA trial received acute percutaneous coronary intervention, and/or thrombolysis (8.3%). At hospital discharge the vast majority (> 90%) of the included patients received guideline adjusted medication including beta-blockers, ACE-inhibitors or ARBs, statins, acetylsalicylic acid, and clopidogrel. Under these conditions, the rates of SCD were unexpectedly low with 1.5 % in both study groups (OR 0.95, 95% CI 0.56-1.60), whereas total mortality was 4.6% in the Ω-3 group and 3.7% in the control group (OR 1.25. 95% CI 0.90-1.72). Furthermore, there was no significant difference between the study groups with regard to sudden cardiac death or total death in any of the predefined subgroups of patients with higher risk (diabetes, age > 70 years, no acute revascularization, ejection fraction < 35%).

Despite these apparently homogeneous results interpretation is limited as the case estimate in the OMEGA-study was based on an overestimation of the rate of SCD in the control group, thereby leading to an underpowering of the study.

Still, according to a secondary competing risk analysis of the OMEGA-study data, treatment with Ω-3 even tended to have a negative effect on all cause death and had a significant negative effect on non-sudden cardiac death (Rauch G et al. 2012).

Two other randomized controlled trials also failed to show a clear beneficial effect of Ω-3 supplementation. In 563 elderly Norwegian men at high cardiovascular risk a non significant tendency to a reduced all-cause mortality after three years could be observed (HR 0.53, 95% CI 0.27-1.04), but the rate of cardiovascular events remained unchanged (HR 0.89; 95% CI 0.55-1.45, Einvik G et al. 2010). In 2,501 patients with a history of myocardial infarction, unstable angina or ischemic stroke supplementation with EPA/DHA was not associated with a significant decrease of major vascular events during a follow-up of 4.7 years (HR 1.08; 95% CI 0.79-1.47; Galan P et al. 2010).

Clinical Trials for Prevention of Arial Fibrillation

Heterogeneity of the effect of Ω-3 supplementation also can be seen with respect to prevention of atrial fibrillation (Table 1). Positive results (Mozaffarian et al. 2004, primary prevention by fish intake in patients > 65 years of age; Calò L et al. 2005, patients undergoing coronary artery surgery; Macchia A et al. 2008, post myocardial infarction patients) were not confirmed in more recent studies and meta-analyses (Kowey PR et al. 2010; Saravanan P et al. 2010; Bianconi L et al. 2011; Farquharson AL et al. 2011; Liu T at al. 2011). In the multinational OPERA trial including 1,516 patients undergoing cardiac surgery perioperative Ω-3 supplementation also did not reduce the risk of postoperative atrial fibrillation (Mozaffarian D et al. 2012). On the other hand supplementation with DHA 1.5 g and EPA 0.3g daily resulted in a prolongation and reduced dispersion of pulmonary venous and left atrial effective refractory periods in patients with paroxysmal atrial fibrillation (Kumar S et al. 2011). Furthermore, in patients with persistent atrial fibrillation on amiodarone and a renin-angiotensin-aldosterone system inhibitor, additional intake of Ω-3 (2g/day) improved the probability of maintaining sinus rhythm after direct current cardioversion (Nodari S et a. 2011). Therefore, prevention of atrial fibrillation also may depend on distinct clinical and pathophysiological conditions and concomitant medication.

Recent Meta-Analyses on the Effect of Ω-3 Supplementation on Major Clinical Events Including SCD

As mentioned at the beginning of this chapter, evaluation of the biological and clinical effects have attracted and still attract the scientific world for almost four decades and - apart from the numerous studies – many meta-analyses tried to summarize the clinical effect of Ω-3 supplementation. Of these two recently published meta-analyses have to be mentioned as they include all the recently published clinical trials that reflect actual medical and interventional treatments of cardiovascular disease.

In the first meta-analysis including 68,680 patients in 20 studies did not show a beneficial effect of Ω-3 supplementation with respect to total mortality (RR 0.96; 95% CI, 0.91-1.02) and sudden cardiac death (RR 0.87; 95% CI, 0.75-1.01) (Rizos EC et al, 2012).

In another meta-analysis published in parallel also 20 trials were included with a total of 62.851 patients. Supplementation with Ω-3 was associated with reduced vascular death (RR0.86; 95% CI, 0.75-0.99), but there neither was a reduction of total mortality (RR 0.95; 95% CI, 0.86-1.04) nor of the composite endpoint of myocardial infarction, stroke and cardiovascular death (RR0.96; 95% CI 0.90-1.03). There also was no effect on arrhythmias (RR 0.99; 95% CI, 0.85-1.16; p=0.92; Kotwal S et al. 2012).

Conclusion

a) Ω-3 clearly interfere with the physiology of myocardial cell membranes through a variety of specific and unspecific pathways, and thereby potentially influence cardiac rhythm. However, these membrane effects of Ω-3 are complex. This complexity makes it difficult to predict the effects of Ω-3 supplementation on cardiac rhythm within the wide variety of conditions that represent clinical practice.
b) The effect of Ω-3supplementation may depend on the background diet and the pre-existent intake of fish oil (Reiffel JA & McDonald A et al. 2006; Mozaffarian D & Rimm EB 2006; Mozaffarian D 2007).
c) With regard to earlier studies, treatment of patients with coronary artery disease, especially treatment of patients with myocardial infarction has improved markedly.

In the GISSI trial (inclusion period October 1993 – September 1995) only 4.4 % of the patients had acute coronary revascularization at baseline, and only 4.7% were on statins at hospital discharge, increasing to only 46% after 42 months of follow-up (GISSI-Prevenzione Investigators 1999). Furthermore, only 43.9% of the patients included in the GISSI trial were on beta-blocker treatment at the start of the study, and this percentage decreased during follow-up. It therefore may be speculated that up-to-date guideline adjusted treatment of AMI (including acute revascularization, medical treatment and support of life style changes) interferes with molecular and cellular Ω-3 interactions thereby weakening or competing with a potential beneficial Ω-3 effect. Although the available data not homogeneously support this hypothesis (Marchioli R et al. 2007), this aspect should strongly be considered in future research. Especially statin treatment may interfere with potential Ω-3 effects (Eussen SR et al 2012; de Lorgeril M et al 2013).

d) The potential clinical anti-arrhythmic effect of Ω-3 may depend on the pathophysiological conditions facilitating arrhythmias. In the case that Ω-3 supplementation especially protects against ischemia induced arrhythmias, prevention of ischemia by modern treatments (i.e. revascularization, beta-blockers, statins, ACE-inhibitors, inhibition of thrombocyte aggregation, physical exercise) could attenuate this beneficial effect of Ω-3. From this background it has to be remembered that beta-blockers are well known to prevent SCD, and even statins could have some anti-arrhythmic effects (Anh D 2004; Lorenz H et al. 2005).
e) In accordance to these considerations potential anti-arrhythmic effects of Ω-3 by augmentation of vagal activity (Mozaffarian D et al. 2005; O`Keefe JH et al. 2006) could be blunted by beta-blocker treatment and increased physical training during cardiac rehabilitation (Nolan JP et al. 2008; Billman GE 2009).

In summary, the anti-arrhythmic effect proven under some well-defined experimental conditions in animal models and suggested in the earlier clinical studies appears to depend on the clinical conditions being studied. These clinical conditions are determined by the type and stage of the underlying myocardial disease and represent a sum of various pathophysiological conditions (including ischemia, reperfusion, ischemic preconditioning, scar tissue, inflammation, congenital defects etc.), the effects of modern medication including beta-blockers, ACE-inhibitors, statins, but also nutrition determining preexisting Ω-3 levels and other interventions potentially interfering with the arrhythmic risk, such as exercise training. For the future it will be necessary to exactly define the clinical conditions in which supplementation with of Ω-3 may be beneficial, and without potentially harmful effects. As long as these conditions are not well defined supplementation with Ω-3 compounds remains questionable and Mediterranean diet including increased fish consumption remains the way to be preferred for cardiovascular disease prevention.

Conflict of Interest Statement

Rauch B.: no conflicts of interest;

Senges J: received honoraria for educational presentations from Trommsdorff GmbH & CoKG Arzneimittel, Alsdorf, Germany and Pronova Biopharma, Lysaker, Norway

References

Abdukeyum, G.G., Owen, A.J., McLennan, P.L., 2004. Dietary (n-3) long-chain polyunsaturated fatty acids inhibit ischemia and reperfusion arrhythmias and infarction in rat heart not enhanced by ischemic preconditioning. *J. Nutr.* 138, 1902-1909.

Albert, C.M., Campos, H., Stampfer, M.J., Ridker, P.M., Manson, J.E., Willett, W.C., Ma, J. (2002). Blood levels of long-chain n-3 fatty acids and the risk of sudden death. *N. Engl. J. Med.* 346, 1113-1118.

Anh, D., Marine, J.E., (2004). Beta bockers as anti-arrhythmic agents. *Heart Fail.Rev.* 9, 139-147.

Bianconi, L., Calò, L., Mennuni, M., Santini, L., Morosetti, P., Azzolini, P., Barbato, G., Biscione, F., Romano, P., Santini, M., (2011). N-3 polyunsaturated fatty acids for the prevention of arrhythmia recurrence after electrical cardioversion of chronic persistent atrial fibrillation: a randomized, double-blind, multicentre study. *Europace* 13, 174-181.

Billman, GE.,(2006). A comprehensive review and analysis of 25 years of data from an in vivo canine model of sudden cardiac death: implications for future anti-arrhythmic drug development. *Pharmacol.& Ther.* 111, 808-835.

Billman, GE.,(2009). Cardiac autonomic neural remodelling and susceptibility to sudden cardiac death: effect of endurance exercise training. *Am. J. Physiol.Heart Circ. Physiol.* 297, H1171-H1193.

Billman, G.E., Carnes, A.C., Adamson, P.B., Vanoli, E., Schwartz, P.J., (2012). Dietary omega-3 fatty acids and susceptibility to ventricular fibrillation. lack of protection and a proarrhytmic effect.*Circ. Arrhythm Electrophysiol.*5, 553-560.

Brouwer, I.A., Zock, P.L., Camm, A.J., Böcker, D., Hauer, R.N.W., Wever, E.F.D., Dullemeijer, C., Ronden, J.E., Katan, M.B., Lubinski, A., Buschler, H., Schouten, E.G., for the SOFA Study Group., (2006). Effect of fish oil on ventricular tachyarrhythmia and death in patients with implantable cardioverter defibrillators. The Study on Omega-3 Fatty Acids and Ventricular Arrhythmia (SOFA) Randomized Trial. *JAMA* 295, 2613-2619.

Bucher, H.C., Hengstler, P., Schindler, C., Meier, G., (2002). N-3 polyunsaturated fatty acids in coronary heart disease: a meta-analysis of randomized controlled trials. *Am. J. Med.* 112, 298-304.

Burr, M.L., Fehily, A.M., Gilbert, J.F., Rogers, S., Holliday, R.M., Sweetnam, P.M., Elwood, P.C., Deadman, N.M., (1989). Effects of changes in fat, fish, and fibre intakes on death and myocardial reinfarction: Diet and Reinfarction Trial (DART). *The Lancet*334, 757-761.

Calò, L., Bianconi, L., Colivicchi, F., Lamberti, F., Loricchio, M.L., de Ruvo, E., Meo, A., Pandozi, C., Staibano, M., Santini, M., (2005). N-3 fatty acids for the prevention of atrial fibrillation after coronary artery bypass surgery: a randomized, controlled trial. *J. Am. Coll. Cardio*l. 45, 1723-1728.

Cheng, J., (2006). Evidences of the gender-related differences in cardiac repolarization and the underlying mechanisms in different animal species and human. *Fundam.Clin.Pharmacol.* 20, 1-8.

Cheng, J.W., Santoni, F., (2008). Omega-3 fatty acid: arole in the management of cardiac arrhythmias? *J. Altern. Complement. Med.*14, 965-974.

Chien, K.R., Reeves, J.P., Buja, L.M., Bonte, F., Parkey, R.W., Willerson, J.T., (1981). Phospholipid alterations in canine ischemic myocardium. Temporal and topographic correlations with Tc-99m-PPi accumulation and an in vitro sarcolemmal Ca2+ permeability defect. *Circ. Res.*48, 711-719.

Coronel, R., Wilms-Schopman, F.J.G., Den Ruijter, H.M., Belterman, C.N., Schumacher, C.A., Opthof, T., Hovenier, R., Lemmens, A.G., Terpstra, A.H.M., Katan, M.B., Zock, P., (2007). Dietary n-3 fatty acids promote arrhythmias during acute regional myocardial ischemia in isolated pig hearts. *Cardiovasc. Res.* 73, 386-394.

Corr, P.B., Gross, R.W., Sobel, B.E., (1984). Amphipathic metabolites and membrane dysfunction in ischemic myocardium. *Circ. Res.* 55, 135-154.

Croset, M., Kinsella, J.E., (1989). Changes in phospholipid fatty acid composition of mouse cardiac organelles after feeding graded amounts of docohexaenoate in presence of high levels of linoleate. Effects on cardiac ATPase activities. *Ann. Nutr. Metab.* 33, 125-191.

Daviglus, M.L., Stamler, J., Orencia, A.J., Dyer, A.R., Liu, K., Greenland, P., Walsh, M.K., Morris, D., Shekelle, R.B., (1997). Fish consumption and the 30-year risk of fatal myocardial infarction. *N. Engl. J. Med.* 336, 1046-1053.

de Lorgeril, M., Renaud, S., Mamelle, N., Salen, P., Martin, J.L., Monjaud,Il, Guidollet,J., Touboul, P., Delaye, J., (1994). Mediteranean alpha-linolenic acid – rich diet in secondary prevention of coronary heart disease. *Lancet* 343, 1454-1459.

de Lorgeril, M., Salen, P., Martin, J.L., Monjaud, I., Boucher, P., Mamelle, N.,(1998). Mediterranean dietary pattern in a randomized trial. *Arch. Intern. Med.* 158, 1181-1187.

de Lorgeril, M., Salen, P., Martin, J.L., Monjaud, I, Delaye, J., Mamelle, N., (1999). Mediterranean diet, traditional risk factors, and the rate of cardiovascular complications after myocardial infarction: final report of the Lyon Diet Heart Study. *Circulation* 99; 779-785.

de Lorgeril, M., Salen, P., Defaye, P., Rabaeus, M., (2013). Recent findings on the health effects of omega-3 fatty acids and statins, and their interactions: do statins inhibit omega-3? *BMC Medicine,* 11, 5.

Den Ruijter, H.M., Verkerk, A.O., Berecki, G., Bakker, D., van Ginneken, A.C.G., Coronel, R., (2006). Dietary fish oil reduces the occurrence of early afterdepolarizations in pig ventricular myocytes. *J. Mol. Cell. Cardiol.* 41, 914-917.

Den Ruijter, H.M., Berecki, G., Opthof, T., Verkerk, A.O., Zock, P.L., Coronel, R., (2007). Pro- and antiarrhythmic properties of a diet rich fish oil. *Cardiovascular Res.* 73, 316-325.

Den Ruijter, H.M., Berecki, G., Verkerk, A.O., Bakker, D., Baartscheer, A., Schumacher, C.A., Belterman, C.N.W., de Jonge, N., Fiolet, J.W.T., Brouwer, I.A., Coronel, R., (2008). Acute administration of fish pills inhibits triggered activity in isolated myocytes from rabbits and patients with heart failure. *Circulation* 117, 536-544.

Den Ruijter, H.M., Verkerk, A.O., Coronel, R., (2010). Incorporated fish oil fatty acids prevent action potential shortening induced by circulating fish oil fatty acids. *Frontiers in Physiology*1, 149.

Dhein, S., Michaelis, B., Mohr, F.W., (2005).Antiarrhythmic and electrophysiological effects of long-chain ω-3 polyunsaturated fatty acids. *Naunyn-Schmiedeberg`s Arch. Pharmacol.* 371, 202-211.

Einvik, G., Klemsdal, T.O., Sandvik, L., Hjerkinn, E.M., (2010). A randomized clinical trial on n-3 polyunsaturated fatty acids supplementation and all-cause mortality in elderly men at high cardiovascular risk. *Eur. J. Cardiovasc. Prev. Rehabil.* 17, 588-592.

Eussen, S.R., Geleijnse, J.M., Giltay, E.J., Rompelberg, C.J., Klungel, O.H., Kromhout, D., (2012). Effects of n-3 fatty acids on major cardiovascular events in statin users and non-users with a history of myocardial infarction. *Eur. Heart J.* 33, 1582-1588.

Farquharson, A.L., Metcalf, R.G., Sanders, P., Stuklis, R., Edwards, J.R., Gibson, R.A., Cleland, L.G., Sullivan, T.R., James, M.J., Young, G.D.., (2011). Effect of dietary fish oil on atrial fibrillation after cardiac surgery. *Am. J. Cardiol.*; 108, 851-856.

Filion, K.B., El Khoury, F., Bielinski, M., Schiller,I., Dendukuri, N., Brophy, J.M., (2010). Omega-3 fatty acids in high-risk cardiovascular patients: a meta-analysis of randomized controlled trials. *BMC Cardiovasc. Disord.* 10, 24. Doi:10.1186/1471-2261-10-24.

Finzi, A.A., Latini, R., Barlera, S., Rossi, M.G., Ruggeri, A., Mezzani, A., Favero, C., Franzosi, M.G., Serra, D., Lucci, D., Bianchini, F., Bernasconi, R., Maggioni, A.P., Nicolosi, G., Porcu, M., Tognoni, G., Tavazzi, L., Marchiolo, R., (2011). Effects of n-3 polyunsaturated fatty acids on malignant ventricular arrhythmias in patients with chronic heart failure and implantable cardioverter-defibrillators: A substudy of the Gruppo Italiano per lo Studio della Sopravvivenza nell` Insufficienza Cardiaca (GISSI-HF) trial. *Am. Heart J.* 161, 338-343.

Galan, P., Kesse-Guyot, E., Czernichow, S., Briancon, S., Blacher, J., Hercberg S., for the SU.FOL.OM3 Collaborative Group., (2010). Effects of B vitamins and omega 3 fatty acids on cardiovascular diseases: a randomized placebo controlled trial. *BMJ* 341, c6273.

GISSI-Prevenzione Investigators (Gruppo Italiano per lo Studio della Sopravvivenza nell` Infarto miocardico)., (1999). Dietary supplementation with n-3 polyunsaturated fatty acids and vitamin E after myocardial infarction: results of the GISSI-Prevenzione trial. *Lancet* 354, 447-455.

Grynberg, A., Fournier, A., Sergiel, J.P., Athias, P.,(1996). Membrane docohexaenoic acid vs. eicosapentaenoic acid and the beating function of the cardiomyocyte and its regulation through the adrenergic receptors. *Lipids* 31, S205-S210.

Hallaq, H., Sellmayer, A., Smith, T.W., Leaf, A.,(1990). Protective effect of eicopentaenoic acid on quabain toxicity in neonatal rat cardiac myocytes. *Proc. Natl. Acad. Sci. USA.* 87, 7834-7838.

Hallaq, H., Smith, T.W., Leaf, A., (1992). Modulation of dihydropyridine-sensitive calcium channels in heart single cells by fish oil fatty acids. *Proc. Natl. Acad. Sci. USA.* 89, 1760-1764.

Honore, E., Barhanin, J., Attali, B., Lesage, F., Lazdunski, M., (1994). External blockade of the major cardiac delayed – rectifier K+ channel (Kv1.5) by polyunsaturated fatty acids. *Proc. Natl. Acad. Sci. USA.* 91, 1937-1941.

Hooper, L., Thompson, R.L., Harrison, R.A., Summerbell, C.D., Ness, A.R., Moore, H.J., Worthington, H.V., Durrington, P.N., Higgins, J.P., Capps, N.E., Riemersma, R.A., Ebrahim, S.B., Davey Smith, G., (2006). Risks and benefits of omega 3 fats for mortality, cardiovascular disease, and cancer: systematic review. *BMJ* 332, 752-760.

Hu, F.B., Bronner, L., Wilett, W.C., Stampfer, M.J., Rexrode, K.M., Albert, C.M., Hunter, D., Manson, J.E., (2002).Fish and omega-3 fatty acid intake and risk of coronary heart disease in woman. *JAMA* 287, 1815-1821.

Janse, M.J., Wit, A.L., (1989). Electrophysiological mechanisms of ventricular arrhythmias resulting from myocardial ischemia and infarction. *Physiol. Rev.* 69, 1049-1169.

Janse, M.J., Coronel, R., Wiilms-Schopman, F.J., de Groot, J.R., (2003). Mechanical effects on arrhythmogenesis: from pipette to patient. *Prog. Biophys. Mol. Biol.* 82, 187-195.

Jenkins, D.J.A., Josse, A.R., Dorian, P., Burr, M.L., LaBelle R., Trangmar, R., Kendall, C.W.C., Cunnane, S.C., (2008). Heterogeneity in Randomized Controlled Trials of Long Chain (Fish) Omega-3 Fatty Acids in Restenosis, Secondary Prevention and Ventricular Arrhythmias. *J. Am. Coll. Nutr.* 27, 367-378.

Jenkins, D.J.A., Josse, A.R., Beyene, J., Dorian, P., Burr, M.L., LaBelle, R., Kendall, C.W.C., Cunnane, S.C., (2008). Fish-oil supplementation in patients with implantable cardioverter defibrillators: a meta-analysis.*CMAJ* 178, 157-164.

Judé, S., Roger, S., Martel, E., Besson, P., Richard, S., Bougnoux, P., Champeroux, P., Le Guennec, J.Y., (2006). Dietary long-chain omega-3 fatty acids of marine origin: a comparison of their protective effects on coronary heart disease and breast cancers. *Prog. Biophys. Mol. Biol.* 90, 299-325.

Kang, J.X., Leaf, A., (1996).Evidence that free polyunsaturated fatty acids modify Na^+ channels by directly binding to the channel proteins. *Proc. Natl. Acad. Sci.* 93, 3542-3546.

Karagueuzian, H.S., Pennec, J.P., Deroubbaix, E., de Leiris, J., Coraboeuf, E., (1982). Effects of excess free fatty acids on the electrophysiological properties of ventricular specialized conducting tissue: a comparative study between the sheep and the dog. *J. Cardiovasc. Pharmacol.* 4, 462-468.

Kinoshita, I., Itoh, K.,Nishida-Nakai, M., Hirota, H., Otsuji, S., Shibata, N., (1994). Antiarrhythmic effects of eicosapentaenoic acid during myocardial infarction-enhanced cardiac microsomal Ca(2+)-Mg(2+)-ATPase activity. *Jpn. Circ. J.* 58, 903-912.

Kotwal, S., Jun, M., Sullivan, D., Perkovic, V., Neal, B., (2012). Omega 3 Fatty Acids and Cardiovascular Outcomes. Systematic Review and Meta-Analysis. *Circ. Cardiovasc. Qual. Outcomes* 5, 808-818.

Kowey, P.R., Reiffel, J.A., Ellenbogen, K.A., Naccarelli, G.V., Pratt, C.M., (2010). Efficacy and safety of prescription omega-3 fatty acids forthe prevention of recurrent symptomatic atrial fibrillation: a randomized controlled trial. *JAMA* 304, 2363-2372.

Kris-Etherton, P.M., Harris, W.S., Appel, L.J., for the AHA Nutrition Committee., (2003). Omega-3 fatty acids and cardiovascular disease: new recommendations from the American Heart Association. *Arterioscler. Thromb. Vasc. Biol.* 23: 151-152.

Kromhout, D., Giltay, E.J., Geleijnse, J.M. for the Alpha Omega Trial Group. , (2010). N-3 Fatty Acids and Cardiovascular Events after Myocardial Infarction. *N. Engl. J. Med.* 363, 2015-2026.

Kromhout, D., Geleijnse, J.M., de Goede, J., Oude Griep, L.M., Mulder, B.J., de Boer, M.J., Deckers, J.W., Boersma, E., Zock, P.L., Giltay, E.J., (2011). N-3 fatty acids, ventricular arrhythmia-related events, and fatal myocardial infarction in postmyocardial infarction patients with diabetes. *Diabetes Care* 34,2515-2520.

Kumar, S., Sutherland, F., The, A.W., Heck, P.M., Lee, G., Garg, M.L., Sparks, P.B., (2011), Effects of chronic omega-3 polyunsaturated fatty acid supplementation on human pulmonary vein and left atrial electrophysiology in paroxysmal atrial fibrillation. *Am. J. Cardiol.* 108, 531-535.

Leaf, A., Kang, J.X., Xiao, Y.F., Billman, G.E., (2003). Clinical prevention of sudden cardiac death by n-3 polyunsaturated fatty acids and mechanism of prevention of arrhythmias by n-3 fish oils. *Circulation* 107: 2646-2652.

Leaf, A., Albert, C.M., Josephson, M., Steinhaus, D., Kluger, J., Kang, J.X., Cox, B., Zhang, H., Schoenfeld, D., for the Fatty Acid Antiarrhythmia Trial Investigators., (2005a). Prevention of fatal arrhythmias in high-risk subjects by fish oil n-3 fatty acid intake. *Circulation* 112, 2762-2768.

Leaf, A., Xiao, Y.F., Kang, J.X., Billman, G.E., (2005b). Membrane effects of the n-3 fish oil fatty acids, which prevent fatal ventricular arrhythmias. *J. Membr. Biol.* 206, 129-139.

Leifert, W.R., McMurchie, E.J., Saint, D.A., (1999). Inhibition of cardiac sodium currents in adult rat myocytes by n-3 polyunsaturated fatty acids. *J. Physiol.* 283, H1688-H1694.

Leifert, W.R., Jahangiri, A., Saint, D.A., McMurchie, E.J., (2000a). Effects of dietary n-3 fatty acids on contractility. Na^+ and K^+ currents in a rat cardiomyocyte model of arrhythmia. *J. Nutr. Biochem.* 11,282-292.

Leifert, W.R., Jahangiri, A., McMurchie, E.J., (2000b).Membrane fluidity changes are associated with antiarrhythmic effects of docohexaenoic acid in adult rat cardiomyocytes. *J. Nutr. Biochem.* 11, 38-44.

Liu, T., Korantzopoulos, P., Shehata, M., Li, G., Wang, X., Kaul, S., (2011).Prevention of atrial fibrillation with omega-3 fatty acids: a meta-analysis of randomized clinical trials. *Heart* 97,1034-1040.

León, H., Shibata, M.C., Sivakumaran, S., Dorgan, M., Chatterley, T., Tsuyuki, R.T., (2008). Effect of fish oil on arrhythmias and mortality: systematic review. *BMJ* 337,a2931.

Lombardi, F., Terranova, P., (2007). Anti-Arrhythmic Properties of N-3 Poly-Unsaturated Fatty Acids (n-3 PUFA). *Curr. Med. Chem.*14, 2070-2080.

Lorenz, H., Jünger, C., Seidl, K., Gitt, A., Schneider, S., Schiele, R., Wienbergen, H., Winkler, R., Gottwik, M., Delius, W., Senges, J., Rauch, B., (2005).Do statins influence the prognostic impact of non – sustained ventricular tachycardia after ST-elevation myocardial infarction? *Eur. Heart J.* 26, 1078-1085.

Macchia, A., Monte, S., Pellegrini, F., Romero, M., Ferrante, D., Doval, H., D`Ettore, A., Maggioni, A.P., Tognoni, G., (2008). Omega-3 fatty acid supplementation reduces one-year risk of atrial fibrillation in patients hospitalized with myocardial infarction. *Eur. J. Clin. Pharmacol.* 64, 627-634.

Marchioli, R., Barzi, F., Bomba, E., Chieffo, C., Gregorio, D.D., Mascio, R.D., Franzosi, M.G., Geraci, E., Levantensi, G., Maggioni, A.P., Mantini, L., Marfisi, R.M., Mastroguiseppe, G., Mininni, N., Nicolosi, G.L., Santini, M., Schweiger, C., Tavazzi, L., Tognoni, G., Tucci, C., Valugassa, F. on behalf of the GISSI-Prevenzione Investigators. (2002). Early protection against sudden death by n-3 polyunsaturated fatty acids after myocardial infarction. Time – course analysis of the results of the Gruppo Italiano per lo Studio della Sopravvivenza nell` Infarto Miocardio (GISSI) –Prevenzione. *Circulation* 105,1897-1903.

Marchioli, R., Marfisi, R.M., Borreli, G., Chieffo, C., Franzosi, M.G., Levantesi, G., Maggioni, A.P., Nicolosi, G.L., Scarano, M., Silletta, M.G., Schweiger, C., Tavazzi, L., Tognoni, G., (2007). Efficacy of n-3 polyunsaturated fatty acids according to clinical characteristics of patients with recent myocardial infarction: Insight from the GISSI-Prevenzione trial. *J. Cardiovasc. Med.* 8, Suppl 1, S34-S37.

Marik, P.E., Varon, J., (2009). Omega-3 dietary supplements and the risk of cardiovascular events: a systematic review. *Clin. Cardiol.* 32, 365-372.

Matthan, N.R., Jordan, H., Chung, M., Lichtenstein, A.H., Lathrop, D.A., Lau, J., (2005). A systematic review and meta-analysis of the impact of ω-3 fatty acids on selected arrhythmia outcomes in animal models. *Metabolism* 54, 1557-1565.

McLennan, P.L., (2001). Myocardial membrane fatty acids and the antiarrhythmic actions of dietary fish oil in animal models. *Lipids* 36: S111-S114.

McLennan, P.L., Abeywardena, M.Y., (2005). Membrane basis for fish oil effects on the heart: linking natural hibernators to prevention of human sudden cardiac death. *J. Membrane Biol.* 206, 85-102.

McMurchie, E.J., Patten, G.S., McLennan, P.L., Charnock, J.S., Nestel, P.J., (1988). The influence of dietary lipid supplementation on cardiac beta-adrenergic receptor adenylate cyclase activity in the marmoset monkey. *Biochim. Biophys. Acta.* 937, 347-358.

Mozaffarian, D., Lemaitre, R.N., Kuller, L.H., Burke, G.L., Tracy, R.P., Siscovick, D.S., (2003). Cardiac benefits of fish consumption may depend on the type of fish meal consumed: the Cardiovascular Health Study. *Circulation* 107,1372-1377.

Mozaffarian, D., Psaty, B.M., Rimm, E.B., Lemaitre, R.N., Burke, G.L., Lyles, M.F., Lefkowitz, D., Siscovick, D.S., (2004). Fish intake and risk of incident atrial fibrillation. *Circulation*110, 368-373.

Mozaffarian, D., Geelen, A., Brouwer, I.A., Geleijense, J.M., Zock, P.L., Katan, M.B., (2005), Effect of fish oil on heart rate in humans: a meta-analysis of randomized controlled trials. *Circulation* 112: 1945-1952.

Mozaffarian, D., Rimm, E.B., (2006). Fish intake, concomitants, and human health: evaluating the risks and the benefits. *JAMA* 296, 1885-1899.

Mozaffarian, D. (2007). JELIS, fish oil, and cardiac events. *Lancet* 369, 1062-1063.

Mozaffarian, D., Jason, H.Y., Wu, J.H.Y., (2011). Omega-3 fatty acids and cardiovascular disease. Effects on risk factors, molecular pathways, and clinical events. *J. Am. Coll. Cardiol.* 58, 2047-2067.

Mozaffarian, D., Marchioli, R., Macchia, A., Silletta, M.G., Ferrazzi, P., Gardner, T., Latini, R., Libby, P., Lombardi, F., O`Gara, P., Page, R., Tavazzi, L., Tognoni, G., (2012).Fish Oil and Postoperative Atrial Fibrillation. The omega-3 fatty acids for prevention of post-operative atrial fibrillation (OPERA) randomized trial. *JAMA* 308, 2001-2011.

Neubronner, J., Schuchardt, J.P., Kressel, G., Merkel, M., von Schacky, C., Hahn, A., (2001).Enhanced increase of omega-3 index in response to long-term n-3 fatty acid supplementation from triacylglycerides versus ethyl esters. *Eur. J. Clin. Nutr.* 65, 247-254.

Niroomand, F., Weinbrenner, C., Weis, A., Bangert, M., Schwencke, C., Marquetant, R., Beyer, T., Strasser, R.H., Kübler, W., Rauch, B., (1995).Impaired Function of Inhibitory G Proteins During Acute Myocardial Ischemia of Canine Hearts and Its Reversal During Reperfusion and a Second Period of Ischemia. Possible Implications for the Protective Mechanism of Ischemic Preconditioning. *Circ. Res.* 76, 861-870.

Nodari, S., Metra, M., Milesis, G., Manerba, A., Cesana, B.M., Gheorghiade, M., Dei Cas, L.(2009). The role of n-3 PUFAs in preventing the arrhythmic risk in patients with idiopathic dilated cardiomyopathy. *Cardiovasc. Drugs Ther.* 23, 5-15.

Nolan, R.P., Jong, P., Barry-Bianchi, S.M., Tanaka, T.H., Floras, J.S., (2008). Effects of drug, biobehavioural and exercise therapies on heart rate variability in coronary artery disease: a systematic review. *Eur. J. Cardiovasc. Prev.Rehabil.* 15, 386-396.

O`Keefe, J.H., Abuissa, H., Sastre, A., Steinhaus, D.M., Harris, W.S., (2006). Effects of Omega-3 Fatty Acids on Resting Heart Rate, Heart Rate Recovery After Exercise, and Heart Rate Variability in Men With Healed Myocardial Infarctions and Depressed Ejection Fractions. *Am. J. Cardiol.* 97, 1127-1130.

Pouleur, A.C., Barkoudah, E., Uno, H., Skali, H., Finn, P.V., Zelenkofske., S.L., Belenkov, Y.N., Mareev, V., Velazquez., E.J., Rouleau, J.L., Maggioni, A.P., Køber, L., Califf, R.M., McMurray, J.J., Pfeffer, M.A., Solomon, S.D., (2010). Pathogenesis of sudden unexpected death in a clinical trial of patients with myocardial infarction and left ventricular dysfunction, heart failure, or both. *Circulation* 122, 597-602.

Raitt, M.H., Connor, W.E., Morris, C., Kron, J., Halperin, B., Chugh, S.S., McClelland, J., Cook, J., MacMurdy, K., Swenson, R., Connor, S.L., Gerhard, G., Kraemer, D.F., Oseran, D., Marchant, C., Calhoun, D., Shnider, R., McAnulty, J., (2005). Fish oil supplementation and risk of ventricular tachycardia and ventricular fibrillation in patients with implantable defibrillators: a randomized controlled trial: *JAMA* 293: 2884-2891.

Rauch, B., Niroomand, F., Messineo, F.C., Weis, A., Kübler, W., Hasselbach, W., (1994).Effect of phospholipid hydrolysis by phospholipase A_2 on the kinetics of antagonist binding to cardiac muscarinic receptors. *Biochem. Pharmacol.* 48: 1289-1296.

Rauch, B., Colvin, R.A., Messineo, F.C., (1989). Inhibition of ^{3}H-Quinuclidinyl Benzylate Binding to Cardiac Muscarinic Receptor by Log Chain Fatty Acids can be Attenuated by Ligand Occupation of the Receptor. *J. Mol. Cell. Cardiol.* 21, 495-506.

Rauch, B., Schiele, R., Schneider, S., Gohlke, H, Diller, F, Gottwik, M, Steinbeck, G, Heer, T, Katus, H, Zimmer, R, Erdogan, A, Pfafferot, C, Senges, J.,fort he OMEGA Study group., (2006).Highly purified omega-3 fatty acids for secondary prevention of sudden cardiac death after myocardial infarction: aims and methods of the OMEGA-study. *Cardiovasc. Drugs Ther.*20, 365-375.

Rauch, B., Schiele, R., Schneider, S., Diller, F., Victor, N., Gohlke, H., Gottwik, M., Steinbeck, G., DelCastillo, U., Sack, R., Worth, H., Katus, H., Spitzer, W., Sabin, G., Senges, J. for the OMEGA Study Group., (2010). OMEGA, a randomized, placebocontrolled trial to test the effect of highly purified omega-3 fatty acids on top of modern guideline-adjusted therapy after myocardial infarction.*Circulation* 122, 2152-2159.

Rauch, B., Senges, J., (2012). The effects of supplementation with omega-3 polyunsaturated fatty acids on cardiac rhythm: anti-arrhythmic, pro-arrhythmic, both or neither? It depends. *Frontiers in Physiology,* doi:10.3389/fphys.2012.00057.

Rauch, G., Kieser, M., Ulrich, S., Doherty, P., Rauch, B., Schneider, S., Riemer, T., Senges, J.(2012).Competing time-to-event endpoints in cardiology trials: A simulation study to illustrate the importance of an adequate statistical analysis. *Eur. J. Prev. Cardiol.* DOI: 10.1177/2047487312460518

Reiffel, J.A., McDonald, A., (2006).Antiarrhythmic Effects of Omega-3 Fatty Acids. *Am. J. Cardiol.* 98(Suppl), 50i-60i.

Rizos, E.C., Ntzani, E.E., Bika, E., Kostapanos, M.S., Elisaf, M.S., (2012).Association between Omega-3 Fatty Acid Supplementation and Risk of Major Cardiovascular Disease Events. A Systematic Review and Meta-analysis. *JAMA* 308, 1024-1033.

Saravanan, P., Bridgewater, B., West, A.L., O`Neill, S.C., Calder, P.C., Davidson, N.C., (2010). Omega-3 fatty acid supplementation does not reduce risk of atrial fibrillation after coronary artery bypass surgery: a randomized, double blind, placebo-controlled clinical trial. *Circ., Arrhythm. and Electrophysiol.* 3, 46-53.

Schuchardt, J.P., Schneider, I., Meyer, H., Neubronner, J., von Schacky, C., Hahn, A., (2011) Incorporation of EPA and DHA into plasma phospholipids in response to different omega-3 fatty acid formulations – a comparative bioavailability study of fish oil vs. krill oil. *Lipids Health Dis.* 10, 145.

Shattock, M.J., Bers, D.M., (1989). Rat vs. rabbit ventricle: Ca flux and intracellular Na assessed by ion-selective microelectrodes. *Am. J. Physiol.* 256, C813-C822.

Siddiqui, R.A., Harvey, K.A., Zaloga, G.P., (2008). Modulation of enzymatic activities by n-3 polyunsaturated fatty acids to support cardiovascular health. *J. Nutr. Biochem.* 19, 417-437.

Siscovick, D.S., Raghunathan, T.E., King, I., Weinmann, S., Wicklund, K.G., Albright, J., Bovbjerg, V., Arbogast, P., Smith, H., Kushi, L.H., Cobb, L.A., Copass, M.K., Psaty, B.M., Lemaitre,R., Retzlaff, B., Childs, M., Knopp, R.H.,.(1995). Dietary intake and cell membrane levels of long-chain n-3 polyunsaturated fatty acids and the risk of primary cardiac arrest. *JAMA* 274, 1363-1367.

Solomon, S.D., Zelenkofske, S., McMurray, J.J., Finn, P.V., Velazquez, E., Ertl, G., Harsanyi, A., Rouleau, J.L., Maggioni, A., Kober, L., White, H., Van de Werf, F., Pieper, K., Califf, R.M., Pfeffer, M.A., (2005). Valsartan in Acute Myocardial Infarction Trial (VALIANT) Investigators. Sudden deaths in patients with myocardial infarction and left ventricular dysfunction, heart failure, or both. *N. Engl. J. Med.* 352, 2581-2588.

Tanaka, H., Komikado, C., Namekata, I., Nakamura, H., Suzuki, M., Tsuneoka, Y., Shigenobu, K., Takahara, A., (2008). Species differences in the contribution of T-type calcium current to cardiac pacemaking as revealed by r(-)-efonidipine. *J. Pharmacol. Sciences* 107, 99-102.

Verkerk, A.O., van Ginneken, A.C.G., Berecki, G., den Ruijter, H.M., Schuhmacher, C.A., Veldkamp, M.W., Baartscheer, A., Casini, S., Opthof, T., Hovenier, R., Fiolet, J.W.T., Zock, P.L, Coronel, R., (2006). Incorporated sarcolemmal fish oil fatty acids shorten pig ventricular action potentials. *Cardiovascular Research* 70, 509-520.

Verkerk, A.O., den Ruijter, H.M., de Jonge, N., Coronel, R., (2009). Fish oil curtails the human action potential dome in a heterogeneous manner: implication for arrhythmogenesis.*Int. J. Cardiol.* 132: 138-140.

Wang, C., Harris, W.S., Chung, M., Lichtenstein, A.H., Balk, E.M., Kupelnick, B., Jordan, H.S., Lau, J., (2006).N-3 fatty acids from fish or fish-oil supplements, but not α-linolenic acid, benefit cardiovascular disease outcomes in primary- and secondary-prevention studies: a systematic review. *Am. J. Clin. Nutr.* 84, 5-17.

Wang, R.X., Li, X.R., Guo, T., Sun, L.P., Guo, S.X., Yang, Z.Y., Yang, X.J., Jiang, W.P., (2010). Docohexaenoic acid has influence on action potentials and transient outward potassium currents of ventricular myocytes. *Lipids Health Dis.* 9, 39.

Whelton, S.P., He, J., Whelton, P.K., Muntner, P., (2004). Meta-analysis of observational studies on fish intake and coronary heart disease. *Am. J. Cardiol.* 2004; 93: 1119-23.

Xiao, Y.F., Kang, J.X., Morgan, J.P., Leaf, A., (1995). Blocking effects of polyunsaturated fatty acids on Na^+ channels of neonatal rat ventricular myocytes. *Proc. Natl. Acad. Sci.* 92, 11000-11004.

Xiao, Y.F., Gomez, A.M., Morgan, J.P., Lederer, W.J., Leaf, A., (1997). Suppression of voltage-gated L-type Ca^{2+} currents by polyunsaturated fatty acids in adult and neonatal rat ventricular myocytes. *Proc. Natl. Acad. Sci.* 94, 4182-4187.

Xiao, Y.F., Wright, S.N., Wang, G.K., Morgan, J.P., Leaf, A., (1998). Fatty acids suppress voltage-gated Na^{+} currents in HEK293t cells transfected with the alpha subunit of the human cardiac sodium channel. *Proc. Natl. Acad. Sci.* 95, 2680-2685.

Yan, G.X., Joshi, A., Guo, D., Hlaing, T., Martin, J., Xu, X., Kowey, P.R., (2004).Phase 2 reentry as a trigger to initiate ventricular fibrillation during early acute myocardial ischemia. *Circulation* 110, 1036-1041.

Yokoyama, M., Origasa, H., Matsuzaki, M., Matsuzawa, Y., Saito, Y., Ishikawa, Y., Oikawa, S,, Sasaki, J., Hishida, H., Itakura, H., Kita, T., Kitabatake, A., Nakaya, N., Sakata, T., Shimada, K., Shirato, K., Japan EPA lipid intervention study (JELIS) Investigators, (2007). Effects of eicosapentaenoic acid on major coronary events in hypercholesterolaemic patients (JELIS): a randomized open-label, blinded endpoint analysis. *Lancet* 369, 1090-1098.

Yzebe, D., Lievre, M., (2004). Fish oils in the care of coronary heart disease patients: a meta-analysis of randomized controlled trials. *Fundam.Cin. Pharmacol.* 18, 581-592.

Zeghichi-Hamri, S., de Lorgeril, M., Salen, P., Chibane, M., de Leiris., J., Boucher, F., Laporte, F., (2010). Protective effect of dietary n-3 polyunsaturated fatty acids on myocardial resistance to ischemia-reperfusion injury in rats. *Nutr. Res.* 30, 849-857.

Zhao, Y.T., Chen, Q., Sun, Y.X., Li, X.B., Zhang, P., Xu, Y., Guo, J.H., (2009). Prevention of sudden cardiac death with omega-3 fatty acids in patients with coronary heart disease: a meta-analysis of randomized controlled trials. *Ann. Med.* 41, 301-310.

In: Omega-3 Fatty Acids
Editor: Md. Wasim Khan
ISBN: 978-1-62948-516-4

Chapter 8

Omega-3 Fatty Acids: Role in Intestinal Adaptation and Recovery

Igor Sukhotnik*
Technion-Israel Institute of Technology, the Ruth and Bruce Rappaport Faculty of Medicine, Dept of Pediatric Surgery and Pathology, Bnai Zion Medical Center, Haifa, Israel

Abstract

Over the past decades, growing evidence from several clinical and experimental studies indicates that omega-3 polyunsaturated fatty acids (n-3PUFA) can reduce not only the activity of inflammatory processes but may also improve the immune system and protect the organism against the negative effect of the systemic inflammatory response syndrome. Dietary n-3PUFAs also decrease the production of different proinflamatory mediators such as IL-1,IL-2, IL-6 and tumor necrosis factor-alpha. The positive anti-inflammatory effect of n-3PUFAs has been described in several inflammatory conditions such as inflammatory bowel disease, rheumatoid arthritis, and hepatitis. In this report, we discuss the use of a fish oil-based lipid emulsion in the treatment of different gastrointestinal disorders including parenteral nutrition-associated liver disease, short bowel syndrome, intestinal ischemia-reperfusion, chemotherapy-induced mucositis, microvillous inclusion disease etc.

The mechanisms of action of n-3PUFAs on intestinal cell turnover (proliferation and apoptosis) as well as the effects of this agent on intestinal re-growth after bowel resection and intestinal recovery following intestinal ischemia or chemotherapy-induced damage in rodent models are discussed. Such knowledge may allow us to exploit omega-3 polyunsaturated fatty acids that act as gut trophic agent for the bowel mucosa and suggest new therapeutic strategies to maintain gut integrity after massive bowel resection or intestinal injury, eliminate the dependence on total parenteral nutrition, and avoid the need for intestinal transplantation.

* E-mail: igor-dr@internet-zahav.net.

Keywords: Omega-3 polyunsaturated fatty acids, intestine, adaptation, short bowel syndrome, intestinal ischemia-reperfusion, chemotherapy-induced mucositis

Introduction

Also known as polyunsaturated fatty acids (PUFAs), omega-3 fatty acids [docosahexaenoic acid (DHA, C22:6) + eicosapentaenoic acid (EPA, C20:5)] play a crucial role in normal growth and development. The broad health benefits of very long chain *n*-3 PUFAs and the pleiotropic effects of dietary fish oil and DHA have been proposed to involve alterations in membrane structure and function, eicosanoid metabolism, gene expression and the formation of lipid peroxidation products, although a comprehensive understanding of the mechanisms of action has yet to be elucidated. *n*-3 PUFAs are fatty acids with a long chain (20 carbons or more), with the first double bond located after the third carbon from the methyl end. They can be found in fish, such as salmon, tuna, and halibut, other seafood including algae and krill, some plants, and nut oils. Research shows that omega-3 fatty acids may help lower risk of chronic diseases such as heart disease, cancer, and arthritis. Since they may reduce the risk of heart disease, the American Heart Association recommends eating fish at least 2 times a week. Fish oil has been shown to lower levels of triglycerides, and to lower the risk of death, heart attack, stroke, and abnormal heart rhythms in people who have already had a heart attack (Sauder et al., 2013; Liosis et al., 2013). Fish oil also appears to help prevent and treat atherosclerosis (hardening of the arteries) by slowing the development of plaque and blood clots, which can clog arteries (Harris et al., 2012; Simopoulos et al., 1999). Omega-3 fatty acids are highly concentrated in the brain and appear to be important for cognitive (brain memory and performance) and behavioral function. Studies show that DHA affects blood-brain barrier functions and neuronal membrane fluidity; it also regulates neurotransmission systems such as serotonergic, dopaminergic, norepinephrinergic, and acetylcholinergic systems (Kidd et al., 2007). In fact, infants who do not get enough omega-3 fatty acids from their mothers during pregnancy are at risk for developing vision and nerve problems (Helland et al.,2003; Larque et al.,2002).. Symptoms of omega-3 fatty acid deficiency include fatigue, poor memory, dry skin, heart problems, mood swings or depression, and poor circulation.

Over the past decades, growing evidence from several clinical and experimental studies indicates that PUFAs can reduce not only the activity of inflammatory processes but may also improve the immune system and protect the organism against the negative effect of the systemic inflammatory response syndrome (Weber et al.,1986). Specifically, n-3PUFAs act to decrease the production of inflammatory eicosanoids, cytokines, reactive oxygen species (ROS) and the expression of inflammatory adhesion molecules. The biologically active essential fatty acids such as arachidonic acid (Omega-6 family) and EPA/DHA (Omega-3 family) play an important role in inflammatory processes as eicosanoid and prostanoid precursors. Arachidonic acid is considered pro-inflammatory and prothrombotic, while EPA/DHA is usually labeled as anti-inflammatory and antithrombotic (Ekema et al., 2008). Recent evidence suggests that converting cellular membranes to an Omega-6/Omega-3 ratio close to 1:1 decreases the substrate availability necessary for production of inflammatory cytokines and replaces them with anti-inflammatory mediators (Alexander et al., 1998).

Omega-3 PUFA reportedly inhibits eicosanoid production by replacing arachidonic acid as a substrate, thus preventing arachidonic acid metabolism. The products formed from n-3 PUFA metabolism, have a much lower pro-inflammatory effect compared to those produced from arachidonic acid metabolism. In addition, n-3PUFAs reduce T-cell receptor expression, phytomitogen response, antigenic stimulation responses by lymphocytes, IgA expression, delayed-type hypersensitivity responses, and accessory cell function. Dietary n-3PUFAs also decreases the production of different proinflamatory mediators such as IL-1,IL-2, IL-6 and tumor necrosis factor-alpha (Meydani et al.,1993). Additionally, the presence of trace amounts of vitamins A and E in fish oil and olive oil may have partially accounted for reduced inflammation via anti-oxidant pathways (Owen et al., 2000).

The positive anti-inflammatory effect of n-3PUFAs has been described in several inflammatory conditions such as inflammatory bowel disease Loeschke et al.,1996; Aslan et al.,1992), rheumatoid arthritis (Kremer et al.,1995; Geusens et al., 1994), and hepatitis (Schmöcker et al., 2007).

Omega-3 Fatty Acids and Intestine

Under normal circumstance, newly formed immature intestinal cells divide in the crypts of Lieberkuhn (finger-like invaginations of the epithelium into the underlying connective tissue) and migrate up to the villus tip, differentiate, mature and become functional, with an important role in absorption of nutrients. The renewal process operates continually, with cells taking two to seven days to make the journey from the site of their final division cycle in the crypt to the point of their exfoliation from the villus tip (Wright et al., 1984). Intestinal stem cells reside near the base of the crypt, providing immature progeny, which continue to divide as they migrate up the crypt until they differentiate into either intestinal absorptive cells (enterocytes) or secretory lineages (goblet, Paneth, and entero-endocrine cells). Absorptive enterocytes, the primary cell type of the epithelial layer, comprise about 90% of the cells in the entire epithelium and are responsible for the terminal digestion and absorption of luminal nutrients. These cells synthesize digestive enzymes called brush border enzymes that become embedded in the plasma membranes around the microvilli. The goblet cells, located throughout the epithelial layer, secrete mucus that helps protect the epithelial layer from digestion, as well as trefoil factor-3, which is ultimately involved in post-injury repair of the small bowel (Beck et al., 2004). The entero-endocrine cells comprise only 1% of the intestinal epithelium and are a part of the enteric endocrine system sense the luminal environment and secrete hormones such as cholecystokinin and gastrin into blood. The overall percentage of secretory lineages within the intestinal epithelium gradually increases from proximal to distal, with the greatest percentage being found in the terminal ileum. The Paneth cells reside at the base of the crypt and are classically believed to have an antimicrobial function in the small intestine (production of lysozyme, an enzyme that destroys bacteria) (Garabedian et al., 1997). The adjacent location of these cells to the intestinal stem cell zone and recent research demonstrating their secretion of cytokines, growth factors, and other products, including soluble WNT proteins known to regulate proliferation, raises the possibility of a potential role for Paneth cells initiating and maintaining the adaptive response (Bjerknes et al., 1981).

Although the mechanism of EPA and DHA action in the small intestine is still not fully defined in molecular terms, it is becoming increasingly clear that these fatty acids are pleiotropic. The challenge for the future is to determine precisely how n-3 PUFA alter membrane microdomain structure and function, eicosanoid metabolism, and the formation and mobilization of reactive oxygen products and intracellular Ca^{2+}. These mechanisms are responsible for the pleiotropic protective effects of n-3 PUFA with respect to colon cancer and inflammatory diseases of the intestine. Dietary intake of EPAandDHA induced changes in gene expression in the small intestine, including many metabolic genes. Especially genes involved in lipid catabolism were upregulated. In contrast, a large cluster of the genes engaged in cholesterol biosynthesis was downregulated. Importantly, all these effects were specifically induced by long-chain n-3 PUFA, EPA and DHA, as compared with their precursor ALA, and could not be detected in the colon for a selected set of genes. Consumption of n-3 PUFA exerted a protective effect at the initiation (DNA adduct formation) and promotional (apoptosis and aberrant crypt foci levels) stages (Davidson et al., 2004). DHA and butyrate work coordinately in the colon to initiate a distinct intrinsic proapoptotic cycle involving the activation of store-operated channels, leading to rapid entry of Ca^{2+} through the plasma membrane and mitochondrial Ca^{2+} loading. This directly or indirectly increases mitochondrial phospholipid hydroperoxides and triggers the opening of the permeability transition pore and release of pro-apoptotic molecules like cytochrome C. These effects culminate in the induction of apoptosis in colonic cells.

Omega-3 Fatty Acids: Role in Intestinal Adaptation after Bowel Resection

Short bowel syndrome (SBS) is defined as intestinal failure following a loss of intestinal length which causes marked maldigestion and malabsorption of dietary nutrients and induces major fecal loss of energy, nitrogen, and fat (Vanderhoof et al., 1996). SBS is a common problem in pediatric surgery andoccurs in newborns and infants suffering from necrotizing enterocolitis, intestinal atresia and volvulus which require massive intestinal resection (Boothet al., 1998). The survival patients after massive small bowel resection is dependant on the ability of the residual bowel to adapt. "Adaptation" is the term applied to the progressive recovery from the intestinal failure which occurs following a loss of intestinal length (Coran et al., 1999; O'Brienet al., 2001). The remaining intestine dilates, elongates and thickens, villous height and crypt depth increase, leading to the mucosal hyperplasia, all of which lead to increased absorptive surface area (structural adaptation). In addition, increased nutrient absorption by enterocytes occurs (functional absorption). Over the past decades, much research has focused on the question of which factors may promote intestinal re-growth in SBS.

Enteral feeding is one of the major trophic factors stimulating intestinal adaptation (Mengeet al., 1975). However, not all nutrients have equal stimulating trophic effects. While traditional dietary management advocates a restricted fat diet containing 50-75% medium-chain triglycerides as a source of fat, there are still some unanswered questions concerning the relative value of high fat vs high-carbohydrate diets for these patients. Among the factors promoting post-resection intestinal adaptation, long-chain fatty acids are considered the most effective. Several experimental studies have demonstrated that bowel re-growth can be altered by changing dietary ratios of different fatty acids. It has been reported that animals with SBS receiving a diet deficient in essential fatty acids have significantly impaired intestinal mucosal hyperplasia (Hart et al., 1988). The fatty Arachidonic acid appears to be a potent stimulator of intestinal adaptation, more than its precursor, linoleic acid (Vanderhoof et al., 1999). On the other hand, dietary linoleic acid has higher stimulating effects on post-resection mucosal hypertrophy compared to palmitic acid (Park et al., 1989). We have recently shown that restriction of palmitic acid in the diet inhibits intestinal adaptation and that a high palmitic acid diet has a mild stimulating effect on intestinal re-growth (Sukhotnik et al., 2008).

Over the past years, several preliminary clinical trials have described the positive effects of omega-3 fatty acids fish-oil based lipid (Omegaven®; Fresenius Kabi, Bad Homburg, Germany) in prevention of parenteral nutrition associated liver disease (PNALD) in children with SBS. In a recent review, Diamond et al discuss the rationale for the use of *n*-3 PUFAs. The authors note that the beneficial effects of substituting *n*-6 PUFAs with *n*-3 PUFAs are likely related to three major mechanisms: 1) the addition of *n*-3 PUFAs results in an improvement in bile flow; 2) *n*-3 PUFAs decrease steatosis; and 3) the addition of *n*-3 PUFAs shifts the eicosanoid profile towards a less inflammatory state. Together, these actions result in decreased cholestasis, hepatitis, and subsequent fibrosis (Diamond et al., 2008). Since liver disease may hinder intestinal adaptation (Weber et al.,2002), in a recent study, the effects of *n*-3 PUFAs fish-oil based lipid (Omegaven®; Fresenius Kabi, Bad Homburg, Germany) on structural intestinal adaptation in a rat model of SBS were evaluated (Sukhotnik et al.,2010). Control or resected animals were treated either with oral or intraperitoneal omegaven. Parameters of intestinal adaptation, overall bowel and mucosal weight, mucosal DNA and protein, villus height and crypt depth, cell proliferation and apoptosis, were determined 14 days following 75% bowel resection. In this study, administration of parenteral omegaven in SBS rats significantly enhanced structural intestinal adaptation. Parenteral omegaven significantly increased overall bowel and mucosal weight, which occurred with a synergistic increase in bowel circumference.

In addition, omegaven significantly increased intestinal mucosal DNA and protein. Omegaven caused also a mild stimulation of cell proliferation and concomitant decrease in cell apoptosis compared to SBS-non-treated animals. Increased mucosal proliferation in functioning intestine as demonstrated by the increased cell proliferation index following IP omegaven administration together with decreased cell death via apoptosis, suggests an activated enterocyte turnover and may be considered as a main mechanism of mucosal hyperplasia in recovering bowel. Bax mRNA levels were down-regulated while bcl-2 mRNA was up-regulated in omegaven treated rats compared to SBS-nontreated animals. In accord, bax/bcl-2 ratio decreased in omegaven treated rats compared to SBS-animals, suggesting increased enterocyte survival.

These results suggest that omegaven may inhibit resection-induced apoptosis by concomitant stimulation of theanti-apoptotic bcl-2 gene and inhibition of the pro-apoptoticbax gene expression. The authors conclude that parenteral omegaven stimulates enterocyte turnover and may be clinically beneficial as an agent to stimulate intestinal adaptation in patients with short bowel syndrome.

Effects of Omega-3 Fatty Acids on Intestinal Recovery Following Chemotherapy-Induced Mucositis

Oral and gastrointestinal mucositis are frequent complications of chemotherapy, in particular with drugs affecting DNA synthesis (such as fluorouracil, methotrexate, and cytarabine), contributing not only to the morbidity of treatment but its cost as well (Sonis et al., 1993; Naidu et al., 2004). Mucositis limits the patient's ability to tolerate chemotherapy or radiation therapy, prolongs hospital stay, increases re-admission rates, compromises the patient's nutritional status, affects the patient's quality of life, and is occasionally fatal. Severe inflammation of the intestinal mucosa plays a significant role in the development of chemotherapy-induced mucositis and is a major characteristic of the condition (Sonis et al., 2004). Different inflammatory mediators and cytokines, such as leukotriene B4and prostaglandin E2, act to amplify signaling cascades, induce apoptosis, and cause further tissue damage. The pathogenesis of chemotherapy induced gastrointestinal mucositis has been described by Sonis et al and includes five phases: initiation by chemotherapy, up-regulation and generation of messenger signals, signaling by pro-inflammatory cytokines and amplify-cation of mucosal injury, ulceration of the mucosa and finally, healing. The initial stages of inflammation in mucositis include increased pro-inflammatory cytokine levels, which act as a homing marker for inflammatory immune cells in the submucosa (Sonis et al., 2004). Since anti-inflammatory properties of n-3PUFAs have been assessed in a number of disease settings, in a recent study the effects of omega-3 fatty acids in reducing the severity of chemotherapy induced mucositis was investigated in a rat model (Koppelmann et al., 2012).

In this study, administration of enteral n-3PUFAs in methotrexate (MTX)-treated rats decreased mucosal inflammation and reversed intestinal mucosal damage. This was evident from the decreased intestinal injury score. Although the mucosa was still severely damaged, MTX- n-3PUFAs rats showed the presence of newly formed crypts and regeneration, less significant villous epithelial atrophy, and less significant polymorphonuclear leukocyte infiltration in the lamina propria (Figure 1).

The mechanisms of this effect may include a decreased generation of pro-inflammatory cytokines (TNF-α, IL-1β, IL-6, IL-8) as well as decreased activation of NFκB signaling, decreased leukocyte chemotaxis and decreased T cell reactivity, which were described in an animal model of inflammatory bowel disease (Calder et al., 2009) may be responsible for the observed protective effects of n-3PUFAs in chemotherapy-induced mucositis. In addition, reduced inflammation was coupled with enhanced intestinal repair. Enteral n-3PUFAs significantly increased overall bowel and mucosal weight, which occurred together with a synergistic increase in mucosal DNA and protein levels, which indicates that the greater intestinal mass is attributed to cellular hyperplasia. n-3PUFAs also caused a significant stimulation of cell proliferation and a concomitant decrease in cell apoptosis compared to MTX-untreated rats, which suggests an activated enterocyte turnover and may be considered as one of the main mechanisms of mucosal hyperplasia in recovering bowel. It has been suggested that incorporation of n-3 PUFAs into the intestinal mucosa may have altered mucosal function and could have contributed to the regulation of growth and transduction signals such as those involving protein kinase C (Calderaro et al., 1994); however, further investigation is required to ascertain the mechanisms by which n-3 PUFA enhance repair in the intestinal mucosa.

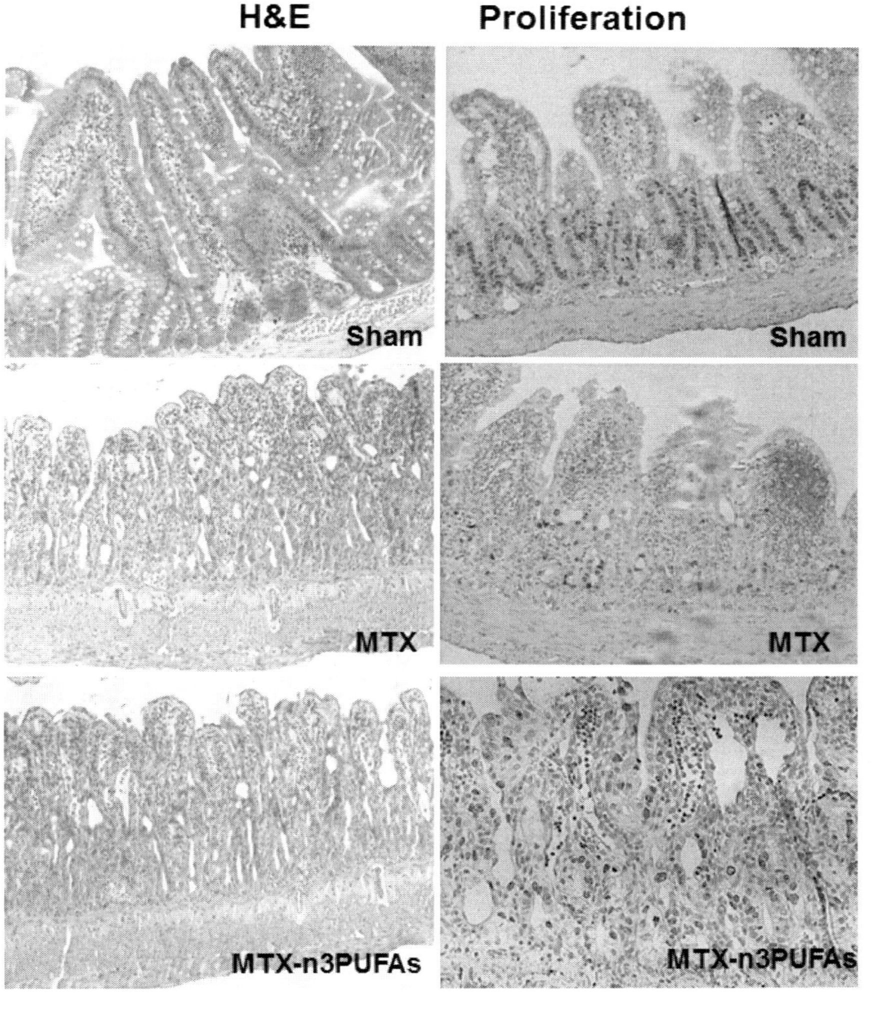

Figure 1. Effect of methotrexate and omega-3 fatty acids on intestinal microscopic appearance and cell proliferation. MTX-methotrexate, n-3PUFA- omega-3 polyunsaturated fatty acids.

An observed increase in enterocyte proliferation rate in MTX-3PUFA animals was accompanied by elevated β-catenin and p-ERK protein levels, which suggests activated stem cell activity and a stimulated MAPK signaling pathway. Treatment with n-3 PUFA also resulted in decreased cell apoptosis. Bax mRNA levels were down-regulated while bcl-2 mRNA was up-regulated in n-3PUFAs treated rats compared to MTX-nontreated animals. Thus, the /bcl-2 ratio decreased in MTX-n-3PUFAs rats compared to MTX-animals, suggesting increased enterocyte survival. Both elevated cell proliferation and decreased cell apoptosis may be considered one of the main mechanisms responsible for compensatory hyperplasia and, thus responsible for increased intestinal cell mass. The authors conclude that enteral omega-3 fatty acids reverse intestinal damage and stimulate intestinal recovery following MTX-induced intestinal damage in a rat. Enhanced cell proliferation and inhibited programmed cell death (through up-regulation of bcl-2 and down-regulation of bax expression) may be responsible for this effect. Enteral omega-3 fatty acids may be clinically beneficial in preventing intestinal damage and in stimulating intestinal recovery in patients with chemotherapy-induced mucositis.

Effects of Omega-3 Fatty Acids on Intestinal Recovery Following Intestinal Ischemia-Reperfusion

Intestinal ischemia-reperfusion (IR) injury remains clinically challenging problem despite decades of research in this area. The causes of mesenteric ischemia are well known and its major complication, gangrenous necrosis of portions of the gastrointestinal tract following necrotizing enterocolitis, midgut volvulus, incarcerated hernia, is well recognized by all pediatric surgeons. In addition, many life-threatening illnesses such as multiple trauma, shock and sepsis may impair intestinal blood flow. Common to most critically ill patients is preservation of blood flow to vital organs, which occurs at the expense of blood flow to the intestine (Haglund et al., 1999).

Ischemia may cause tissue injury directly as a result of decreased oxygen delivery, depletion of cellular energy stores and accumulation of toxic metabolites, resulting in cell necrosis (Khanna et al., 2003). However, reperfusion of ischemic tissue, although necessary as a reparative mechanism, has been shown to exacerbate acute ischemic injury through to the generation of reactive oxygen and nitrogen species. A production of these agents initiates a cascade of events, including activation of mast cells and neutrophils and production of systemic inflammatory mediators and cytokines which lead to neutrophils increased bowel permeability, increased bacterial translocation, a systemic inflammatory reaction and multiple organ failure (Granger et al., 1986).

Therapy of IR injury should include the pharmacological agents that can prevent ischemic injury and can prevent production of reactive oxygen species or agents in order to improve intestinal rehabilitation following the IR event. The *n*-3PUFAs have been shown to be efficaciousin treating and preventing ischemia-reperfusion injury of various organs. *n*-3PUFAs decrease cardiac-related death following myocardial ischemia (Marchioli et al.,2009), lower the risk of death or end-stage renal disease in patients with renal ischemia (Hassan et al.,2009), and have important roles in neuroprotection and in pro-survival signaling cascades after brain ischemia-reperfusion (Niemoller et al.,2010). Recent experimental study addressed the questionof whether parenteral omega-3 PUFA intervention confers intestinal protectionin the ischemic intestine in a rat model (Sukhotnik et al.,2011). The results of this study have shown that parenteral Omegaven exerts a protective effects on intestinal mucosa and ameliorates tissue damage associated with IR. Intestine from *n*-3PUFAs treated rats exhibited reversal of the loss of and injury to villus epithelium, and a decrease in the Park injury score.

The intestinal mucosa was indistinguishable from normal bowel, with evidence of minimal epithelial disruption. In addition, the mucosal hypoplasia caused by IR was ameliorated in part by Omegaven administration.

A comparison of morphometric parameters revealed intestinal hyperplasia to be a predominant feature of mucosal recovery. Increase in mucosal DNA along with hypertrophy of the individual cells, which we demonstrated morphometrically, is characteristic of tissues undergoing increased cell proliferation or repair. Histologically, IR-*n*-3PUFAs rats showed increased villus height compared to IR animals, suggesting increased absorptive surface area (Figure 2).

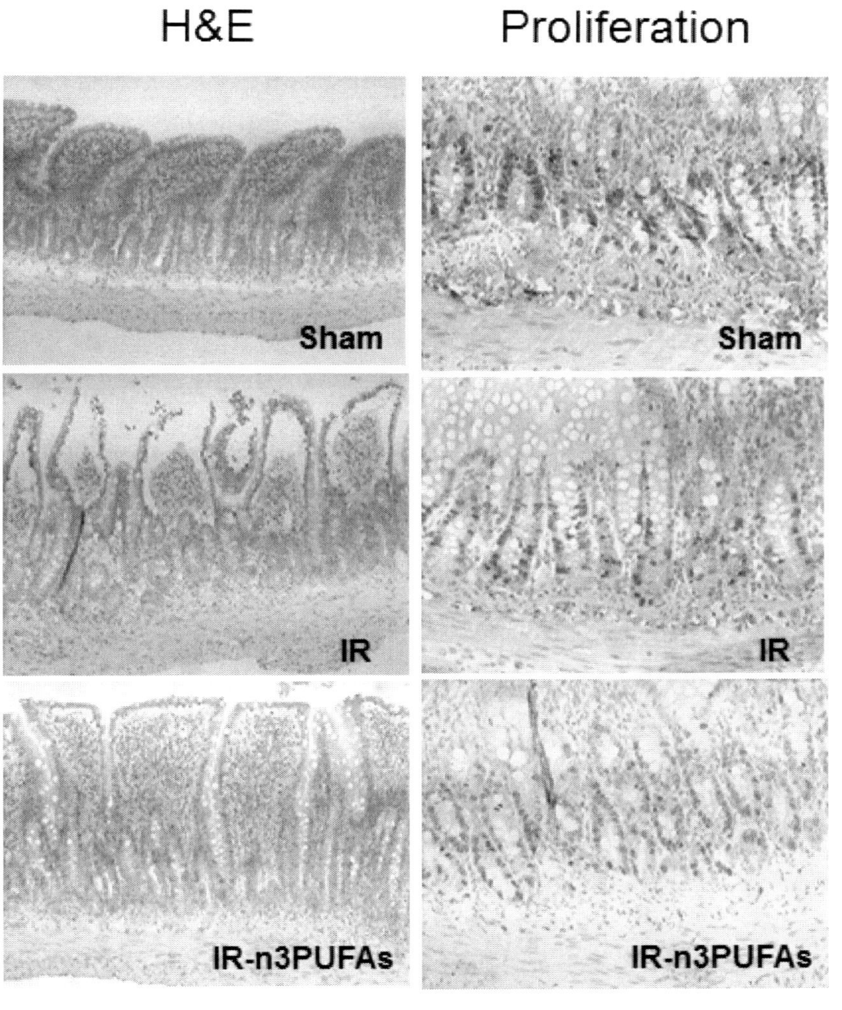

Figure 2. Effect of intestinal ischemia-reperfusion (IR) and omega-3 fatty acids (n3PUFAs) on intestinal microscopic appearance and cell proliferation.

It is interesting that omegaven did not change significantly proliferation rates, but decreased rates of cell apoptosis in both jejunum and ileum. Bax mRNA levels were down-regulated while Bcl-2 mRNA was up-regulated in omegaven treated rats compared to IRnontreated animals. In accord, Bax/Bcl-2 ratio decreased in this group, suggesting increased enterocyte survival. These results suggest that Omegaven may inhibit IR-induced apoptosis by concomitant stimulation of the anti-apoptotic Bcl-2 gene and inhibition of the pro-apoptotic Bax gene expression. The authors conclude that parenteral omega-3 fatty acids administration attenuates the intestinal mucosal injury caused by IR and enhances intestinal recovery.This benefit correlates with a decreased cell death via apoptosis. In the other experiment, intravenous administration of omega-3 lipid emulsion prior to ischemia-reperfusion injury in a rat reduced the oxidative stress and severity of tissue damage (Arisue et al.,2012). The authors conclude that modification of membrane fatty acids may serve as the mechanism underlying this reduction of tissue damage.

In conclusion, the rationale for supplementation with n-3 fatty acids to promote the health of the gastrointestinal tract lies in the anti-inflammatory effects of these lipid compounds, although there are controversial results that may reside in the different study designs used as well as in the various formulations and dosages used. The health benefits of PUFAs are diverse and a plethora of nutritional studiescontinue to demonstrate important benefits from the consumption of omega-3 enriched oils.The beneficial effects of omega-3 lipid emulsions have been extensively studied in various gastrointestinal disorders, including short bowel syndrome, intestinal ischemia–reperfusion injury, and chemotherapy-induced mucositis. It is both appropriate and timely, to precisely determine how PUFAs modulate cell signaling networks and reduces intestinal damage.

References

Alexander, J. W., (1998). Immunonutrition: the role of the omega-3 fatty acids. *Nutrition,* 14:627-33.

Arisue, A., Shimojima, N., Tomiya, M., Shimizu, T., Harada, D., Nakayama, M., Tomita, H., Shinoda, M., Tanabe, M., Maruyama, I., Mizuno, M., Kuroda, T., Wakabayashi, G., Morikawa, Y., (2012).Effect of an omega-3 lipid emulsion in reducing oxidative stress in a rat model of intestinal ischemia-reperfusion injury. *Pediatr. Surg. Int.*, 28:913-18.

Aslan, A., Triadafilopoulos, G., (1992). Fish oil fatty acid supplementation in active ulcerative colitis: a double-blind, placebo-controlled, crossover study. *Am. J. Gastroenterol.,* 87: 432-7.

Beck, P. L., Wong, J. F., Li, Y., Swaminathan, S., Xavier, R. J., Devaney, K. L., Podolsky, D. K., (2004). Chemotherapy- and radiotherapy-induced intestinal damage is regulated by intestinal trefoil factor. *Gastroenterology,* 126:796-808.

Bjerknes, M., Cheng, H., (1981). The stem-cell zone of the small intestinal epithelium. I. Evidence from Paneth cells in the adult mouse. *Am. J. Anat.,* 160: 51-63.

Booth, I. W., Lander, A. D., (1998). Short bowel syndrome. *Bailliere's Clinical. Gastroenterology,* 12: 739-72.

Calder, P. C., (2009). Fatty Acids and Immune Function: Relevance to Inflammatory Bowel Diseases. *International Reviews Immunology,* 28:506-34.

Calderaro, V., Parrillo, C., Balestrieri, M. L., Giovane, A., Filippelli, A., Rossi, F., (1994). Docosahexaenoic acid and signaling pathways in rabbit colon. *Mol. Pharmacol.*, 45: 737-46.

Coran, A. G., Spivak, D., Teitelbaum, D. H., (1999). An analysis of the morbidity and mortality of short bowel syndrome in the pediatric age group. *Eur. J. Pediatr. Surg.,* 9: 228-30.

Davidson, L. A., Nguyen, D. V., Hokanson, R. M., Callaway, E. S., Isett, R. B., Turner, N. D., Dougherty, E. R., Lupton, J. R., Carroll, R. J., Chapkin, R. S., (2004). Chemopreventive n-3 polyunsaturated fatty acids reprogram genetic signatures during colon cancer initiation and progression in the rat. *Cancer Res*., 64:6797-804.

Diamond, I. R., Sterescu, A., Pencharz, P. B., Wales, P. W., (2008). The rationale for the use of parenteral omega-3 lipids in children with short bowel syndrome and liver disease. *Pediatr.Surg. Int.,* 24:773-8.

Ekema, G., Falchetti, D., Boroni, G., Tanca A.R., Altana, C., Righetti, L., Ridella, M., Gambarotti, M., Berchich, L., (2008). Reversal of severe parenteral nutrition-associated liver disease in an infant with short bowel syndrome using parenteral fish oil (Omega-3 fatty acids). *J.Pediatr.Surg.,* 43, 1191-5.

Garabedian, E. M., Roberts, L. J. J., McNevin, M. S., Gordon, J. I., (1997). Examining the role of Paneth cells in the small intestine by lineage ablation in transgenic mice. *J. Biol. Chem.,* 272: 23729-32.

Geusens, P., Wouters, C., Nijs, J., (1994).. Long term effect of omega-3 fatty acid supplementation in active rheumatoid arthritis: a 12 month, double blind,controlled study. *Arthritis. Rheum.,* 37: 824-31.

Granger, D. N., Hollwarth, M. E., Park, D. A., (1986). Ishemia reperfusion injury: role of oxygen-derived free radicals. *Acta. Physiol. Scand.,* (Suppl).548: 47-56.

Haglund, U., Bergqvist, D., (1999). Intestinal ischemia-the basics. *Langenbecks. Arch. Surg.,* 384:233-8.

Harris, W. S., Pottala, J. V., Lacey, S. M., Vasan, R. S., Larson., Robins, S. J., (2012).Clinical correlates and heritability of erythrocyte eicosapentaenoic and docosahexaenoic acid content in the Framingham Heart Study. *Atherosclerosis,* 225:425-31.

Hart, M. H., Grandjean, C. J., Park, J. H. Y., Erdman, S. H., Vanderhoof, J. A., (1988). Essential fatty acid deficiency and postresection mucosal adaptation in the rats. *Gastroenterology*, 94:682–7.

Hassan, I. R., Gronert, K., (2009). Acute changes in dietary omega-3 and omega-6 polyunsaturated fatty acids have a pronounced impact on survival following ischemic renal injury and formation of renoprotective docosahexaenoic acid-derived protectin D1. *J. Immunol.,* 182:3223-32.

Helland, I. B., Smith, L., Saarem, K., Saugstad, O. D., Drevon, C. A., (2003). Maternal supplementation with very-long-chain n-3 fatty acids during pregnancy and lactation augments children's IQ at 4 years of age. *Pediatrics,* 111:e39-e44.

Kidd, P. M., (2007). Omega-3 DHA and EPA for cognition, behavior, and mood: clinical findings and structural-functional synergies with cell membrane phospholipids. *Alternative Medicine Review,* 12:207-27.

Khanna, A., Rossman, J., Caty, M.G., Fung, H. L., (2003). Beneficial effects of intraluminal nitroglycerin in intestinal ischemia-reperfusion injury in rats. *J. Surg. Res*., 114: 15-24.

Koppelmann, T., Pollak, Y., Mogilner, J., Bejar, J., Coran, A. G., Sukhotnik, I., (2012). Reversal of severe MTX- induced intestinal damage using enteral Omega-3 fatty acids. *Br. J. Nutr.,* 28:1-10.

Kremer, J. M., Lawrence, D. A., Petrillo, G. F., (1995). Effects of high-dose fish oil on rheumatoid arthritis after stopping nonsteroidal antiinflammatory drugs: clinical and immune correlates. *Arthritis. Rheum.*, 38: 1107-12.

Larque, E., Demmelmair, H., Koletzko, B., (2002). Perinatal supply and metabolism of long-chain polyunsaturated fatty acids: importance for the early development of the nervous system. *Ann.New.York. Acad. Sci.,* 967:299-310.

Liosis, S., Bauer, T., Schiele, R., Gohlke, H., Gottwik, M., Katus, H., Sabin, G., Zahn, R., Schneider, S., Rauch, B., Senges, J., Zeymer, U., (2013). Predictors of 1-year mortality *Clin. Res. Cardiol.*, (in press).

Loeschke, K., Ueberschaer, B., Pietsch, A., (1996). n-3 fatty acids only delay early relapse of ulcerative colitis in remission. *Dig. Dis. Sci.,* 41: 2087-95.

Marchioli, R., Levantesi, G., Silletta, M. G., (2009). Effect of n-3 polyunsaturated fatty acids and rosuvastatin in patients with heart failure: results of the GISSI-HF trial. *Expert. Rev. Cardiovasc. Ther.,* 7:735-48.

Menge, H., Grafe, M., Lorenz-Meyer, H., Riecken, E. O., (1975). The influence of food intake on the development of structural and functional adaptation following ileal resection in t he rat. *Gut,* 16:468-72.

Meydani, S. N., Lichtenstein, A. H., Cornwall, S., (1993) Immunologic effects of national cholesterol education panel step-2 diets with and without fish-derived N-3 fatty acid enrichment. *J. Clin. Invest.,* 92: 105-12.

Naidu, M. U. R., Ramana, G. V., Rani, P. U., (2004). Chemotherapy-induced and/or radiation therapy-induced oral mucositis? Complicating the treatment of cancer. *Neoplasia,* 6: 423-31.

Niemoller, T. D., Bazan, N. G., (2010). Docosahexaenoic acid neurolipidomics. *Prostaglandins. Other. Lipid. Mediat.,* 91:85-89.

O'Brien, D. P., Nelson, L. A., Huang, F. S., (2001). Intestinal adaptation: structure, function, and regulation. *Sem. Pediatr. Surg.,* 10: 55-64.

Owen, R., Giancosa, A., Hull, W., Haubner, R., Spiegelhalder, B., Bartsch, H., (2000). The antioxidant/anticancer potential of phenolic compounds isolated from olive oil. *Eur. J. Cancer,* 36:1235-47.

Park, J. H. Y., Grandjean, C. J., Hart, M. H., Vanderhoof, J. A., (1989). Effects of dietary linoleic acid on mucosal adaptation after small bowel resection. *Digestion,* 44: 57-65.

Sauder, K. A., Skulas-Ray, A. C., Campbell, T. S., Johnson, J. A., Kris-Etherton, P. M., West, S. G., (2013). Effects of omega-3 fatty acid supplementation on heart rate variability at rest and during acute stress in adults with moderate hypertriglyceridemia. *Psychosom. Med.*, 75:382-9.

Schmöcker, C., Weylandt, K. H., Kahlke, L., (2007). Omega-3 fatty acidsalleviate chemically induced acute hepatitis by suppression of cytokines. *Hepatology,* 45:864-9.

Simopoulos., (1999). Essential fatty acids in health and chronic disease. *Am. J. Clin. Nutr.,* 70(suppl): 560S-9S.

Sonis, S. T. (1993). Complications of cancer and their treatment: oral complications. In: Holland JF, Frei E, Bast RC, eds. Cancer Medicine, 3rd edn. Philadelphia: Lea and Febiger, pp: 2381-8.

Sonis, S. T., Elting, L. S., Keefe, D. M. K., Peterson, D. E., Schubert, M., Hauer-Jensen, M., Bekele, B. N.,Raber-Durlacher, J., Donnelly, J. P., Rubenstein, E. B., (2004).Perspectives on cancer therapy-induced mucosal injury: Pathogenesis, measurement, epidemiology, and consequences for patients.*Cancer,* 100:1995-2025.

Sukhotnik, I., Hayari, L., Bashenko, Y., Chemodanov, E., Mogilner, J., Shamir, R., Shaoul, R., Coran, A. G., (2008). Dietary palmitic acid modulates intestinal re-growth after massive small bowel resection in a rat. *Pediatr. Surg. Int.,* 24:1313-21.

Sukhotnik, I., Shany, A., Bashenko, Y., Hayari, L., Chemodanov, E., Mogilner, J., Coran, A. G., Shaoul, R., (2010). Parenteral but not enteral omega-3 fatty acids (omegaven) modulate intestinal re-growth after massive small bowel resection in a rat. *JPEN. J. Parenter. Enteral. Nutr.*, 34:503-12.

Sukhotnik, I., Slijper, N., Chemodanov, E., Bashenko, Y., Shaoul, R., Coran, A. G., Mogilner, J. G., (2011).Parenteral omega-3 fatty acids (Omegaven) modulate intestinal recovery following intestinal ischemia-reperfusion in a rat. *J. Pediatr. Surg.,* 46:1353-60.

Vanderhoof, J. A., (1996). Short bowel syndrome. *Neonatal. Gastroenterol.,* 23:377-86.

Vanderhoof, J. A., Kollman, K. A.,Lien, M., (1999).The role of eicosanoid synthesis in intestinal adaptation. *J. Pediatr. Gastroenterol. Nutr.,* 28:596-603.

Weber, P. C., Fischer, S., von Schacky, C., Lorenz, R., Strasser, T.,(1986). Dietary omega-3 polyunsaturated fatty acids and eicosanoid formation in man. In: Simopoulos A. P., Kifer R. R., Martin R. E., eds. Health effects of polyunsaturated fatty acids in seafoods. Orlando, FL: Academic Press, 49–60.

Weber, T. R., Keller, M. S., (2002). Adverse effects of liver dysfunction and portal hypertension on intestinal adaptation in short bowel syndrome in children. *Am. J. Surg.*, 184:582-6.

Wright, N., Alison, M., (1984). The biology of epithelial cell populations. Clarendon, Oxford. L.T.

In: Omega-3 Fatty Acids
Editor: Md. Wasim Khan

ISBN: 978-1-62948-516-4

Chapter 9

Gastroprotective and Antinociceptive Effects of Omega-3 Fatty Acids: Potential Mechanisms Underlying These Effects

***J. M. Jiménez-Andrade*[1] *and A. E. Chávez-Piña*[2,*]**
[1]Unidad Académica Multidisciplinaria Reynosa Aztlán,
Universidad Autónoma de Tamaulipas, Reynosa, Tamaulipas, México
[2]Laboratorio de Farmacología, Programa Institucional de Biomedicina Molecular,
Escuela Nacional de Medicina y Homeopatía del Instituto Politécnico Nacional,
México, D. F., México

Abstract

Chronic treatment with fish oil has exhibited gastroprotective effects during several insults such as ethanol, aspirin, pyloric ligation, cold-restrain stress and indomethacin. These actions have been attributed to omega-3 long-chain polyunsaturated fatty acids such as docosahexaenoic acid (DHA) and eicosapentaenoic acid (EPA), which are the main compounds present in fish oil. In support with this, DHA has recently exhibited a gastroprotective effect in the gastric damage-induced by indomethacin mediated by a blockade of gastric LTB_4. The gastroprotective properties of EPA have not been explored yet. Additionally other benefits have been attributed to DHA and EPA. Among those effects, antinociceptive effects by DHA and EPA have been reported in different experimental acute pain models. Mechanisms underlying these antinociceptive effects are related to modulation of synthesis of resolvins from the D and E series (agents that stop inflammatory signaling), as well as activation of opioid receptors, blockade of temperature sensitive transient receptors potential ion channels (thermo TRPs), acid sensing ion channels (ASICs) and increased release of endogenous opioids. Likewise, a reduction of proinflammatory mediators (e.g.eicosanoids and cytokines) and/or a change

[*] Tel: 11(+52)57296000 ext. 55583. E-mail:achavezp@ipn.mx, arapina@yahoo.com.

in the cell membrane phospholipid fatty acid composition have been reported to be involved in the analgesic effect of these omega-3 long-chain polyunsaturated fatty acids.

Keywords: omega-3, fatty acids, gastroprotection, pain

1. Introduction

Gastric damage is a disruption of the mucosal integrity due to stress and chronic ingestion of irritants such as caffeine, alcohol, nicotine and drugs like non-steroidal anti-inflammatory drugs (NSAIDs)(Wallace, 1997). NSAIDs are a group of drugs that are worldwide used for the treatment of pain, inflammation and fever. However their use is limited as their chronic administration is strongly associated to gastric and cardiac adverse effects. While, NSAIDs-induced gastric damage has been diminished with the treatment of proton-pump inhibitor (a kind of drug whose main action is a reduction of gastric acid production, they are the most potent inhibitors of acid secretion available), however recently it has been published that proton-pump inhibitor induces dysbiosis (Wallace et al., 2011). In light of this, there is a medical need to find new alternatives for the treatment of pain and inflammation without inducing significantly negative side effects.

Docosahexaenoic acid (DHA), an omega-3 polyunsaturated fatty acid and one of the main components of fish oil, recently has demonstrated to induce an antinociceptive effect during several acute pain experimental models. In addition, our group of work has recently reported a DHA´s gastroprotective effect against indomethacin-induced gastric injury. In this chapter is outlined the antinociceptive and gastroprotective mechanism of DHA. This knowledge may help to find new alternatives for the treatment of pain and prevention of gastric injury induced by NSAIDs.

2. Gastroprotection

2.1. Definition

Gastroprotection is a term attributed to the ability of the gastric tissue to prevent gastric injury. The stomach secretes a juice that can digest various foods that we ingest, but it seldom digests itself (Wallace, 2008). The prevention of damage mainly consists on several factors such as secretion of mucus-alkaline, mucosal phospholipids layer, increased gastric blood flow and release of autacoids like prostaglandins, lipoxins and gaseous mediators (nitric oxide and hydrogen sulfide)(Chávez-Piña and Navarrete, 2009).

2.2. Gastric Mucosal Defense Mechanism

a) Mucus-Bicarbonate-Phospholipid "Barrier"

The mucus-bicarbonate-phospholipid “barrier” constitutes the first line of mucosal defense, this barrier is formed by mucus gel, bicarbonate, and surfactant phospholipids (Laine et al., 2008). The mucus gel contains phospholipids and its luminal surface is coated with a film of surfactant phospholipids with strong hydrophobic properties (Lichtenber, 1999). Mucus gel is secreted by apical expulsion from surface epithelial cells and contains 95% water and 5% mucin glycoproteins. Mucus secretion is stimulated by gastrointestinal hormones, including gastrin and secretin, as well as PGE_2 and cholinergic agents (Allen and Flemström, 2005). Furthermore, the mucus that is secreted onto the surface of the stomach acts both as a lubricant (to reduce physical damage to the epithelium by ingested materials), and as a trap for bacteria (Wallace, 2008). Mucus performs an important structural role in creating an unstirred layer on the mucosal surface, which supports maintenance of a near-neutral pH at the surface as well as acting as a physical barrier against luminal pepsin (Allen and Flemström, 2005).

The secretion of bicarbonate in a mucus layer creates a pH closer to neutrality at the epithelial surface in the stomach and duodenum and provides the first line of mucosal defense against luminal acid (Laine et al., 2008). Mucus has been suggested to retard the diffusion of protons (Lichtenber, 1999). Besides, "alkaline tide" occurs due to stimulated parietal cells secreting hydrochloric acid into the gastric gland lumen concurrently secrete bicarbonate into the interstitium and lumen of adjacent capillary blood vessels. This bicarbonate is transported upward to the base of the surface epithelial cells and to the gastric lumen, where it contributes to the unstirred layer of mucus and bicarbonate (Laine et al., 2008). The mucus bicarbonate barrier is the only preepithelial barrier between lumen and epithelium (Laine et al., 2008). When the mucosal barrier is overwhelmed or broken down by injury, there are other mechanisms that could be executed in order to defend the mucosa.

b) Epithelium

It is known that apical gastric epithelial cells remain intact and functional when they continuously are exposed to high concentrations of acid. The epithelium secretes mucus and bicarbonate as well as generates prostaglandins (PGs) and heat shock proteins to protect itself from acid diffusion (Laine et al., 2008 ; Wallace, 2008). Epithelium is interconnected by tight junctions, the surface epithelial cells form a barrier preventing back diffusion of acid and pepsin (Werther, 2000). Heat shock proteins are generated by gastric epithelial cells in response to stress, such as increased temperature, oxidative stress, and cytotoxic agents. Integrity of mucosa is another feature that makes gastric epithelium resilient to injury. This integrity is maintained by continuous cell renewal, in fact, gastric epithelium is renewed every 2-4 days and complete replacement occurs after 3-7 days. Cell proliferation of progenitor cells is controlled by growth factors (Laine et al., 2008).

c) Mucosal Microcirculation

Mucosal microcirculation is essential for delivery of oxygen and nutrients and removal of toxic substances. With adequate vascular perfusion, epithelial damage does not generally progress to necrosis of deeper layers of the mucosa.

At the level of the muscularis mucosae, most gastric arteries branch into capillaries, which enter the lamina propia and travel upward in proximity to gastric glandular epithelial cells. At the base of surface epithelial cells, capillaries converge into collecting venules (Gannon et al., 1984).

In the rat, the entire luminal epithelium can be destroyed and there is a little macroscopic evidence of the injury, other than extensive mucus release. This event is due to restitution, and it involves the migration of healthy epithelial cells from the gastric pits over the denuded basement membrane (Wallace et al., 1982; Wallace, 2008). The mucus released from damaged epithelial cells and plasma exuding from the mucosal vasculature coalesces to form a protective layer over the denuded region that has been termed mucoid cap. Even in the presence of very high levels of hydrochloric acid in the stomach, the pH within the mucoid cap can be maintained at close to neutrality. When the gastric mucosa is exposed to an irritant or when acid-back diffusion occurs, a marked and rapid increased in mucosal blood flow occurs.

This increase allows removal and/or dilution of the back-diffusing acid and/or noxious agents. If blood flow to the stomach is interrupted, then the pH within the mucoid cap drops precipitously and hemorrhagic lesions form (Wallace, 2008). Mucosal blood flow increases by influence of endogenous substances such as PGs, nitric oxide and hydrogen sulfide and activation of sensory afferent nerves, this event is essential for gastric mucosal defense participation against noxious stimulus (Wallace and Devchand, 2005 ; Zanardo et al., 2006; Wallace, 2008).

d) Inflammation

Superficial injury to the gastric mucosa also triggers an acute inflammatory response, characterized by the above-mentioned increase in blood flow, as well as by plasma exudation and recruitment into the mucosa of leukocytes. The objective of this response is to minimize tissue injury, facilitate repair of damaged tissue, and prevent entry into the systemic circulation of foreign substances, including microbes and microbial products. For example, leukocytes adhere to gastric endothelium which permits at the same time the liberation of oxygen-derived free radicals and proteases. Leukocyte adherence to vascular endothelium could obstruct capillaries, resulting in a reduction in gastric mucosal blood flow and thereby predisposing the mucosa to injury (Wallace andGranger, 1996; Wallace, 1997). However, it is important to note that while the acute inflammatory response is aimed at reducing mucosal injury, there are circumstances in which this response can be deregulated and can contribute to mucosal injury. Asako studied the role of leukocytes and leukotriene B_4 (LTB_4) after indomethacin treatment, where the inhibition of LTB_4 during indomethacin treatment decreased gastric injury (Asako et al., 1992) (Table 1).

3. Lipid Mediators and Gastroprotective Effect

3.1. Prostaglandins

Prostaglandins are a type of eicosanoids derived from arachidonic acid by the initial action of cyclooxygenase (PG endoperoxide G/H synthase)(O'Neill andFord-Hutchinson, 1993). Arachidonic acid is a fatty acid derived from dietary sources or is synthetized in the body from an essential fatty acid, linoleic acid. It is stored in lipid bilayers of cell membranes and is esterified predominantly to phospholipids such as phosphatidylcholine, phosphatidylethanolamine and phosphatidylinositol by an enzyme named phospholipase A_2 (PLA_2)(Serhan, 2001). Prostaglandins are divided into series that differ in the oxygen substitution in the cyclopentane ring and coded by a letter (PGD, PGE, PGF, PGG and PGH).

		GASTROPROTECTIVE EFFECT		
PUFAs	*Ulcer model*	*Dose of PUFA*	*Mechanism*	*Reference*
Fish oil	Aspirin	200 mg/kg p.o.	Non elucidated	Bhattacharya *et al.*, 2006
		twice daily for 5 days		
Fish oil	Cold restrain stress	200 mg/kg p.o.	Antioxidant activity (attenuation of decreased on	Bhattacharya *et al.*, 2006
		twice daily for 5 days	CAT and GPX, and increased on SOD and LPO)	
Fish oil	Ethanol	200 mg/kg p.o.	Non elucidated	Bhattacharya *et al.*, 2006
		twice daily for 5 days		
Fish oil	Pylorus ligated	200 mg/kg p.o.	Decreased of acid and pepsin concentration on	Bhattacharya *et al.*, 2006
		twice daily for 5 days	gastric juice	
Fish oil	Dexamethasone	40 µL/day during 10 days	Antioxidant activity	Manjari and Das, 2000
DHA	Indomethacin	30 mg/kg p.o.	Decreased on LTB_4	Pineda-Peña *et al.*, 2012

SOD (superoxidedismutase), CAT (catalase), GPX (glutathioneperoxidase), LPO (lipidperoxidation), LTB_4 (leukotriene B_4)

The subscript numeral in PG nomenclature indicates the number of double bonds present in the compound (Serhan, 2001).

Cyclooxygenase (COX) is a heme-containing enzyme that is most abundant in the endoplasmic reticulum: cyclization of arachidonic acid to form PGG_2 and hydroperoxidation of PGG_2 to yield PGH_2 (Li et al., 1998). The latter is a relatively unstable compound that has a half-life of seconds and is a common intermediate that is converted to biologically active products such as thromboxane (TXA_2), prostacyclin (PGI_2), PGD_2, PGE_2 and $PGF_{1\alpha}$, by thromboxane synthase, prostacyclin synthase, PGD_2 isomerase, PGE_2 isomerase and PGF reductase respectively. The synthesis of each kind of prostaglandin depends on the cell and enzyme present in each tissue (Serhan, 2001).

In gastric tissue, PGE_2 and PGI_2 are synthetized. Prostaglandins, derived from COX-1 in gastric mucosa mediate many of the components of gastric mucosa defense, such as the maintenance of gastric blood flow by PGI_2 and bicarbonate and mucus secretion by PGE_2. Additionally, prostaglandins retard the ability of acid and pepsin to penetrate mucus (Wallace and Granger, 1996; Wallace, 2008). In contrats, COX-2 is enrolled in the synthesis of prostaglandins that are made out when the mucosa is challenged and promote ulcer healing, decrease leukocyte adherence and induces re-epithelization of gastric cells (Wallace and Devchand, 2005).

PGE_2 and PGI_2 are potent vasodilators producing this effect in the stomach through EP2/EP4 and IP receptors, respectively (Araki et al., 2000). Vasodilatory effect of prostaglandins facilitate epithelial restitution by contributing to the creation of a relatively neutral pH microenvironment within the mucoid cap that forms over sites of epithelial damage (Wallace and McKnight, 1990). Inhibition of prostaglandins by NSAIDs contributes to gastropathy. There is a correlation between time and dose dependent manner of suppression of gastric prostaglandin synthesis by NSAIDs and their ability to induce gastric ulcers (Lanza, 1989). The most common alterations caused by NSAIDs in gastric area are hemorrhagic gastric erosions. Prostaglandins downregulate the release of a number of other inflammatory mediators that have been suggested to contribute to the generation of mucosa injury in certain circumstances such as tumor necrosis factor-α (TNF- α), histamine, the release of LTB_4 from neutrophils, platelet-activating factor and interleukin-1 from macrophages.

3.2. Lipoxins

Lipoxins are trihydroxytetraene-containing lipid mediators that result from the sequential oxygenation of arachidonic acid at the carbon-15 and the carbon-5 position by 15- and 5-lipooxygenase, respectively. Lipoxins are characterized structurally by the presence of four conjugated double bonds (Serhan, 2001). Lipoxins are formed during cell-cell interactions and are predominantly counter regulators of some mediators (Fierro and Serhan, 2001).

Lipoxins are known to be generated in humans by one of at least three biosynthetic routes working either independently or in concert, in particular biological settings or tissues. The first route occurs between airway epithelial cells or monocytes with neutrophils. The second pathway is induced by interactions predominantly within the vasculature between 5-LO, present in myeloid cells, and 12-LO, present in platelets. The third route involves aspirin and the action of COX-2 and 5-LO (Claria and Serhan, 1995). Aspirin covalently modifies, through acetylation, a serine residue near the active site of COX. In the case of COX-1, this acetylation occurs in a serine residue near the active site (Ser530) which induces a conformational change in the enzyme, and then it can no longer oxidize arachidonic acid (Shimokawa andSmith, 1992). While with COX-2 does not occur the same, aspirin also acetylates a serine residue (Ser516), which induces a conformational change in the enzyme, but COX-2 remains able to metabolize arachidonic acid to 15-(R)-HETE (15-*R*-hydroxyepitetraenoic acid). Then this compound is released from endothelial and epithelial cells and synthetized by leukocyte 5-LO to 15-epimer lipoxin A_4 or aspirin-triggered lipoxins (ATL) (Mancini et al., 1994).

Those lipid mediators such as lipoxins exert potent protective actions on gastric mucosa. It has been reported that lipoxins reduce the severity of aspirin to induce gastric injury, inhibit responses in neutrophils, increase gastric mucosal blood flow and diminish leukocyte adherence in gastric endothelium (Claria and Serhan, 1995 ; Fiorucci et al., 2002).

3.3. Omega-3 Polyunsaturated Fatty Acids (PUFAs)

3.3.1. Definition

Omega-3 polyunsaturated fatty acids are made out of a hydrocarbonated chain of variable length with several double bonds. The position of the first double bond is called omega, counting the methyl carbon as carbon one (Calder, 2013). This class of fatty acids are said to be essential because they cannot be synthetized by the organism and have to be taken from nutrition. The simplest omega-3 fatty acid is α-linoleic acid (18:3 n-3), this is produced in plants, green leaves and some plant oils, nuts and soya beans. EPA (22:5 n-3) and DHA (22:6 n-3) are found in significant quantities in flesh of both lean and oily marine fish (Calder, 2013).

In this chapter is briefly reviewed the gastroprotective, antinociceptive and anti-inflammatory effects of PUFAs due to several diseases such as gastric injury and pain share an inflammatory component.

3.3.2. Anti-Inflammatory Effect of PUFAs

Inflammation is a component of the mechanisms of normal host defense that provides protection from infection and other insults. Inflammation initiates the killing of pathogens and is involved in the processes of tissue repair so aiding restoration of homeostasis at infected or damaged sites. The inflammatory response involves interactions amongst many cell types and the production of a number of chemical mediators. These mediators damage to pathogens but may also cause damage to host tissues. The influx of cells into the site of inflammatory activity and the synthesis and release of the inflammatory mediators are the major mechanism underlying the cardinal signs of inflammation: redness, swelling, heat, pain and loss of function (Calder, 2013).

The first evidence for the important role of dietary intake of PUFAs in inflammation was derived from epidemiological observations of the low incidence of autoimmune and inflammatory disorders, such as psoriasis, asthma and type-1 diabetes, as well as the complete absence of multiple sclerosis, in a population of Greenland Eskimos compared with gender- and age-matched groups living in Denmark (Kromann and Green, 1980). Native Greenland

Eskimos and Japanese have a high dietary intake of long chain omega-3 PUFA from seafood and a low incidence of myocardial infarction and chronic inflammatory autoimmune disorders, even when compared to their Westernized ethnic counterpart (Simopoulos, 2002).

3.3.3. Gastroprotective Role of PUFAs

Gastroprotective effects of fish oil have been reported in gastric ulcers induced by ethanol (Faust et al., 1989; Leung, 1994), aspirin (Faust et al., 1990; Bhattacharya et al., 2006), indomethacin (Güzel et al., 1995), dexamethasone (Manjari and Das, 2000), cold-restrain stress (Ulak et al., 1995) and pyloric ligation (Bhattacharya et al., 2006).

Several mechanisms have been suggested to be involved in the gastroprotective effect of fish oil, for example Bhattacharya in 2006, demonstrated a gastroprotective effect by omega-3 in different experimental models, where aeitiopathogenesis of ulceration are different. The ulcerogenic activity of aspirin involves enhanced acid secretion, increased back diffusion of H^+ ions, and breaking up of defensive mucosal barrier, by inhibiting cytoprotective prostaglandin synthesis. In addition, alcohol-induced gastric ulcerations are associated with breakdown of superficial mucosal cells by induction of chronic gastritis. Furthermore, alcohol appears to stimulate mucosal mast cells to release vasoactive mediators including histamine, PAF and leukotrienes. However, in this study the mechanism of action on aspirin and alcohol-induced gastric injury was not studied.

Furthermore, fish oil administration decreased acid and pepsin levels during pylorus ligated rats and increased antioxidant enzymes on cold-restraint stress (CRS) (Bhattacharya et al., 2006). Pylorus ligation-induced gastric ulcerations are caused by enhanced acid-pepsin secretion leading to auto digestion of the gastric mucosa and breakdown of mucosal barrier (Goel and Bhattacharya, 1991). In this study, it is suggested that protection of gastric mucosa against offensive-pepsin assault is mediated through release of prostaglandins (PGs), where PGI_2 appears to play a major role (Bhattacharya et al., 2006). In contrast, our group of work has recently published that docosahexaenoic acid (DHA), one of the main compounds in fish oil, exhibits gastroprotective effect against-indomethacin induced gastric injury. However, this effect is not through release of prostaglandins as DHA was not able to reduce indomethacin-induced prostaglandin inhibition (Pineda-Pena et al., 2012).

Furthermore, omega-3 fatty acids have been reported to reduce the production of leukotriene. LTB_4 induces inflammation and promotes leukocyte chemotaxis and adherence (Wallace, 1997). Leukocyte adherence contributes to the pathogenesis of gastric mucosal injury in two ways. First, the effects of LTB_4 on gastric and mesenteric microcirculation lead to an increase in leukocyte adherence, and simultaneously allow the liberation of oxygen-derived free radicals and proteases. Furthermore, an increase in the LTB_4-dependent neutrophil adherence to the vascular endothelium could obstruct capillaries, resulting in a reduction in gastric mucosal blood flow and thereby predisposing the mucosa to gastric injury (Wallace, 1997). Asako studied the role of LTB_4 in leukocyte adherence after indomethacin administration. On this study, the increment on leukocyte adherence induced by indomethacin was abolished by a pre-treatment of LTB_4 antagonist (Asako et al., 1992). Additionally, this effect appear to be also mediated by an increment in the expression of intercellular adhesion molecule 1 (ICAM-1) due to pretreatment with an antibody against ICAM-1 reduced leukocyte adherence and susceptibility to indomethacin-induced gastropathy (McCafferty et al., 1995).

It has been shown that NSAIDs-induced gastric injury, such as from treatment with indomethacin, results in an increase on LTB_4-dependent leukocyte adherence (Asako et al., 1992). Our results reveal that the gastroprotective effect of DHA was associated with a significant reduction of gastric LTB_4 (Pineda-Pena et al., 2012). Decrement of LTB_4 by omega-3 polyunsaturated fatty acids has been evidenced in another animal model such as dextran sulfate sodium-induced colitis. In this study, incorporation of n-3 PUFAs into an olive oil diet decreased colonic TNF-α and LTB_4 (Camuesco et al., 2005).

Another possible mechanism of DHA´s gastroprotective effect is that DHA *in vitro* generates a new family of pro-resolution mediators named resolvins. DHA is enzymatically converted to resolvin D1 (RvD1) in a multistep reaction that proceeds through a transcellular biosynthetic route during endothelial cell-neutrophil interactions (Serhan et al., 2000). RvD1 stops neutrophil recruitment, regulates cytokine and chemokine expression and decreases LTB_4 functions (Norling et al., 2012). RvD1´s effect has been explored in dextran sulfate sodium-induced colitis, where RvD1 reduced colonic cytokine levels (Bento et al., 2011). RvD1 binds to the lipoxin A_4 and annexin A1 receptor FPR2/ALX (receptors involved in gastric resistance to aspirin-induced gastric injury) (Fiorucci et al., 2002) and binds also to RvD1 receptor (also known as GPR32) (Krishnamoorthy et al., 2010). RvD1 may also contribute to DHA gastroprotective effects; however, there are no studies performed to determine whether the gastroprotective effect of acute DHA is associated with changes in the synthesis of RvD1.

On another side, omega-3 fatty acids could alter cell membrane phospholipid fatty acid composition. It is well known that phospholipid of cells involved in inflammation typically contain a high proportion of arachidonic acid and much less, often little, EPA or DHA. However, including fish oil in the diet of experimental animals or healthy human volunteers results in increased accumulation of EPA and DHA in these cells, with is associated with a decreased content of arachidonic acid (Calder et al., 1994 ; Yaqoob et al., 2000). With those events, PUFAs offer stability to the cell membrane and changes in signaling (Calder, 2012, 2013) (Table 2).

4. Pain and Nociception

Pain is a multidimensional sensory experience that is intrinsically unpleasant and associated with hurting and soreness. It is essentially a sensation, pain has strong cognitive and emotional components, it is linked to, or described in terms of, suffering (Clifford, 2004). In contrast, nociception is just the perception of noxious stimuli. Nociception is initiated by noxious stimuli that activate the peripheral terminals of primary afferent neurons called nociceptors. Primary afferent sensory neurons are the gateway by which sensory information from peripheral tissues is transmitted to the spinal cord and brain and these sensory neurons innervate every organ of the body with the exception of the brain (Clifford, 2004). To accomplish this, nociceptors express an extremely diverse repertoire of receptors and transduction molecules that can sense forms of noxious stimuli including thermal, mechanical and chemical albeit with varying degrees of sensitivity. Adaptative pain contributes to survival by protecting the organism from injury or promoting healing when injury has occurred.

		ANTINOCICEPTIVE EFFECT		
PUFAs	*Pronociceptive model*	*Dose of PUFA*	*Mechanism*	*Reference*
DHA	Tail flick test	25 mmol/kg, p.o.	Opioid receptors-mediated pain control activity	Nakamoto *et al.*, 2010
	Formalin test			
	Acetic acid writhing test			
DHA	Acetic acid writhing test	150 nmol/mouse, i.c.v.	Release of β-endorphins	Nakamoto *et al.*, 2011
DHA	Formalin test	50 μg/mouse, i.c.v.	Release of β-endorphins	Nakamoto *et al.*, 2012
			and activation of GPR40 receptor	
RvD1	TNBS-induced mechanical	100 ng/kg, i.t.	Blocking expression of cytokines (TNF-α, IL-1β, IL-1)	Quan-Xin *et al.*, 2012
	allodinya			
RvD1	Capsaicin-induced pain	10 ng, i.t.	Inhibitor of TRPV1 and TRPA1	
RvD2	Capsaicin-induced pain	10 ng, i.t.	Potent inhibitor of TRPV1 and TRPA1	
RvE1	Capsaicin-induced pain	10 ng, i.t.	Inhibitor of TRPV1 and TRPA1	Park *et al.*, 2011
DHA	Capsaicin-induced pain	10 μg, i.t.		
EPA	Capsaicin-induced pain	10 μg, i.t.		
RvD2	CFA	0.01 ng, i.t.		
Fish oil	Thermal	Concentrate, 6 months	Decrease in ion channels expression (TRPV1, ASIC1a	Veigas *et al.*, 2011
			and ASIC and cFos)	

Adaptive pain is divided into nociceptive and inflammatory. Nociceptive pain system is a key early warning device, an alarm system that announces the presence of a potentially damaging stimulus, is a vital physiologic sensation. If tissue damage occurs despite the nociceptive defensive system, inflammatory pain is used to accomplished this goal. In this state, sensitivity is increased such that stimuli to the affected part that would normally not cause pain now do so. As a result, we prevent contact with or movement of the injured part until repair is complete, minimizing further damage (Clifford, 2004).

4.1. Antinociception and Omega-3 Fatty Acids

Conventionally, nonopiod analgesics such as NSAIDs, opioid analgesics such as morphine, and supplementary analgesics have been used to relieve pain. However, the use of these analgesics is limited by significant adverse side effects. For example, non-selective NSAIDS can cause gastric injury or tromboembolic problems. Opioids are effective in attenuating inflammatory pain but are frequently accompanied by side effects such as constipation, sedation, nausea, vomiting, and respiratory depression. Therefore novel analgesics and/or supplemental analgesics are needed that fully controlling chronic inflammatory pain.

DHA´s antinociceptive effect has been reported in several experimental pain models of thermal and chemical nociception (tail flick test, acetic acid writhing test and formalin test in mice) (Nakamoto et al., 2010). n-3 PUFAs probably reduce pain by inhibition of the production of proinflammatory eicosanoids and cytokines (Zaloga and P., 2001) or blocking cationic channels such as voltage-gated Na^{+} channels (Xiao et al., 1995), L-type Ca^{2+} and transient vanilloid 1 (TRPV1) receptors (Matta et al., 2007). Recently, it was elucidated that one of the actions contributing to the antinociceptive mechanism of DHA is not performed directly on the opiod receptor, instead DHA indirectly through the release of an endogenous opioid peptide β-endorphin (Nakamoto et al., 2011 ; Nishinaka et al., 2013). Another mechanism involved in the antinociceptive effect of PUFAs is that fatty acids can bind to several G-protein coupled cell membrane receptors such as GPR40 and GPR120 (Calder, 2013). GPR40 is preferentially expressed on β pancreatic cells, where it mediates insulin secretion (Feng et al., 2012 ; Zhao et al., 2013). Additionally, GPR40 receptor is expressed in the olfactory, bulb, striatum, hippocampus, midbrain, hypothalamus, medulla oblongata, cerebellum and cerebral cortex in the brain as well as in the the spinal cord. Thus, it has been hyptothetized that DHA could active a signal to trigger GPR40 in the presinaptic neuron to liberate endorphins (Nishinaka et al., 2013).

Furthermore, recently EPA and DHA have been involved with peroxisome proliferator-activated receptor gamma (PPARγ) activation. PPARs are transcription factors that regulate gene expression. PPARγ is also expressed in inflammatory cells, where it is involved in regulating the production of inflammatory mediators, having an anti-inflammatory action (Szanto and Nagy, 2008). PPARs are activated by noncovalent binding of ligands that include (n-3) PUFA and various eicosanoids mediators (Forman et al., 1997 ; Kliewer et al., 1997). Through activation of PPARs, (n-3) PUFAs are able to regulate inflammatory cytokines such us TNF-α and IL-6 upon endotoxin stimulation (Kong et al., 2011). Thus, over activation of PPAR, (n-3) PUFAs are able to regulate metabolism and other cell and tissue responses, including inflammation (Calder, 2012). In support with this, recently it has been evidenced that pharmacological activation of PPARγ in the brain rapidly inhibits local edema and the spinal transmission of noxious inflammatory signals during inflammatory pain (Morgenweck et al., 2010).

Another possible mechanism involved is that during inflammatory pain, prostaglandins play a pivotal role in the persistance of nociceptive effect. EPA, an omega-3 fatty acid, present in fish oil competes with arachidonic acid, an omega-6 fatty acid present in membranes, for prostaglandin and leukotriene synthesis at the cyclooxygenase and lipooxygenase level (Lee et al., 1985 ; Sperling et al., 1993). Due to EPA is a 20-carbon fatty acid highly unsaturated, it is also a substrate for the COX, LOX and cytochrome P450 enzymes that produce eicosanoids. However, the mediators produced have a different structure from those made out from arachidonic acid. EPA generates the synthesis of PGI_3 and LTB_5, the last one a weak inducer of inflammation and a weak chemotactic agent. Then, EPA results in a decreased production of potent eicosanoids and an increased production of weak eicosanoids from arachidonic acid (Calder, 2013). Besides, EPA and DHA *in vitro* generate a new family of pro-resolution mediators named resolvins (Rv). EPA and DHA are enzymatically converted to RvE1 and RvD1/AT-RvD1 in a multi-step reaction that proceeds through a transcellular biosynthetic route during endothelial/neutrophil cell-cell interactions (Serhan et al., 2000). RvE1 (5*S*,12*R*,18*R*-trihydroxy-6*Z*,8*E*,10*E*,14*Z*,16*E*-eicosapentaenoic acid) requires several enzymes such as COX-2 or cytochrome P450, and 5-lipoxygenase (5-LOX) for its biosynthesis from EPA (Oh et al., 2011). By comparison, RvD1 (7*S*,8*R*,17*R*-trihydroxy-4*Z*,9*E*,11*E*,13*Z*,15*E*-*19*Z-docosahexaenoic acid) utilizes 15-LOX and 5-LOX for its biosynthesis from DHA (Sun et al., 2011). It is important to point out that consumed fish oil will not automatically be metabolized to resolvins under normal conditions, partly because the biosynthetic enzymes have low expression levels in most tissues in the non-injured states (Ji et al., 2011).

RvD1 produces anti-inflammatory and antinociceptive effects in several animal models (Ji et al., 2011). RvD1 stops neutrophil recruitment, regulates cytokine and chemokine expression and decreases LTB_4 functions (Norling et al., 2012). Resolvins derived from the E and D series have shown antinociceptive effect in several pain models such as chronic pancreatitis-induced pain (Quan-Xin et al., 2012), inflammatory pain (Xu et al., 2010 ; Park et al., 2011), arthritis-associated inflammatory pain (Xu andJi, 2011). Recently, it has been revealed RvD2 such as a potent endogenous inhibitor for transient receptor potential subtype vanilloid 1 (TRPV1) and TRP ankyryn 1 (TRPA1); TRPV1/TRPA1 contribute importantly to the genesis of inflammatory pain via both peripheral mechanisms (peripheral sensitization) and spinal cord mechanisms (central sensitization) (Park et al., 2011).

Conclusion and Future Studies

Several pre-clinical studies have demonstrated that PUFAs have anti-inflammatory and anti-nociceptive effects. Furthermore, PUFAS have been reported to be useful in reducing the gastric injury induced by NSAIDs. Additionally, recent pre-clinical studies have elucidated some of the mechanisms underlying these beneficial effects by PUFAs. Thus, vigorous and clinical studies are needed to test the efficacy and safety of PUFAs in inflammatory chronic pain and/or the beneficial use of concomitant administration of PUFAs with NSAIDs to reduce gastric injury induced by NSAIDs.

Acknowledgment

Authors acknowledge the support of National Polytechnic Institute with the project SIP-20131157 and SIP-SNI-2011/01 and to the National Council for Science and Technology with the project CONACyT178027.

References

Allen, A., Flemström, G., (2005). Gastroduodenal mucus bicarbonate barrier: protection against acid and pepsin. *Am. J. Physiol. Cell Physiol.,* 288,1-19.

Araki, H., Ukawa, H., Sugawa, Y., Yagi, K., Suzuki, K., Takeuchi, K., (2000). The roles of prostaglandin E receptor subtypes in the cytoprotective action of prostaglandin E2 in rat stomach. *Aliment Pharmacol. Ther.,* 14 (1), 116-124.

Asako, H., Kubes, P., Wallace, J., Gaginella, T., Wolf, R. E. , Granger, D. N., (1992). Indomethacin-induced leukocyte adhesion in mesenteric venules: role of lipoxygenase products. *Am. J. Physiol.,* 262 (5), 903-908.

Bento, A. F., Claudino, R. F., Dutra, R. C., Marcon, R., Calixto, J. B., (2011). Omega-3 fatty acid-derived mediators 17(R)-hydroxydocosahexaenoic acid, aspirin-triggered resolvin D1 and resolvin D2 prevent experimental colitis in mice. *J. Immunol.,* 187, 1957-1969.

Bhattacharya, A., Ghosal, S., Bhattacharya, S. K. (2006). Effect of fish oil on offensive and defensive factors in gastric ulceration in rats. *Prostaglandins Leukot Essent Fatty Acids,* 74, 1 09-116.

Calder, P. C., (2012). Mechanisms of action of (n-3) fatty acids. *J. Nutri.,* 142 (3), 592S-599S.

Calder, P. C., (2013). Omega-3 polyunsaturated fatty acids and inflammatory processes: nutrition or pharmacology? *Br. J. Clin. Pharmacol.,* 75 (3), 7.

Calder, P. C., Yaqoob, P., Harvey, D. J., Watts, A., Newsholme, E. A., (1994). Incorporation of fatty acids by concanavalin A-stimulated lymphocytes and the effect on fatty acid composition and membrane fluidity. *Biochem. J.,* 300 (2), 508-518.

Camuesco, D., Gálvez, J., Nieto, A., Comalada, M., Rodríguez-Cabezas, M. E., Concha, A., Xaus, J., Zarzuelo, A., (2005). Dietary olive oil supplemented with fish oil, rich in EPA and DHA (n-3) polyunsaturated fatty acids, attenuates colonic inflammation in rats with DSS-induced colitis. *J. Nutri.,* 135 (4), 687-694.

Claria, J., Serhan, C. N., (1995). Aspirin triggers previously undescribed bioactive eicosanoids by human endothelial cell-leukocyte interactions. *Proc. Natl. Acad. Sci.,* 92,9475-9479.

Clifford, J. W., (2004). Pain: moving from symptom contol toward mechanism-specific pharmacologic management. *Ann. Intern. Med.,* 140, 441-451.

Chávez-Piña, A. E., Navarrete, A. (2009). Gastroprotective Triterpenoids: Pharmacological Mechanism In A. I. Varela, J. (Ed.), Medicinal Plants: Classification, Biosynthesis and Pharmacology (pp. 97-138). New York: Nova Science Publishers Inc.

Faust, T., Redfern, J. S., Lee, E., Feldman, M., (1989). Effects of fish oil on gastric mucosal 6-keto-PGF1 alpha synthesis and ethanol-induced injury. *Am. J. Physiol., 257*,9-13.

Faust, T. W., Redfern, J. S., Podolsky, I. , Lee, E. , Grundy, S. M., Feldman, M. (1990). Effects of aspirin on gastric mucosal prostaglandin E2 and F2 alpha content and on gastric mucosal injury in humans receiving fish oil or olive oil. *Gastroenterology,* 98, 586-591.

Feng, X. T., Leng, J., Xie, Z., Li, S.L., Zhao, W., Tang, Q. L., (2012). GPR40: a therapeutic target for mediating insulin secretion (review). *Int. J. Mol. Med.,* 30 (6), 1261-1266.

Fierro, I. M., Serhan, C. N., (2001). Mechanism in anti-inflammation and resolution: the role of lipoxins and aspirin-triggered lipoxins. *Braz. J. Med. Biol. Res.,* 34 (5), 555-566.

Fiorucci, S., de Lima, O. M., Mencarelli, A., Palazzetti, B., Distrutti, E., McKnight, W., Dicay, M., Ma, L., Romano, M., Morelli, A., Wallace, J. L., (2002). Cyclooxygenase-2–derived lipoxin A4 increases gastric resistance to aspirin-induced damage. *Gastroenterology,* 123 (5), 1598-1606.

Forman, B. M., Chen, J., R. M., Evans., (1997). Hypolipidemic drugs, polyunsaturated fatty acids, and eicosanoids are ligands for peroxisome proliferator-activated receptors alpha and delta. *Proc. Natl. Acad. Sci. U S A,* 94 (9), 4312-4317.

Gannon, B., Browning, J., O'Brien, P., Rogers, P., (1984). Mucosal microvascular architecture of the fundus and body of human stomach. *Gastroenterology,* 1984 (86), 866-875.

Goel, R. K., Bhattacharya, S. K., (1991). Gastroduodenal mucosal defense and mucosal protective agents. *Indian J. Exp. Biol.,* 29, 701-714.

Güzel, C., Ulak, G., Sermet, A., Cicek, R., Ulak, M., (1995). Effect of fish oil on indomethacin-induced gastric lesions in rats. *Arzneimittelforschung,* 45, 1172-1173.

Ji, R. R., Xu, Z. Z., Strichartz, G., Serhan, C. N., (2011). Emerging roles of resolvins in the resolution of inflammation and pain. *Trends Neurosci.,* 34 (11), 599-609.

Kliewer, S. A., Sundseth, S. S., Jones, S. A., Brown, P. J., Wisely, G. B., Koble, C. S., Devchand, P., Wahli, W., Willson, T. M., Lenhard, J. M., Lehmann, J. M., (1997). Fatty acids and eicosanoids regulate gene expression through direct interactions with peroxisome proliferator-activated receptors alpha and gamma. *Proc. Natl. Acad. Sci. U S A,* 94 (9), 4318-4323.

Kong, W., Yen, J. H., Ganea, D., (2011). Docosahexaenoic acid prevents dendritic cell maturation, inhibits antigen-specific Th1/Th17 differentiation and suppresses experimental autoimmune encephalomyelitis. *Brain Behav. Immun.,* 25 (5), 872-882.

Krishnamoorthy, S., Recchiuti, A., Chiang, N., Yacoubian, S., Lee, C. H., Yang, R., Petasis, N. A., Serhan, C. N., (2010). Resolvin D1 binds human phagocytes with evidence for proresolving receptors. *Proc. Natl. Acad. Sci.,* 107 (4), 1660-1665.

Kromann, N., Green, A., (1980). Epidemiological studies in the Upernavik district, Greenland. Incidence of some chronic diseases. *Acta. Med. Scand.,* 208,401-406.

Laine, L., Takeuchi, K., Tarnawski, A., (2008). Gastric mucosal defense and cytoprotection: bench to bedside. *Gastroenterology,* 135 (1), 41-60.

Lanza, F. L., (1989). A review of gastric ulcer and gastroduodenal injury in normal volunteers receiving aspirin and other nonsteroidal anti-inflammatory drugs. *Scand. J. Gastroenterology,* 24 (16), 24-31.

Lee, T. H., Hoover, R. L., Williams, J. D., Sperling, R. I., Ravalese, J., Spur, B. W., Robinson, D. R., Corey, E. J., Lewis, R. A., Austen, K. F., (1985). Effect of dietary enrichment with eicosapentaenoic and docosahexaenoic acids on in vitro neutrophil and monocyte leukotriene generation and neutrophil function. *N. Engl. J. Med.,* 312 (19), 1217.

Leung, F. W., (1994). Prostaglandins mediate fish oil protection against ethanol-induced gastric mucosal injury in rats. *Dig. Dis. Sci.,* 39, 893.

Li, Y., Smith, T. , Gasbski, S., (1998). The membrane associaton sequences of prostaglandin endoperoxide synthases-1 and 2 isozymes. *J. Biol. Chem.,* 273, 29830-29837.

Lichtenber, L. M., (1999). Gastroduodenal mucosal defense. *Curr. Opin. Gastroenterol.,* 15,463-472.

Mancini, J. A., O'Neill, G. P., Bayly, C., Vickers, P. J., (1994). Mutation of serine-516 in human prostaglandin G/H synthase-2 to methionine or aspirin acetylation of this residue stimulates 15-R-HETE synthesis. *FEBS Lett.,* 342, 33-37.

Manjari, V., Das, U. N. (2000). Effect of polyunsaturated fatty acids on dexamethasone-induced gastric mucosal damage. *Prostaglandins Leukot. Essent. Fatty Acids,* 62, 85-96.

Matta, J. A., Miyares, R. L., Ahern, G. P., (2007). TRPV1 is a novel target for omega-3 polyunsaturated fatty acids. *J. Physiol.,* 578 (2), 397-411.

McCafferty, D. M., Granger, D. N., Wallace, J. L., (1995). Indomethacin-induced gastric injury and leukocyte adherence in arthritic versus healthy rats. *Gastroenterology,* 109 (4), 1173-1180.

Morgenweck, J., Abdel-Aleem, O. S., McNamara, K. C., Donahue, R. R., Badr, M. Z., Taylor, B. K., (2010). Activation of peroxisome proliferator-activated receptor gamma in brain inhibits inflammatory pain, dorsal horn expression of Fos, and local edema. *Neuropharmacology,* 58 (2), 337-345.

Nakamoto, K., Nishinaka, T., Ambo, A., Mankura, M., Kasuya, F., Tokuyama, S., (2011). Possible involvement of beta-endorphin in docosahexaenoic acid-induced antinociception. *Eur. J. Pharmacol.,* 666 (1-3), 100-104.

Nakamoto, K., Nishinaka, T., Mankura, M., Fujita-Hamabe, W., Tokuyama, S., (2010). Antinociceptive effects of docosahexaenoic acid against various pain stimuli in mice. *Biol. Pharm. Bull.,* 33 (6), 1070-1072.

Nishinaka, T., Matsumoto, K., Nakamoto, K., Anbo, A., Mankura, M., Koyama, Y., Tokuyama, S., (2013). Elucidation of Mechanisms Underlying Docosahexaenoic Acid-induced Antinociception. *Yakugaku Zasshi,* 133 (5), 493-499.

Norling, L. V., Dalli, J., Flower, R. J., Serhan, C. N., Perreti, M. (2012). Resolvin D1 limits polymorphonuclear leukocyte recruitment to inflammatory loci: receptor-dependent actions. *Arterioscler. Thromb Vasc. Biol.,* 32, 1970-1978.

O'Neill, G. P., Ford-Hutchinson, A. W. (1993). Expression of mRNA for cyclooxygenase-1 and cyclooxygenase-2 in human tissues. *FEBS Lett.,* 330, 156-160.

Oh, S. F., Pillai, P. S., Recchiuti, A., Yang, R., Serhan, C. N., (2011). Pro-resolving actions and stereoselective biosynthesis of 18S E-series resolvins in human leukocytes and murine inflammation. *J. Clin. Invest.,* 121 (2), 569.

Park, C. K., Xu, Z. Z., Liu, T., Lu, N., Serhan, C. N., Ji, R. R., (2011). Resolvin D2 is a potent endogenous inhibitor for transient receptor potential subtype V1/A1, inflammatory pain, and spinal cord synaptic plasticity in mice: distinct roles of resolvin D1, D2, and E1. *J. Neurosci.,* 31 (50), 18433-18438.

Pineda-Pena, E. A., Jimenez-Andrade, J. M., Castaneda-Hernandez, G., Chavez-Pina, A. E., (2012). Docosahexaenoic acid, an omega-3 polyunsaturated acid protects against indomethacin-induced gastric injury. *Eur. J. Pharmacol.,* 697 (1-3), 139-143.

Quan-Xin, F., Fan, F., Xiang-Ying, F., Shu-Jun, L., Shi-Qi, W., Zhao-Xu, L., Xu-Jie, Z., Qing-Chuan, Z., Wei, W., (2012). Resolvin D1 reverses chronic pancreatitis-induced mechanical allodynia, phosphorylation of NMDA receptors, and cytokines expression in the thoracic spinal dorsal horn. *BMC Gastroenterol.,* 12 (148).

Serhan, C. N. (2001). Eicosanoids. In W. J. Koopman (Ed.), Arthritis and allied conditions. Birmingham, Alabama, USA: Lippincott Williams and Wilkins.

Serhan, C. N., Clish, C. B., Brannon, J., Colgan, S. P., Chiang, N., Gronert, K., (2000). Novel functional sets of lipid-derived mediators with antiinflammatory actions generated from omega-3 fatty acids via cyclooxigenase 2-nonsteroidal antiinflammatory drugs and transcellular processing. *J. Exp. Med.,* 192, 1197-1204.

Shimokawa, T., Smith, T. L.,(1992).Prostaglandin endoperoxide synthase. The aspirin acetylation region. *J. Biol. Chem.,* 267, 12387-12392.

Simopoulos, A. P.,(2002).Omega-3 fatty acids in inflammation and autoimmune diseases. *J. Am. Coll. Nutr.,* 21 (6), 495-505.

Sperling, R. I., Benincaso, A. I., Knoell, C. T., Larkin, J. K, Austen, K. F., Robinson, D. R., (1993). Dietary omega-3 polyunsaturated fatty acids inhibit phosphoinositide formation and chemotaxis in neutrophils. *J. Clin. Invest.,* 91 (2), 651-660.

Sun, Y. P., Oh, S. F., Uddin, J., Yang, R., Gotlinger, K., Campbell, E., Colgan, S. P., Petasis, N. A., Serhan, C. N., (2011). Resolvin D1 and its aspirin-triggered 17R epimer. Stereochemical assignments, anti-inflammatory properties, and enzymatic inactivation. *J. Biol. Chem.,* 282 (13), 9323-9334.

Szanto, A., Nagy, L., (2008). The many faces of PPARgamma: anti-inflammatory by any means? *Immunobiology,* 213 (9-10), 789-803.

Ulak, G., Cicek, R., Sermet, A., C., Güzel, Ulak, M., Denli, O.,(1995).Protective effect of fish oil against stress-induced gastric injury in rats. *Arzneimittelforschung,* 45, 1174-1175.

Wallace, J. L., (2008). Prostaglandins, NSAIDs, and gastric mucosal protection: why doesn't the stomach digest itself? *Physiol. Rev.,* 88 (4), 19.

Wallace, J. L., Devchand, P. R., (2005). Emerging roles for cyclooxygenase-2 in gastrointestinal mucosal defense. *Br. J. Pharmacol.,* 145 (3), 275-282.

Wallace, J. L., Granger, D. N., (1996). The cellular and molecuar basis for gastrodueodenal mucosal defense. *FASEB J.,* 10, 731-740.

Wallace, J. L., (1997). Nonsteroidal anti-inflammatory drugs and gastroenteropathy: the second hundred years. *Gastroenterology,* 112, 1000-1016.

Wallace, J. L., McKnight, W., (1990). The mucoid cap over superficial gastric damage in the rat. A high-pH microenviroment dissipated by nonsteroidal anti-inflammatory drugs and endothelin. *Gastroenterology,* 99, 295-304.

Wallace, J. L., Morris, G. P., Krausse, E. J., Greaves, S. E., (1982). Reduction by cytoprotective agents of ethanol-induced damage to the rat gastric mucosa: a correlated morphological and physiological study. *Can. J. Physiol. Pharmacol.,* 60 (12), 1686-1699.

Wallace, J. L., Syer, S., Denou, E., de Palma, G., Vong, L., McKnight, W., Jury, J., Bolla, M., Bercik, P., Collins, S. M., Verdu, E., Ongini, E.,(2011).Proton pump inhibitors exacerbate NSAID-induced small intestinal injury by inducing dysbiosis. *Gastroenterology,* 141 (4), 1314-1322.

Werther, J. L. (2000). The gastric mucosal barrier. *Mt. Sinai. J. Med.,* 67, 41-53.

Xiao, Y. F., Kang, J. X., Morgan, J. P., Leaf, A., (1995). Blocking effects of polyunsaturated fatty acids on Na+ channels of neonatal rat ventricular myocytes. *Proc. Natl. Acad. Sci.,* 21 (24), 11000-11004.

Xu, Z. Z., Zhang, L., Liu, T., Park, J. Y., Berta, T., Yang, R., Serhan, C. N., Ji, R. R., (2010). Resolvins RvE1 and RvD1 attenuate inflammatory pain via central and peripheral actions. *Nat. Med.,* 16 (5), 592-597, 591p following 597.

Xu, Z. Z., Ji, R. R., (2011). Resolvins are potent analgesics for arthritic pain. *Br. J. Pharmacol.,* 164 (2), 274-277.

Yaqoob, P., Pala, H. S., Cortina-Borja, M., Newsholme, E. A., Calder, P. C., (2000). Encapsulated fish oil enriched in alpha-tocopherol alters plasma phospholipid and mononuclear cell fatty acid compositions but not mononuclear cell functions. *Eur. J. Clin. Invest.,* 30 (3),260-274.

Zaloga, G. P., P., Marik., (2001). Lipid modulation and systemic inflammation. *Crit. Care Clin.,* 17 (1), 201-207.

Zanardo, R. C., Brancaleone, V., Distrutti, E., Fiorucci, S., Cirino, G., Wallace, J. L., (2006). Hydrogen sulfide is an endogenous modulator of leukocyte-mediated inflammation. *FASEB J.,* 20 (12), 2118-2120.

Zhao, Y. F., Wang, L., Zha, D., Qiao, L., Lu, L., Yu, J., Qu, P., Sun, Q., Qiu, J., Chen, C., (2013). GW9508 inhibits insulin secretion by activating KATP channels in rat pancreatic beta-cells. *J. Mol. Endocrinol.,* 51(1),69-77.

In: Omega-3 Fatty Acids
Editor: Md. Wasim Khan

ISBN: 978-1-62948-516-4

Chapter 10

Omega-3 Dietary Intake: Review on Supplementation, Health Benefits and Resources Sustainability

Ivone Almeida, Telmo J. R. Fernandes, Bruno M. R. Guimarães and M. Beatriz P. P. Oliveira*

REQUIMTE, Department of Chemical Sciences, Faculty of Pharmacy, University of Porto, Portugal

Abstract

Fish, particularly fatty fish, is the primary food source of omega-3 (*n*-3) long-chain polyunsaturated fatty acids (LCPUFA). Due to different circumstances, it is not always possible to ingest adequate amounts of *n*-3 fatty acids (FA) through fish consumption, which has resulted in an increased interest in omega-3 dietary supplements (mainly fish oil supplements). On the potential beneficial health effects of omega-3 supplements, many epidemiologic and clinical studies have been conducted.

These studies have shown that dietary supplementation with *n*-3 PUFA help to prevent a large variety of medical conditions from occurring or progressing, such as cardiovascular disease, cancer depression, maternal/fetal well-being, and Alzheimer's disease. Omega-3 dietary supplements are available in several retail outlets, health stores and pharmacies and provide *n*-3 PUFA mostly as ethyl esters (EE) or as re-esterified triglycerides (rTG) forms. Different formulations are available, with diverse amounts of EPA and DHA, depending on the omega-3 dietary requirements, and based in various recommendations by scientific, governmental and regulatory bodies, regarding *n*-3 PUFA intake.

Since marine resources are the most important readily available and edible source of *n*-3 PUFA, is crucial to understand whether they're being or not over-exploited. Recent data suggest that wild-fish catch may be in decline, while the human population is still growing. Aquaculture is probably the most effective way to maintain global consumption

* Corresponding author: beatoliv@ff.up.pt.

of *n*-3 PUFA. One of the major efforts of aquaculture industry is to replace aquafeed marine-derived ingredients for plant sources, since there is a notorious lack of sustainability of the wild-harvest resources that composes fishmeal (FM) and fish oil (FO).

1. Introduction

Since the ability of human body to produce polyunsaturated fatty acids of the omega-3 series (*n*-3 PUFA) from their precursors is limited, their intake is of vital importance. The abundance of *n*-3 PUFA in seafood, particularly of eicosapentaenoic acid (20:5 *n*-3, EPA) and docosahexaenoic acid (22:6 *n*-3, DHA), has enhanced the consumer's acceptance of fish and fish-based products in recent years. In addition to providing essential marine-derived *n*-3 fatty acids (FA), fish is also a good source of high quality protein, B-complex vitamins and essential trace minerals, including potassium, phosphorus, and selenium (Ariño et al., 2013).

The consumption of fish has increased in recent years mainly due to the wide range of health benefits related to the PUFA content of this kind of food. The first studies of possible health benefits related with the consumption of high amounts of fish fat were undertaken in Eskimos of Greenland. It was shown that this population had a lower risk of cardiovascular disease, mainly due to the high *n*-3 PUFA contents of their marine diet (Dyerberg et al., 1975).

Since then, and during the past few years, research has shown that *n*-3 PUFA, primarily EPA and DHA, possess a wide range of biological effects and health benefits (Riediger et al., 2009). *n*-3 PUFA can be found in fatty fish such as herring, tuna or salmon, and in some vegetable sources. Nevertheless, it is not always possible to consume adequate amounts of *n*-3 PUFA through the diet, which has been leading to an increasing interest in omega-3 supplements (Opperman et al., 2011). For individuals who cannot meet current recommendations for *n*-3 FA through fish consumption, omega-3 dietary supplements (mainly fish oil supplements) are an effective alternative to increase *n*-3 LCPUFA intake and status.

Omega-3 resources scarcity is nowadays an example of the recurrent conflicting relationships between human and nature. Large-scale consumption of fish and fish oils may be good for human health but negative for biosphere's health and therefore, ultimately, may also affect human health (BenDor et al., 2009; McMichael & Butler, 2005). It is clear that the availability of *n*-3 PUFA in human diet is crucial and the primary dietary sources of these FA are fish and seafood.

Recent data suggest that wild-fish catch may be in decline, while the human population is still growing resulting in an over-exploitation of marine resources. In this context, aquaculture has emerged as the probable most effective way to maintain global consumption of *n-3* PUFA but also new perspectives of providing EPA and DHA in foods and different pathways to synthesize these LC PUFA must be pointed out (FAO, 2009; Kang, 2011).

2. Omega-3 Supplementation

2.1. *n-3* LCPUFA Dietary Supplements

Data available from different countries suggests that populations are not consuming enough *n*-3 PUFA, with the estimated intake of marine food products and EPA plus DHA being much lower than the international recommendations (Abreu, et al., 2012; Lucas, et al., 2009). The consumption of fatty fish is generally considered the preferred way to increase *n*-3 LC PUFA intake. However, for individuals who are unable or unwilling to consume sufficient *n*-3 FA from fish, because of taste preferences, healthy, ethical, or ideological reasons, and environmental sustainability concerns, supplementation is an alternative. Fish oils, krill oils, and single cell oils, added to foods or consumed as food supplements, offer an effective way of increasing *n*-3 LC PUFA intake and status. *n*-3 PUFAs are available on prescription and as dietary supplements preparations, being marketed through retail outlets, health stores and pharmacies.

Unlike medicines, which must be proven both safe and effective through clinical trials, the manufacturer of a dietary supplement or dietary ingredient is responsible for ensuring that the product is safe before it is marketed (EC, 2013). Thus, sometimes these products may lack consistency in dose and quality. Tatarczyk et al., (2007), in an analysis of nine *n*-3 FA supplements marketed in Austria, found that preparations differed widely from each other regarding the percentage of EPA and DHA. When comparing the combined EPA and DHA content, the results for the majority of the examined products did not differ significantly from labeled levels. Similar results were reported by Kolanowski et al. (2010) for fish body oil and fish liver oil capsules available in Poland. The contents of *n*-3 LCPUFA for the majority of the 19 brands analyzed were in compliance with label claims. More recently, 45 *n*-3 fish oil supplements, commercially available on the South African market, were analyzed to determine the *n*-3 FA content. More than 50% of the supplements failed to meet the EPA and DHA levels stated on the label (Opperman et al., 2011).

Fish oil supplements are usually considered safe, free of detectable traces of mercury (Foran et al., 2003) and are very low in polychlorinated biphenyls (PCBs) (Hilbert et al., 1998; Melanson et al., 2005). Thus, those concerned with environmental toxins from fish may also opt for *n*-3 PUFA supplementation.

2.2. Omega-3 Fish Oil Supplements: Triglycerides vs. Ethyl Esters

Based on the original observations of Dyerberg and co-workers (1975) in Greenland Eskimos concerning an association between dietary intake of *n-3* PUFA and biological and cellular functions, many heart associations and governmental bodies have recommended to increase the intake of fish oil and the use of *n-3* PUFA supplements for prevention of coronary heart disease (Kris-Etherton et al., 2002).

Initially, the omega-3 supplements used in controlled studies contained deodorized fish oils composed by *n-3* PUFA in the natural triglyceride form. Triglycerides (TG) are molecular structures combining three fatty acids esterified (bonded) to a glycerol backbone. Since free fatty acids (FFA) are rapidly oxidized, the TG structure offers greater stability, preventing the breakdown and oxidation of the FA (Segura, 1988). However, and due to the relative low concentration of *n-3* PUFA in these supplements (which can make it impractical to obtain higher doses), problems with patient compliance were reported.

Thus, the development of more concentrated compounds was stimulated (Ackman, 1992). Currently, *n-3* PUFA are mainly present in the formulations as ethyl esters (EE) or as re-esterified TG (rTG), although some supplements deliver *n-3* PUFA as FFA. FA ethyl esters are a class of lipids that derived from the reaction of FFA with ethanol in a process called trans-esterification, which involves a reaction whereby the glycerol backbone of a TG is removed and replaced by ethanol (Mogelson et al., 1984). The resulting EE allow for the selective concentration, by distillation of the LC FA at lower temperatures, such as EPA and DHA, to levels greater than those found naturally in fish. rTG could be manufactured adding one step to the production process of EE. Using food grade enzymes and by glycerolysis, the ethanol molecule of EE is removed, as well as short chain and saturated fatty acids (SFA). EPA and DHA can then be re-esterified to a glycerol backbone (Shahidi & Wanasundara, 1998), enabling the formation of a final product more concentrated, without contaminants and, unlike EE, a natural occurring form.

Numerous studies have assessed the absorption and bioavailability of the different fish oil formulations with mixed results. Although few studies have found that the absorption rate is similar between the two types of oils (Krokan et al., 1993; Luley et al., 1990; Nordøy et al., 1991), the overall evidence suggests that TG fish oils are better absorbed in comparison to EE. Natural TG fish oil results in 50% more plasma EPA and DHA after absorption in comparison to EE oils (Beckermann et al., 1990). TG forms of EPA and DHA were shown to be 48 % and 36 % better absorbed than EE forms (Lawson & Hughes, 1988). Moreover, EPA incorporation into plasma lipids was found to be considerably smaller and longer when administered as an EE (El Boustani et al., 1987), and in the rat, DHA TG supplementation led to higher plasma and erythrocyte DHA content than did DHA EE (Valenzuela et al., 2005).

On the subject of omega-3 bioavailability, two major long-term clinical trials have been published. On the study by Dyerberg and colleagues (Dyerberg et al., 2010) the bioavailability of EPA+DHA from rTG was higher (124%) when compared to natural fish oil, whereas the bioavailability from EE was inferior (73%). In another controlled experiment, the bioavailability of a moderate dose of *n-3* PUFA, administered as EE or rTG, was compared to a placebo. This study used as a measure of bioavailability the omega-3 index, which represents the percentage of EPA+DHA in red blood cell membranes. The results showed a significant incorporation of EPA+DHA in the red blood cell membranes via rTG over EE by over 25% (Neubronner et al., 2011).

n-3 FA in the form of EE are much less stable than those in the natural TG form and readily oxidize. The oxidation kinetics of DHA as an EE or as a TG has already been assessed by measuring the concentration of oxygen found in the headspace of a reaction vessel with both TG and EE forms. The EE form of DHA was more reactive, and quickly oxidized, demonstrating that EEs are far less stable and can more readily produce harmful oxidation products (Yoshii et al., 2002). Another study analyzed the stability of DHA in EE and TG form during a 10-week period and found that DHA in EE form decayed 33% more rapidly (Song et al., 1997).

Fish oil supplements in the natural TG form offer many advantages when compared to those in the EE form: oils in a TG form are completely safe to consume, are naturally occurring, provide increased absorption and bioavailability, and are much more stable.

Therefore, since TG fish oil can be more effectively absorbed then it can be potentially better at reaching therapeutic benefits in comparison to EE fish oils. However, omega-3 concentrates in the TG form are usually more expensive than the EE form.

2.3. Health Benefits Studies

The evidence of the benefit srelated to the ingestion of omega-3 FA resulted from studies conducted in populations of Eskimos of Greenland and Alaska, and coastal fishermen of Japan (Dyerberg et al., 1975). These studies showed that the risk of cardiovascular disease (CVD) was higher in Caucasian populations, compared with the studied population of Eskimos. The same studies also revealed that the Eskimos had a lower level of cholesterol, TGand very low-density lipoprotein (VLDL), and a high level of high-density lipoprotein (HDL) in both genders, despite the ingestion of a greater amount of fat from aquatic mammals and fish. The major cause for these differences was attributed not to the genetic profile of the population, but to the high content of *n*-3 PUFA intake in the traditional diet of the Eskimos compared with the typical Caucasian population (Barrow et al., 2009; Dyerberg et al., 1975). Later epidemiological studies confirmed that the fats consumed by Eskimos were rich in EPA and DHA, which was associated with a cardioprotective role against the risk of cardiovascular disease in healthy subjects (Albert et al., 2005; Barrow et al., 2009).

These early observations led to an increase in research, examining the beneficial and/or preventive effects of *n-3*FA on several debilitating and common conditions, including CVD, rheumatoid arthritis and asthma (Riediger et al., 2009).

The Diet and Reinfarction Trial (DART), in 1989, was the first randomized controlled trial to test the hypothesis that fish oil confers protection against coronary heart disease (CHD). It showed a reduction in mortality during the 2 years after myocardial infarction (MI) amongst men who were instructed to eat about 300g of oily fish per week, or who consume fish oil supplements at an equivalent amount of *n*-3 FA (Burr et al., 1989). The GISSI-Prevenzione trial was a secondary prevention study with 11324 patients that experienced a recent MI (≤3 months). Patients were randomly assigned into different groups to determine the impact of a daily dose of 1 capsule of 850 mg/day of EPA + DHA with or without vitamin E over 42 months. In both cases there was a 15% reduction in death by heart attack or stroke. The overall mortality decreased by 21% and sudden death caused by heart disease decreased 45% (GISSI, 1999).

In 2002, a meta-analysis of past studies on *n*-3FA (dietary and supplemental intake) in patients with CHD was conducted, comparing these interventions to a control diet or placebo (Bucher et al., 2002). The meta-analysis demonstrated that dietary or supplemental consumption of *n*-3 FA were more effective than a controlled diet or placebo for decreasing fatal myocardial infarction (27% reduction), sudden death (31% reduction), and overall mortality (19% reduction).

Recently, a prospective cohort study investigated associations of plasma phospholipid EPA, DHA, DPA (docosapentaenoic acid, 22:5 *n*-3), and total *n*-3 PUFA levels with total and cause-specific mortality among healthy older adults not receiving supplements. The research team examined 16 years of data from 2692 adults aged 74 years (±5 years) without prevalent CHD, stroke, or heart failure. DHA was most strongly related to lower risk of CHD death

(40% lower risk), especially CHD death due to arrhythmias. DPA was strongly associated with lower risk of stoke death, and EPA most strongly linked with lower risk of nonfatal heart attack. Overall, study participants with the highest levels of all three fatty acids had a 27% lower risk of total mortality due to all causes (Mozaffarian et al., 2013).

Originally the cardiovascular benefits associated with the consumption of *n*-3 PUFA were associated with antithrombotic effects (Dyerberg et al., 1975). However, recent studies point to the fact that the benefits of *n*-3 PUFA intake may be associated with favorable antiarrhythmic effects. It was found that supplementation with *n-3* PUFA in patients who have suffered myocardial infarction correlated with decreased risk of mortality associated with cardiac arrhythmias (Wang et al., 2006). Other proposed and potential cardiovascular benefits of *n*-3 PUFA may include anti-inflammatory properties (de Caterina et al., 2000), reduction in platelet aggregation, decreased heart rate, blood pressure reduction, and stabilization of existing arterial plaques (Grimsgaard et al., 1998; Harris, 2004; Kris-Etherton et al., 2003). Regardless of the mechanism of action, it is clear that in past and recent epidemiologic and randomized clinical studies, these FA have a beneficial cardiovascular effect.

n-3 FA contribute to various membrane properties such as fluidity, flexibility and permeability, and are essential to the proper functioning of tissues (Schmitz & Ecker, 2008; Simopoulos, 2002). Increased consumption of EPA and DHA results in their incorporation into inflammatory cell phospholipids in a time and dose-dependent manner (Healy et al., 2000; Rees et al., 2006). They are incorporated partly at the expense of arachidonic acid (20:4 *n*-6, ARA), but also of other *n-6* FA. Both EPA and ARA compete for the common cyclooxygenase and lipoxygenase enzymes and compared to EPA, ARA produces more potent inflammatory and pro-aggregatory eicosanoids. Thus, being incorporated into inflammatory cells phospholipids at the expense of ARA, EPA and DHA help reduce the substrate for the synthesis of ARA-derived pro-inflammatory eicosanoids including PGE2, thromboxane B2, LTB4 and LTE4 by inflammatory cells (Kelley et al., 1999; Trebble et al., 2003). In addition, EPA is also capable to act as a substrate for both cyclooxygenase and 5-lipoxygenase, giving rise to eicosanoids with a different structure to those formed from ARA. Thus, fish oil supplementation has been shown to result in increased production of LTB5, LTE5 and 5-hydroxyeicosapentaenoic acid by inflammatory cells (Lee et al., 1985; Sperling et al., 1993; Von Schacky et al., 1993). Moreover, diet supplementation with *n-3* rich fish oil is very important and appears to have a beneficial effect, as demonstrated by several authors, in various inflammatory diseases such as rheumatoid arthritis, Chron's disease, cystic fibrosis, obesity and atherosclerosis (Calder, 2006; He et al., 2007; Moyad, 2005; Munro & Garg, 2013). Recently a new family of EPA and DHA-derived lipid mediators called resolvins has been described. These molecules appear to have potent anti-inflammatory and inflammation resolving properties in neutrophils, macrophages, dendritic cells and T cells (Serhan et al., 2000a; Serhan et al., 2000b; Serhan et al., 2008). It seems likely that these mediators will explain many of the anti-inflammatory actions of *n*-3 FA that have been described (Hudert et al., 2006; Ishida et al., 2010).

One of the more exciting potential impacts of *n*-3 FA is their potential to reduce the risk of chronic neurodegenerative diseases and mental illness such as Alzheimer's disease, attention-deficit hyperactivity disorder (ADHD) and depression. Several epidemiologic studies have reported low plasma DHA status in individuals with schizophrenia, ADHD, dyslexia, personality disorder, depression, and bipolar disorder (Edwards et al., 1998; Peet &

Stokes, 2005; Richardson, 2006; Timonen et al., 2004; Yao et al., 1994). ADHD is associated with low red blood cell DHA and high *n*-6 FA levels (Colter et al., 2008). Stevens and colleagues (Stevens et al., 2003) found that EPA levels of red blood cells were positively associated with reduced disruptive behavior in children with ADHD receiving 480 mg/day DHA and 80 mg/day EPA for 4 months.

Kamphuis and associates (Kamphuis et al., 2006) stated that every 50 mg/day increase in *n-3*FA intake was associated with a 7% risk reduction of depressive symptoms in elderly men. Likewise, Su and colleagues (Su et al., 2003) found that following four weeks of intervention, subjects treated with EPA/DHA displayed significantly lower depression scores as compared to controls. Nemets and colleagues (Nemets et al., 2006) reported improvements in depression score in children with major depressive disorder receiving 400 mg/day and 200 mg/day of EPA and DHA, respectively, for 4 months. Recently, a double-blind placebo-controlled study confirmed the positive effects of *n*-3 PUFA supplementation in the treatment of elderly depression (Rizzo et al., 2012).

Dementia, including both vascular dementia and Alzheimer's disease, afflicts more than 30% of elderly persons older than 85 years (Ott et al., 1995). Since *n*-3 FA have anti-inflammatory properties and inflammatory markers have been found in the brain of patients with Alzheimer's disease, it seems reasonable to suggest that *n*-3 FA may delay the onset of Alzheimer's disease by reducing brain inflammatory state (Honig, 2000). This may be one of the reasons behind the prevention of Alzheimer's disease/dementia by adequate DHA/EPA intake suggested by the Framingham heart study (Schaefer et al., 2006). A randomized double-blind placebo controlled study (Freund-Levi et al., 2006) showed that in patients with very mild Alzheimer's disease, *n*-3 PUFA supplementation was associated with a significantly higher mean Mini Mental State Examination scores (i.e., less severe Alzheimer's disease symptoms) at 6 months compared to placebo-treated group. However in the group with mild to moderate Alzheimer disease, the decline in cognitive function did not differ when compared with the placebo-treated group. This may suggest that supplementation with *n*-3 PUFA may be more effective in the treatment of early stage ofAlzheimer disease.

Animal studies suggest that EPA and DHA can reduce the risk of osteoporosis and fractures (Sun et al., 2003; Sun et al., 2004). These results are consistent with other studies that show that reducing the *n*-6/*n*-3 ratio could result in increased bone marrow cellularity and bone strength (Atkinson et al., 1997; Kokkinos et al., 1993). Studies in humans and experimental studies in animals have shown that EPA and DHA promoted calcium absorption, reduced calcium excretion and has overall positive effects on calcium content and bone density (Kruger & Schollum, 2005). A recent systematic review reports that conclusions regarding *n*-3 PUFA and osteoporosis are somewhat limited because of the small number and inadequate sample sizes of randomized controlled trials. The same study also relates that the potential benefit of *n*-3 PUFA on skeletal health may be enhanced by simultaneous administration of calcium (Orchard et al., 2012).

n-3 PUFA are required for normal conception, growth, and development of an embryo. Throughout the third trimester, approximately 50 to 60 mg/day of maternal DHA stores are transferred to the fetus via the placenta (Clandinin et al., 1980). A positive association has been observed between red blood cell DHA content and improved visual acuity as well as other indexes of brain development in human infants (Lauritzen et al., 2004). Numerous clinical studies have also described improved visual acuity (Birch et al., 2000; Decsi et al., 2005; Hoffman et al., 2003) and hand-eye coordination (Dunstan et al., 2006) in infants receiving greater DHA in the womb and/or during lactation. Evidence from randomized controlled trials indicates that enhancement of maternal intake of *n*-3 PUFA prolongs the duration of gestation in low-risk pregnancies and may also contribute to the prevention of early preterm birth in both low-risk and high-risk pregnancies (Decsi, 2009). Given the importance of *n*-3 PUFA on pre-natal, child and maternal health and the potential contamination of fish with mercury and other compounds accompanied by the lack of a high concentration of beneficial *n*-3 PUFA in numerous fish and shellfish products, the use of omega-3 supplements to enrich pregnant and lactating women diets should be encouraged to ensure the optimal health of their infants, especially during the third trimester when *n*-3 PUFA intake may be most effective in achieving optimal infant development.

Age-related macular degeneration (AMD) is the leading cause of blindness in industrialized countries. This condition has no cure and the treatments do not reestablish vision, but only prevents the progression of the disease (Lim et al., 2012). High dietary intakes of *n*-3 PUFA and fish have been consistently associated with a reduced risk for AMD (Merle & Delyfer, 2013). *n*-3 PUFA have important protective and structural roles in the retina, and DHA reaches its highest concentration in the membranes of photoreceptors being fundamental for its differentiation and survival (SanGiovanni & Chew, 2005). A meta-analysis (Chong et al., 2008) of 9 epidemiological studies showed that the risk for AMD is greatly reduced in participants with high consumption of *n*-3 PUFA and fish. However, these results are still inconclusive, since some authors had reported different outcomes (Hodge et al., 2006; Ouchi et al., 2002; Sanders et al., 1993). While there is some clinical evidence for protection of AMD from *n*-3 PUFAs, more studies are still needed, especially prospective cohort designs and randomized clinical trials.

Dietary fats have been known to play an important role in the etiology of cancer (Cohen, 1992). According to Newman (Newman, 1990) certain PUFAs help to inhibit the growth of carcinoma cells and melanoma. Later, another study confirmed the anti-melanoma effects of *n*-3 PUFAs (Xia et al., 2006). More recently *n*-3 PUFA was shown to have preventive properties against various cancers and their complications including colon and breast cancer (Cockbain et al., 2012; Patterson et al., 2011). It is important to mention that the evidence for the effectiveness of *n*-3 PUFA against cancer has mainly come from epidemiological studies and animal experimentation and to really establish the potential anticancer effects of *n*-3 PUFA, well-designed, large-scale clinical trials are required.

2.4. Dietary Recommendations for Fish and *n-3* PUFA Consumption

Over the last decades, there's been a marked progress regarding the establishment of recommendations and dietary guidelines for *n*-3 PUFA. Worldwide, health and nutrition organizations have issued recommendations for *n*-3 fatty acids (including ALA, EPA, and DHA) to prevent deficiency and for the prevention and the treatment of cardiovascular *disease* (CVD) (EFSA, 2010; Lichtenstein et al., 2006; FAO/WHO, 2010).

α-linolenic acid (18:3 *n*-3, ALA) is an essential fatty acid that mammals cannot synthesize *de novo* and therefore, it has to be obtained through the diet. An Adequate Intake (AI) of 0.5% of total daily energy intake has been set for ALA as the minimum dose *required*to*prevent deficiency symptoms (*EFSA 2010). Dietary recommendations for *n*-3 LC PUFA are based on cardiovascular disease (CVD) risk reduction. Current dietary recommendations from national and international bodies for *n*-3 LC PUFAs (mostly as EPA and DHA) range from 200 mg to >600 mg/day for adults without documented CVD (EFSA, 2010). This level can be achieved by the consumption of at least 2 servings of fatty fish per week (where each serving is defined as providing 200-500 mg/day of DHA and EPA) (Lichtenstein et al., 2006; EFSA, 2010). For patients with preexisting CVD, the American Heart Association (AHA) and the European Society for Cardiology (ESC) recommend the intake of about 1 g/day of EPA and DHA preferably from fatty fish, although supplementation can be an alternative (Lichtenstein et al., 2006; van de Werf et al., 2003). For triglyceride-lowering purposes in patients with hypertriglyceridemia, intakes of 2 - 4g/day of combined EPA and DHA are recommended, and should be *provided under a physician's supervision* (Lichtenstein et al., 2006; EFSA, 2012). There is insufficient evidence to set a Tolerable Upper Intake Level (UL) for *n*-3 LC PUFA (individually or combined) for any population group. However, the EFSA Panel on Dietetic Products Nutrition and Allergies (NDA) has ruled that supplementation up to 5 grams of EPA plus DHA do not increase the risk of adverse effects in healthy adults (EFSA, 2012). This decision was based on reported adverse effects associated with high intakes of *n*-3 LCPUFA on immune function, on LDL cholesterol, bleeding tendencies, and on lipid and glucose metabolism.

n-3 PUFA supplements are generally well tolerated, but some mild side effects can occur. The most commonly observed adverse effects include eructation and a fishy taste, gastrointestinal upset and nausea, and intolerance to the capsules (*Sacks et al., 1995;* GISSI, 1999).

Studies *suggest*that humans have evolved on a diet *with* a *ratio* of *n-6* to *n*-3 essential fatty acids of about 1–2:1 (Simopoulos, 2003). Over the course of human evolution there has been a dramatic change in the ratio of *n*-6/*n*-3 fats consumed in the diet. Available evidence indicates that high dietary intakes of *n*-6 PUFA, characteristic of the Western diet, may increase the synthesis of proinflammatory eicosanoids (Simopoulos, 2003). On the contrary, higher intakes of *n*-3 LCPUFA are associated with production of less inflammatory or even anti-inflammatory eicosanoids (Calder, 2006).

The opposing roles of *n-3* and *n-6* PUFA in the regulation of the inflammatory process suggest the importance of the *n-6*/*n-3* ratio in the development and severity of diseases of inflammatory nature. This is of particular importance when considering the profusion of *n*-6 FA and the lack of *n*-3 FA in actual human diet (Simopoulos, 2003). However, more experimental evidence is necessary in order to recommend a specific ratio of *n*-6 to *n*-3 FA (EFSA, 2010; Harris, 2006).

3. Omega-3 Resources Sustainability

3.1. Natural Omega-3 Resources Over-exploitation

There are three main different sources for humans to obtain *n*-3 PUFA: marine phytoplankton (microalgae), which are an aquatic animal feed or can be directly used for food products; fish and seafood (the most important dietary source for humans) which consume phytoplankton directly or eat other animals that feed on phytoplankton; and livestock which are fed meal that may include *n*-3 PUFA (Kang, 2011).

Marine phytoplankton (microalgae) represent the basis of the food web for aquatic creatures and provide large amounts of LC PUFA, thus being particularly valuable in maintaining the availability of *n*-3 PUFA in human diet. In fact, freshwater ecosystems are often relatively rich in EPA, although DHA is generally more limited (Kang, 2011, Tocher, 2009). Recent studies alerted about the vulnerability of phytoplankton to global climatic and environmental changes. Ultraviolet-B radiation clearly reduces the overall phytoplankton growth and their PUFA content. Together with increased carbon dioxide levels itmay also decrease PUFA synthesis or destroy already synthesized PUFA. Generally, global climate changes could significantly affect the overall availability of *n*-3 PUFA and therefore diminishing the omega-3 content of food chain products: primarily,of fish/marine products; animal feed/livestock; fish oils and nutritional supplements, resulting in a lower*n*-3/*n*-6 ratio (Carvalho & Malcata, 2005; Leu et al., 2007; Schindell et al., 1998).

Fish and seafood are the most important readily available and edible source of *n*-3 PUFA, since EPA and DHA accumulate in the fish from the lipids obtained from microalgae (Brunner et al., 2009). Despite the greatest importance for consumers of these FA, fish also provide a source of high quality and easily digested protein (about 20% of the total animal protein consumed), minerals (such as iodine and selenium) and vitamins (A, D and E) (McMichael & Butler, 2005; Tocher, 2009). Particularly, the poorest countries are highly dependent on fish to fulfill their dietary needs and the over-exploitation of the world's fish resources may represent a threat to these communities (Brunner et al., 2009).

The main reason to the lack of marine resources sustainability is that fish and seafood catch has reached a plateau phase (may even be in decline), while the size of human population is expanding and therefore the demand for fish and seafood is continuingly increasing (BenDor et al., 2009; Napier & Sayanova, 2005; Tocher, 2009; Villasante et al., 2012). In fact, FAO recently estimated that a quarter of the world's fish resources are over-exploited and in decline and half are fully exploited (FAO, 2009), which may be also related to the unsustainable fishing practices and rapid improvements in fishery technology in the last few decades (BenDor et al., 2009).

There is an urgent need to provide sustainable alternatives to fulfill the human dietary intake of omega-3 and aquaculture has been the apparent solution to fill this supply gap (Napier et al., 2005; Tocher, 2009).

3.2. Aquaculture

Aquaculture is probably the most effective way to maintain global consumption of *n*-3 PUFA. It is seen as the major source of future supplies of fish and seafood taking into account the rising demand and declining wild species and stocks. Globally, the culture of all aquatic species has grown at an average rate of about 10% a year for the last three decades providing nowadays about 50% of the global fish and seafood supplies (fish, crustaceans, mollusks and plants) destined for direct human consumption (FAO, 2009; McMichael & Butler, 2005; Tocher, 2009). One of the major advantages of aquaculture is the better control of possible contaminants, which may result inlower health risks regarding farmed fish consumption (Tocher, 2009).

However, there are some environmental impacts of aquaculture, which have begun to raise concern among industry and the scientific community. Typically, there is a notorious dependence of aquaculture on marine-derived sources (Glencross, 2009). Two of the basic ingredients for aquafeeds production are fishmeal (FM) and fish oil (FO), which provide a high digestibility and sufficient content of essential FA to farmed fish, in particular *n*-3 PUFA. In 2010, 75% of the global fish oil production was used for fish feeds. At the moment, aquaculture uses about 40% and 60% of the world's production of fishmeal and fish oil, respectively. Predictably, in the next ten years, fish oil production might not be enough for aquafeeds manufacture, highlighting the need for marine-derived compounds substitution (Nasopoulou & Zabekatis, 2012). Thus, one of the major efforts of aquaculture industry is to reduce the dependence on marine-derived ingredients trough a better management, for example by its replacement for plant sources, justified by a clear lack of sustainability of the wild-harvest resources that composes FM and FO (Fernandes et al., 2012; Matos et al., 2012; Pratoomyot et al., 2010).

Significant breakthroughs have occurred over the past few years in replacing FM and FO. Plant protein (PP) and vegetable oils (VO) seem to be the strongest alternatives in aquafeeds (Bell et al., 2004; Hardy, 2010; Cabral et al., 2013). Including VO in aquafeeds, the costs are reduced due to the fact that these oils have steadily increasing production, high availability and better economic value. The most common vegetable oils used for aquafeeds production are soybean, linseed, rapeseed, sunflower, palm and olive oil (Glencross, 2009; Nasopoulou & Zabetakis, 2012).

However, the substitution of high levels of dietary FM and FO for PP and VO might substantially change the FA composition of fish tissues, including flesh, the edible portion of fish. The FA composition of fish fed individual treatments usually reflects the FA profile of the ingested lipids. The partial replacement by plant sources can minimize the impact of the marine ingredients reduction on the nutritional quality of flesh to human consumers. On the other hand, the use of FO "finishing"dietary treatments during the final stage of growth helps to restore the levels of LC PUFA on fish muscle, namely EPA and DHA, while the C_{18} FA contribution by PP and VO is diluted (Nasopoulou & Zabetakis, 2012; Tocher, 2009).

3.3. Other Alternatives

Additionally to the major efforts of aquaculture industry to reduce its reliance on marine fish stocks, further knowledge has provided sustainable alternative ways of maintaining human omega-3 intake. Possible alternative marine sources from lower trophic levels have been used, specifically zooplankton, such as krill. However, krill is at the base of a food chain that includes whales and penguins, which is a major opposition to its use. Livestock is also being fed with *n*-3 PUFA enriched feed in order to increase the content of these FA in goods such as meat, milk and eggs (Kang, 2011; Tocher, 2009).

Biotechnology can also provide viable solutions for the nutritional demands since the use of transgenic plants genetically engineered to synthesize *n*-3 PUFA has recently been demonstrated as a sustainable resource for the future (Napier & Sayanova, 2005). Animals and plants could be able to efficiently convert *n*-6 PUFA, which are highly abundant, to *n*-3 PUFA so that they and their derived products would contain higher omega-3contents (Kang, 2011). The possibility of "reverse engineering" LC PUFA biosynthesis into transgenic plants has been successfully demonstrated for EPA and ARA, although the production and accumulation of DHA in transgenic oil seeds may be some years further away. In consequence, the omega-3-rich food products would provide a more accessible land sources and reduce the reliance on marine fishing (Kang, 2011, Napier & Sayanova, 2005). Since aquaculture is also dependent on fish oils, transgenic-derived plant oils enriched with *n*-3 PUFA may also be able to sustainably replace this additional demand.

Conclusion

Recent data suggest that populations are not consuming enough *n*-3 PUFA, which along with the described health benefits have been leading to an increased interest in omega-3 dietary supplements in recent years. This review has focused on the omega-3 dietary intake, specifically on how *n*-3 PUFA supplementation can improve human health and the possible ways to maintain the global intake of these crucial FA, whereas fish stocks are currently over-exploited. Through the various controlled studies regarding the multiple health benefits of fish oil consumption (cardiovascular protection, anti-inflammatory effect, improving visual function, neuropsychiatry disorders) that were undertaken over the last decades, it was possible to place some dietary advice and guidance regarding the adequate amounts of *n*-3 PUFA intake to the consumer. In order to meet the current demand for omega-3 sources and at the same time to reduce the pressure on wild fish stocks, aquaculture industry is increasingly turning to vegetable resources as an alternative to feed farmed fish. Moreover, biotechnology has been assumed as a real sustainable alternative for the future, providing pathways to synthesize *n*-3 PUFA.

References

Abreu, D., Guessous, I., Vaucher, J., Preisig, M., Waeber, G., Vollenweider, P., & Marques-Vidal, P. (2012). Low compliance with dietary recommendations for food intake among adults. *Clinical Nutrition,* 1-6.

Ariño, A., Beltrán, J. A., Herrera, Roncalés, & A. P. (2013). *Fish and seafood: Nutritional Value.* In Lindsay H. Allen, Andrew Prentice, & Benjamin Caballero, Eds., Encyclopedia of Human Nutrition. (3rd ed., pp. 254-261). Waltham, Academic Press.

Ackman, R. G. (1992). The absorption of fish oils and concentrates. *Lipids*, 27(11), 858-862.

Albert, C. M., Oh, K., Whang, W., Manson, J. E., Chae, C. U., Stampfer, M. J., Willett, W. C., & Hu F. B. (2005). Dietary alpha-linolenic acid intake and risk of sudden cardiac death and coronary heart disease. *Circulation*, 112(21), 3232-3238.

Atkinson, T. G., Barker, H. J., & Meckling-Gill, K. A. (1997). Incorporation of long-chain *n*-3 fatty acids in tissues and enhanced bone marrow cellularity with docosahexaenoic acid feeding in post-weanling Fischer 344 rats. *Lipids*, 32(3), 293-302.

Barrow, C. J., Nolan, C., & Holub, B. J. (2009). Bioequivalence of encapsulated and microencapsulated fish-oil supplementation. *Journal of Functional Foods*, 1(1), 38-43.

Beckermann, B., Beneke, M., & Seitz, I. (1990). Comparative bioavailability of eicosapentaenoic acid and docasahexaenoic acid from triglycerides, free fatty acids and ethyl esters in volunteers. *Arzneimittel Forschung*, 40, 700-704.

Bell, J., Henderson, R., Tocher, D., & Sargent, J. (2004). Replacement of dietary fish oil with increasing levels of linseed oil: Modification of flesh fatty acid compositions in Atlantic salmon (*Salmo salar*) using a fish oil finishing diet. *Lipids*, 39, 223-232.

BenDor, T., Schefran, J., & Hannon, B. (2009). Ecological and economic sustainability in fishery management: A multi-agent model for understanding competition and cooperation. *Ecological Economics*, 68, 1061-1073.

Birch, E. E., Garfield, S., Hoffman, D. R., Uauy, R., & Birch, D. G. (2000). A randomized controlled trial of early dietary supply of long-chain polyunsaturated fatty acids and mental development in term infants. *Developmental Medicine and Child Neurology*, 42(3), 174-181.

Brunner, E. J., Jones, P. J. S., Friel, S., & Bartley, M. (2009). Fish, human health and marine ecosystem health: policies in collision. *International Journal of Epidemiology*, 38, 93-100.

Bucher, H. C., Hengstler, P., Schindler, C., & Meier, G. (2002). *n*-3 polyunsaturated fatty acids in coronary heart disease: a meta-analysis of randomized controlled trials. *The American Journal of Medicine*, 112(4), 298-304.

Burr, M. L., Fehily, A. M., Rogers, S., Welsby, E., King, S., & Sandham, S. (1989). Diet and reinfarction trial (DART): design, recruitment, and compliance. *European Heart Journal*, 10, 558-567.

Cabral, E. M., Fernandes, T. J. R., Campos, S. D., Castro-Cunha, M., Oliveira, M. B. P. P., Cunha, L. M., & Valente, L. M. P. (2013). Replacement of fish meal by plant protein sources up to 75% induces good growth performance without affecting flesh quality in ongrowing Senegalese sole. *Aquaculture*, 380-383, 130-138

Calder, P. C. (2006). Polyunsaturated fatty acids and inflammation. *Prostaglandines, Leukotrienes and Essential Fatty Acids,* 75(3), 197-202.

Carvalho, A. P., & Malcata, F. X. (2005). Optimization of ω-3 fatty acid production by microalgae: crossover effects of CO2 and light intensity under batch and continuous cultivation modes. *Journal of Marine Biotechnology*, 7, 381-388.

Chong E. T., Wong T. Y., Simpson J. A., Guymer R. H., & Kreis, A. J. (2008). Dietary ω-3 fatty acid and fish intake in the primary prevention of age-related macular degeneration: A systematic review and meta-analysis. *Archives of Ophthalmology*, 126(6), 826-833.

Clandinin, M. T., Chappell, J. E., Leong, S., Heim, T., Swyer, P. R., & Chance, G. W. (1980). Intrauterine fatty acid accretion rates in human brain: implications for fatty acid requirements. *Early Human Development*, 4(2), 121-129.

Cockbain, A. J., Toogood, G. J., & Hull, M. A. (2012). Omega-3 polyunsaturated fatty acids for the treatment and prevention of colorectal cancer. *Gut*, 61(1), 135-149.

Cohen, L. (1992). Lipids in cancer: an introduction. *Lipids*. 27(10), 791-792.

Colter, A. L., Cutler, C., & Meckling, K. A. (2008). Fatty acid status and behavioral symptoms of Attention Deficit Hyperactivity Disorder in adolescents: A case-control study. *Nutrition Journal*, 7:8.

de Caterina, R., Liao, J. K., & Libby, P. (2000). Fatty acid modulation of endothelial activation. *The American Journal of Clinical Nutrition*, 71(1), 213–223.

van de Werf, F., Bax, J., Betriu, A., Blomstrom-Lundqvist, C., Crea, F., Falk, V., Filippatos, G., Fox, K., Huber, K., Kastrati, A., Rosengren, A., Steg, P. G., Tubaro, M., Verheugt, F., Weidinger, F., & Weis, M. (2003). Management of acute myocardial infarction in patients presenting with persistent ST-segment elevation: the Task Force on the Management of ST-Segment Elevation Acute Myocardial Infarction of the European Society of Cardiology. *European Heart Journal*, 29, 2909 -2945.

Decsi, T. (2009). Effects of supplementing LCPUFA to the diet of pregnant women: data from RCT. *Advances in experimental medicine and biology*, 646, 65-69.

Decsi, T., Campoy, C., & Koletzko, B. (2005). Effect of *n*-3 polyunsaturated fatty acid supplementation in pregnancy: the Nuheal trial. *Advances in experimental medicine and biology*, 569, 109-113.

Dunstan, J. A., Simmer, K., Dixon, G., & Prescott, S. L. (2006). Cognitive assessment at 2 1/2 years following fish oil supplementation in pregnancy: a randomized controlled trial. *Arch Dis Child Fetal Neonat ED*, 93, 45-50.

Dyerberg, J., Bang, H. O., & Hjorne, N. (1975). Fatty acid composition of the plasma lipids in Greenland Eskimos. *The American journal of clinical nutrition*, 28(9), 958-966.

Dyerberg, J., Madsen, P., Møller, J. M., Aardestrup, I., & Schmidt, E. B. (2010). Bioavailability of marine *n*-3 fatty acid formulations. *Prostaglandins leukotrienes and essential fatty acids*, 83(3), 137-141.

Edwards, R., Peet, M., Shay, J., & Horrobin, D. (1998). Omega-3 polyunsaturated fatty acid levels in the diet and in red blood cell membranes of depressed patients. *Journal of Affective Disorders*, 48(2-3), 149-155.

EFSA (2010). Scientific Opinion on Dietary Reference Values for fats, including saturated fatty acids, polyunsaturated fatty acids, monounsaturated fatty acids, *trans* fatty acids, and cholesterol. *EFSA Journal*, 8(3), 1461. Available at: http://www.efsa.europa.eu/en/efsajournal/pub/1461.htm.

EFSA (2010). Scientific Opinion related to the Tolerable Upper Intake Level of eicosapentaenoic acid (EPA), docosahexaenoic acid (DHA) and docosapentaenoic acid (DPA). *EFSA Journal*, 10(7), 2815.

European Commission (2013). *Directorate General for Health and Consumers. Food supplements.* Available at: http://ec.europa.eu/food/food/ labellingnutrition/supplements/index_en.htm. Accessed May 14, 2013.

El Boustani, S., Colette, C., Monnier, L., Descomps, B., Crastes de Paulet, A., & Mendy, F. (1987). Enteral absorption in man of eicosapentaenoic acid in different chemical forms. *Lipids*, 22(10), 711-714.

FAO (2009). *The state of world fisheries and aquaculture.* FAO Fisheries and Aquaculture Department. Food and Agriculture Organization of the United Nations, Rome.

Fernandes, T. J. R., Alves, R. C., Souza, T., Silva, J. M. G., Castro-Cunha, M., Valente, L. M. P., & Oliveira, M. B. P. P. (2012). Lipid content and fatty acid profile of Senegalese sole (*Solea senegalensis* Kaup, 1858) juveniles as affected by feed containing different amounts of plant protein sources. *Food Chemistry*, 134, 1337-1342.

Foran, S. E., Flood, J. G., & Lewandrowski, K. B. (2003). Measurement of Mercury Levels in Concentrated Over-the-Counter Fish Oil Preparations: Is Fish Oil Healthier Than Fish? *Archives of Pathology & Laboratory Medicine*, 127 (12), 1603-1605.

Freund-Levi, Y., Eriksdotter-Jönhagen, M., Cederholm, T., Basun, H., Faxén-Irving, G., Garlind, A., Vedin, I., Vessby, B., Wahlund, L. O., & Palmblad, J., (2006). Omega-3 fatty acid treatment in 174 patients with mild to moderate Alzheimer disease: OmegAD study: a randomized double-blind trial. *Archives of Neurology*, 63(10), 1402-1408.

GISSI (1999). Dietary supplementation with *n*-3 polyunsaturated fatty acids and vitamin E after myocardial infarction: results of the GISSI-Prevenzione trial. *The Lancet*, 354, 447-455.

Glencross, B. D. (2009). Exploring the nutritional demand for essential fatty acids by aquaculture species. *Reviews in Aquaculture*, 1, 71-124.

Grimsgaard, S., Bonaa, K. H., Hansen, J. B., & Myhre, E. S. (1998). Effects of highly purified eicosapentaenoic acid and docosahexaenoic acid on hemodynamics in humans. *The American Journal of Clinical Nutrition,* 68, 52-59.

Hallahan, B., Hibbeln, J. R., Davis, J. M., & Garland, M. R. (2007). Omega-3 fatty acid supplementation in patients with recurrent self-harm. Single-centre double-blind randomised controlled trial, *The British Journal of Psychiatry*, 190, 118-122.

Hardy, R. W. (2010). Utilization of plant proteins in fish diets: effects of global demand and supplies of fishmeal. *Aquaculture Research*, 41, 770-776.

Harris, S. (2004). Fish oil supplementation: evidence for health benefits. *Cleveland Clinic Journal of Medicine*, 71(3), 208-210, 212, 215-218.

He, X., Yang, X., & Guo, Y. (2007). Effects of different dietary oil sources on immune function in cyclophosphamide immunosuppressed chickens. *Animal Feed Science and Technology*, 139(3-4), 186-200.

Healy, D. A., Wallace, F. A., Miles, E. A., Calder, P. C., & Newsholm, P. (2000). Effect of low-to-moderate amounts of dietary fish oil on neutrophil lipid composition and function. *Lipids*, 35, 763-768.

Hilbert, G., Lillemark, L., Balchen, S., & Højskov, C. S. (1998). Reduction of organochlorine contaminants from fish oil during refining. *Chemosphere*, 37, 1241-1252.

Hodge, W. G., Schachter, H. M., Barnes, D., Pan, Y., Lowcock, E. C., Zhang, L., Sampson, M., Morrison, A., Tran, K., Miguelez, M., & Lewin, G. (2006). Efficacy of omega-3 fatty acids in preventing age-related macular degeneration: a systematic review. *Ophthalmology*, 113(7), 1165-1172; 1172-1173, 1178.

Hoffman, D. R., Birch, E. E., Castañeda, Y. S., Fawcett, S. L., Wheaton, D. H., Birch, D. G., & Uauy, R. (2003). Visual function in breast-fed term infants weaned to formula with or without long-chain polyunsaturates at 4 to 6 months: a randomized clinical trial. *The Journal of pediatrics,* 142, 669-677.

Honig, L. (2000). Inflammation in neurodegenerative disease: Good, bad, or irrelevant? *Archives of Neurology*, 57(6), 786-788.

Hudert, C. A., Weylandt, K. H., Lu, Y., Wang, J., Hong, S., Dignass, A., Serhan, C. N., & Kang, J. X. (2006). Transgenic mice rich in endogenous omega-3 fatty acids are protected from colitis. *Proceedings of the National Academy of Sciences of the United States of America*, 103(30), 11276-11281.

Ishida, T., Yoshida, M., Arita, M., Nishitani, Y., Nishiumi, S., Masuda, A., Mizuno, S., Takagawa, T., Morita, Y., Kutsumi, H., Inokuchi, H., Serhan, C. N., Blumberg, R. S., & Azuma, T. (2010). Resolvin E1, an endogenous lipid mediator derived from eicosapentaenoic acid, prevents dextran sulfate sodium-induced colitis. *Inflammatory Bowel Diseases*, 16(1), 87-95.

FAO/WHO (2010). Fats and Fatty Acids in Human Nutrition. Report of an expert consultation. Geneva, Switzerland: Food and Agriculture Organization/World Health Organization. *FAO Food and Nutrition Paper,* 91, 1–166.

Kalmijn, S. (2000). Fatty acid intake and the risk of dementia and cognitive decline: a review of clinical and epidemiological studies. *The journal of nutrition health aging*, 4(4), 202-207.

Kamphuis, M. H., Geerlings, M. I., Tijhuis, M. A. R., Kalmijn, S., Grobbee, D. E., & Kromhout, D. (2006). Depression and cardiovascular mortality: a role for *n*-3 fatty acids?, *The American Journal of Clinical Nutrition*, 84(6), 1513-1517.

Kang, J. X. (2011). Omega-3: A link between global climate change and human health. *Biotechnology Advances*, 29, 388-390.

Kelley, D. S., Taylor, P. C., Nelson, G. J., Schmidt, P. C., Ferretti, A., Erickson, K. L., Yu, R., Chandra, R. K., & Mackey, B. E. (1999). Docosahexaenoic acid ingestion inhibits natural killer cell activity and production of inflammatory mediators in young healthy men. *Lipids,* 34, 317-324.

Kokkinos, P. P., Shaye, R., Alam, B. S., & Alam, S. Q. (1993). Dietary lipids, prostaglandin E2 levels, and tooth movement in alveolar bone of rats. *Calcified Tissue International*, 53 (5), 333-337.

Kolanowski, W. (2010). Omega-3 LC PUFA Contents and Oxidative Stability of Encapsulated Fish Oil Dietary Supplements. *International Journal of Food Properties*, 13 (3), 498-511.

Kris-Etherton, P. M., Harris, W. S., & Appel, L. J. (2002). Fish consumption, fish oil, omega--3 fatty acids, and cardiovascular disease. *Circulation*, 106(21), 2747-2757.

Kris-Etherton, P. M., Harris, W. S., & Appel, L. J., For The Aha Nutrition Committee. (2003). Omega-3 Fatty Acids and Cardiovascular disease. *Arteriosclerosis, Thrombosis, and Vascular Biology,* 23: 151-152.

Krokan, H. E., Bjerve, K. S., & Mørk, E. (1993). The enteral bioavailability of eicosapentaenoic acid and docosahexaenoic acid is as good from ethyl esters as from glyceryl esters in spite of lower hydrolytic rates by pancreatic lipase *in vitro*. *Biochimica et Biophysica Acta (BBA) - Lipids and Lipid Metabolism*, 1168(1), 59-67.

Kruger, M. C., & Schollum, L. M. (2005). Is docosahexaenoic acid more effective than eicosapentaenoic acid for increasing calcium bioavailability? *Prostaglandins leukotrienes and essential fatty acids*, 73(5), 327-334.

Lawson, L. D., & Hughes, B. G. (1988). Human absorption of fish oil fatty acids as triacylglycerols, free acids, or ethyl esters. *Biochemical and Biophysical Research Communications*, 152(1), 328-335.

Lauritzen, L., Jørgensen, M. H., Mikkelsen, T. B., Skovgaard, L. M., Straarup, E. M., Olsen, S. F., Høy, C. E., & Michaelsen, K. F. (2004). Maternal fish oil supplementation in lactation: effect on visual acuity and *n*-3 fatty acid content of infant erythrocytes. *Lipids,* 39, 195-206.

Lee, T. H., Hoover, R. L., Williams, J. D., Sperling, R. I., Ravalese, J., Spur, B. W., Robinson, D. R., Corey, E. J., Lewis, R. A., & Austen, K. F. (1985). Effect of dietary enrichment with eicosapentaenoic and docosahexaenoic acids on *in vitro* neutrophil and monocyte leukotriene generation and neutrophil function. *The New England Journal of Medicine*, 312(19), 1217-1224.

Leu, E., Falk-Petersonm S., & Hessen D. O. (2007). Ultraviolet radiation negatively affects growth but not food quality of Artic diatoms. *Limnology and Oceanography*, 53, 787-797.

Lichtenstein, A. H., Appel, L. J., & Brands, M. (2006). Diet and lifestyle recommendations revision 2006: a scientific statement from the American Heart Association Nutrition Committee. *Circulation*, 114, 82-96.

Lim, L. S., Mitchell, P., Seddon, J. M., Holz, F. G., & Wong, T. Y. (2012). Age-related macular degeneration. *Lancet*, 379(9827), 1728–1738.

Lucas, M., Asselin, G., Plourde, M., Cunnane, S. C., Dewailly, E., & Dodin, S. (2009). *n*-3 Fatty acid intake from marine food products among Quebecers: comparison to worldwide recommendations. *Public Health Nutrition*, 13(1), 63–70.

Luley, C., Wieland, H., & Gruwald, J. (1990). Bioavailability of omega-3 fatty acids: ethyl ester preparations are as suitable as triglyceride preparations. *Akt Ernahrungsmed*, 15, 122-125.

Matos, E., Gonçalves, A., Bandarra, N., Colen, R., Nunes, M. L., Valente, L. M. P., Dinis, M. T., & Dias, J. (2012). Plant proteins and vegetable oil do not have detrimental effects on post-morten muscle instrumental texture, sensory properties and nutritional value of gilthead seabream. *Aquaculture*, 358-359, 205-212.

McMichael, A. J., & Butler, C. D. (2005). Fish, Health and Sustainability. *American Journal of Preventive Medicine*, 29(4), 322-323.

Melanson, S. F, Lewandrowski, E. L, Flood, J. G., & Lewandrowski, K. B. (2005). Measurement of organochlorines in commercial over-the-counter fish oil preparations: implications for dietary and therapeutic recommendations for omega-3 fatty acids and a review of the literature. *Archives of Pathology & Laboratory Medicine*, 129, 74-77.

Merle, B., & Delyfer, M. (2013). High Concentrations of Plasma n3 Fatty Acids Are Associated with Decreased Risk for Late Age-Related Macular Degeneration. *The Journal of Nutrition*, 505-511.

Mogelson, S., Pieper, S. J., & Lange, L. G. (1984). Thermodynamic bases for fatty acid ethyl ester synthase catalyzed esterification of free fatty acid with ethanol and accumulation of fatty acid ethyl esters. *Biochemistry*, 23(18), 4082-4087.

Moyad, M. (2005). An introduction to dietary/supplemental omega-3 fatty acids for general health and prevention: part I. *Urologic oncology*, 23(1), 28-35.

Mozaffarian, D., Lemaitre, R. N., King, I. B., Song, X., Huang, H., Sacks, F. M., Rimm, E. B., Wang, M., & Siscovick, D. S. (2013). Plasma Phospholipid Long-Chain ω-3 Fatty Acids and Total and Cause-Specific Mortality in Older Adults. A Cohort Study. *Annals of internal Medicine*, 158(7), 515-525.

Munro, I. A. & Garg, M. L. (2013). Prior supplementation with long chain omega-3 polyunsaturated fatty acids promotes weight loss in obese adults: a double-blinded randomised controlled trial. *Food & function*, 25(4), 650-658.

Napier, J. A. & Sayanova, O. (2005). The production of very-long chain PUFA biosynthesis in transgenic plants: towards a sustainable source of fish oils. *Proceedings of the Nutrition Society*, 64, 387-393.

Nasopoulou, C. & Zabetakis (2012). Benefits of fish oil replacement by plant originated oils in compounded fish feeds: A review. *LWT - Food Science and Technology*, 47, 217-224.

Nemets, H., Nemets, B., Apter, A., Bracha, Z., & Belmaker, R. H. (2006). Omega-3 treatment of childhood depression: a controlled, double-blind pilot study. *The American Journal of Psychiatry*, 163(6), 1098-1100.

Neubronner, J., Schuchardt, J. P., Kressel, G., Merkel, M., von Schacky, C., & Hahn, A. (2011). Enhanced increase of omega-3 index in response to long-term *n*-3 fatty acid supplementation from triacylglycerides versus ethyl esters. *European journal of clinical nutrition*, 65(2), 247-54.

Newman, M. J. (1990). Inhibition of carcinoma and melanoma cell growth by type 1 transforming growth factor beta is dependent on the presence of polyunsaturated fatty acids. *Proceedings of the National Academy of Sciences of the United States of America*, 87(14), 5543-5547.

Opperman, M., Marais, D. W., & Spinnler Benade, A. J. (2011). Analysis of omega-3 fatty acid content of South African fish oil supplements. *Cardiovascular Journal of Africa*, 22(6), 324-329.

Orchard, T., Pan, X., Cheek, F., Ing, S. W., & Jackson, R. D. (2012). A systematic review of omega-3 fatty acids and osteoporosis. *British Journal of Nutrition*, 107(2), 253-260.

Ott, A., Breteler, M. M., van Harskamp, F., Claus, J. J., van der Cammen, T. J., Grobbee, D. E., & Hofman, A. (1995). Prevalence of Alzheimer's disease and vascular dementia: association with education. The Rotterdam study. *British Medical Journal*, 310, 970-973.

Ouchi, M., Ikeda, T., Nakamura, K., Harino, S., & Kinoshita, S. (2002). A novel relation of fatty acid with age-related macular degeneration. *Ophthalmologica,* 216(5), 363-367.

Patterson, R. E., Flatt, S. W., Newman, V. A., Natarajan, L., Rock, C. L., Thomson, C. A., Caan, B. J., Parker, B. A., & Pierce, J. P. (2011). Marine fatty acid intake is associated with breast cancer prognosis. *The Journal of nutrition*, 141(2), 201-206.

Peet, M., & Stokes, C. (2005). Omega-3 fatty acids in the treatment of psychiatric disorders. *Drugs*, 65(8), 1051-1059.

Pratoomyot, J., Bendiksen, E. Ä., Bell, J. G., & Tocher, D. R. (2010). Effects of increasing replacement of dietary fishmeal with plant protein sources on growth performance and body lipid composition of Atlantic salmon (*Salmo salar* L.). *Aquaculture*, 305, 124-132.

Rees, D., Miles, E.A., Banerjee, T., Wells, S. J., Roynette, C. E., Wahle, K. W. J., & Calder, P. C. (2006). Dose-related effects of eicosapentaenoic acid on innate immune function in healthy humans: a comparison of young and older men. *The American Journal of Clinical Nutrition*, 83(2), 331-342.

Richardson, A. J. (2006). Omega-3 fatty acids in ADHD and related neurodevelopmental disorders. *International review of psychiatry Abingdon England*, 18(2), 155-172.

Riediger, N. D., Othman, R., Suh, M., & Moghadasian, M. H. (2009). A systemic review of the roles of *n*-3 fatty acids in health and disease. *Journal of the American Dietetic Association*, 109(4), 668-679.

Rizzo, A. M., Corsetto, P. A., Montorfano, G., Opizzi, A., Faliva, M., Giacosa, A., Ricevuti, G., Pelucchi, C., Berra, B., & Rondanelli, M. (2012). Comparison between the AA/EPA ratio in depressed and non-depressed elderly females: omega-3 fatty acid supplementation correlates with improved symptoms but does not change immunological parameters. *Nutrition Journal*, 11:82.

Sacks, F. M., Stone, P. H., Gibson, C. M., Silverman, D. I., Rosner, B., & Pasternak, R. C. (1995). Controlled trial of fish oil for regression of human coronary atherosclerosis. HARP Research Group. *Journal of the American College of Cardiology*, 25,1492-*1498*.

Sanders, T. A, Haines, A. P., Wormald, R., Wright, L. A. & Obeid, O. (1993). Essential fatty acids, plasma cholesterol, and fat-soluble vitamins in subjects with age-related maculopathy and matched control subjects. *The American journal of clinical nutrition*, 57(3), 428-433.

SanGiovanni, J. P., & Chew, E. Y. (2005). The role of omega-3 long-chain polyunsaturated fatty acids in health and disease of the retina. *Progress in Retinal and Eye Research*, 24(1), 87-138.

Schaefer, E. J., Bongard, V., Beiser, A. S., Lamon-Fava, S., Robins, S. J., Au, R., Tucker, K. L., Kyle, D. J., Wilson, P. W., & Wolf, P. A. (2006). Plasma phosphatidylcholine docosahexaenoic acid content and risk of dementia and Alzheimer disease: the Framingham Heart Study. *Archives of Neurology*, 63, 1545-1550.

Schmitz, G., & Ecker, J. (2008). The opposing effects of *n*-3 and *n*-6 fatty acids. *Progress in Lipid Research*, 47(2), 147-155.

Sechindell, D. T. Rind, D., & Lonergan, P. (1998). Increased polar stratospheric ozone losses and delayed eventual recovery owing to increasing greenhouse-gas concentrations. *Nature*, 392, 589-592.

Segura, R. (1988). Preparation of fatty acid methyl esters by direct transesterification of lipids with aluminium chloride-methanol. *Journal of Chromatography*, 441(1), 99-113.

Serhan, C. N, Chiang, N., & van Dyke, T. E. (2008). Resolving inflammation: dual anti-inflammatory and pro-resolution lipid mediators. *Nature Reviews Immunology*, 8(5), 349-361.

Serhan, C. N., Clish, C. B., Brannon, J., Colgan, S. P., Gronert, K., & Chiang, N. (2000a). Anti-inflammatory lipid signals generated from dietary *n*-3 fatty acids via cyclooxygenase-2 and transcellular processing: a novel mechanism for NSAID and *n*-3 PUFA therapeutic actions. *Journal of physiology and pharmacology an official journal of the Polish Physiological Society*, 51, 643-654.

Serhan, C. N, Clish, C. B., Brannon, J., Colgan, S. P., Chiang, N., & Gronert, K. (2000b). Novel functional sets of lipid-derived mediators with antiinflammatory actions generated from omega-3 fatty acids via cyclooxygenase 2-nonsteroidal antiinflammatory drugs and transcellular processing. *The Journal of Experimental Medicine*, 192(8), 1197-1204.

Shahidi, F., & Wanasundara, U. N. (1998). Omega-3 fatty acid concentrates: nutritional aspects and production technologies. *Trends in Food Science & Technology*, 9(6), 230-240.

Simopoulos, A. P. (2002). Omega-3 fatty acids in inflammation and autoimmune diseases. *Journal of the American College of Nutrition*, 21(6), 495-505.

Simopoulos, A. P. (2003). Essential fatty acids in health and chronic diseases. *Forum of nutrition*, 56(3), 67-70.

Song, J. H., Inoue, Y., & Miyazawa, T. (1997). Oxidative stability of docosahexaenoic acid-containing oils in the form of phospholipids, triacylglycerols, and ethyl esters. *Bioscience, biotechnology and biochemistry*, 61(12), 2085-2088.

Sperling, R. I., Benincaso, A. I., Knoell, C. T., Larkin, J. K., Austen, K. F., & Robinson, D. R. (1993). Dietary omega-3 polyunsaturated fatty acids inhibit phosphoinositide formation and chemotaxis in neutrophils. *Journal of Clinical Investigation*, 91(2), 651-660.

Stevens, L., Zhang, W., Peck, L., Kuczek, T., Grevstad, N., Mahon, A., Zentall, S. S., Arnold, L. E., & Burgess, J. R. (2003). EFA supplementation in children with inattention, hyperactivity, and other disruptive behaviors. *Lipids,* 38, 1007-1021.

Su, K.-P., Huang, S.-Y., Chiu, C.-C., & Shen, W. W. (2003). Omega-3 fatty acids in major depressive disorder. A preliminary double-blind, placebo-controlled trial. *European neuropsychopharmacology the journal of the European College of Neuropsychopharmacology*, 13, 267-271.

Sun, D., Krishnan, A., Zaman, K., Lawrence, R., Bhattacharya, A., & Fernandes, G. (2003). Dietary *n*-3 fatty acids decrease osteoclastogenesis and loss of bone mass in ovariectomized mice. *Journal of bone and mineral research*, 18(7), 1206-1216.

Sun, L., Tamaki, H., Ishimaru, T., Teruya, T., Ohta, Y., Katsuyama, N., & Chinen, I. (2004). Inhibition of osteoporosis due to restricted food intake by the fish oils DHA and EPA and perilla oil in the rat. *Bioscience biotechnology and biochemistry*, 68(12), 2613-2615.

Tatarczyk, T., Engl, J., Ciardi, C., Laimer, M., Kaser, S., Salzmann, K., Lenners, R., Patsch, J. R., & Ebenbichler, C. F. (2007). Analysis of long-chain ω-3 fatty acid content in fish-oil supplements.*Wien Klin Wochenschr*, 119(13-14), 417-422.

Timonen, M., Horrobin, D., Jokelainen, J., Laitinen, J., Herva, A., & Räsänen, P. (2004). Fish consumption and depression: the Northern Finland 1966 birth cohort study. *Journal of Affective Disorders*, 82(3), 447-452.

Tocher, D. R. (2009). Issues surrounding fish as a source of omega-3 long-chain polyunsaturated fatty acids. *Lipid Technology*, 21(1), 13-16.

Trebble, T. M., Wootton, S. A., Miles, E. A., Mullee, M., Arden, N. K., Ballinger, A. B., Stroud, M. A., Burdge, G. C., & Calder, P. C. (2003). Prostaglandin E2 production and T cell function after fish-oil supplementation: response to antioxidant cosupplementation. *The American Journal of Clinical Nutrition*, 78(3), 376-382.

Uauy, R., & Dangour, A. D. (2006). Nutrition in Brain Development and Aging: Role of Essential Fatty Acids. *Nutrition*, 64, 24-33.

Valenzuela, A., Valenzuela, V., Sanhueza, J., & Nieto, S. (2005). Effect of supplementation with docosahexaenoic acid ethyl ester and sn-2 docosahexaenyl monoacylglyceride on plasma and erythrocyte fatty acids in rats. *Annals of nutrition metabolism*, 49(1), 49-53.

Villasante, S., Morato, T., Rodriguez-Gonzalez, D., Altelo, M., Osterblom, H., Watling, L., Nouvian, C., Gianni, M., & Macho, G. (2012). Sustainability of deep-sea fish species under the European Union Common Fisheries Policy. *Ocean & Coastal Management*, 70, 31-37.

von Schacky, C., Kiefl, R., Jendraschak, E., & Kaminski, W. E. (1993). *n*-3 fatty acids and cysteinyl-leukotriene formation in humans *in vitro*, *ex vivo*, and *in vivo*. *The Journal of laboratory and clinical medicine*, 121, 302-309.

Wang, C., Harris, W. S., Chung, M., Lichtenstein, A. H., Balk, E. M., Kupelnick, B., Jordan, H. S., & Lau, J. (2006). *n*-3 Fatty acids from fish or fish-oil supplements, but not alpha-linolenic acid, benefit cardiovascular disease outcomes in primary- and secondary-prevention studies: a systematic review. *The American Journal of Clinical Nutrition*, 84(1), 5-17.

Yoshii, H., Furuta, T., Siga, H., Moriyama, S., Baba, T., Maruyama, K., Misawa, Y., Hata, N., & Linko, P (2002). Autoxidation kinetic analysis of docosahexaenoic acid ethyl ester and docosahexaenoic triglyceride with oxygen sensor. *Bioscience biotechnology and biochemistry*, 66(4), 749-753.

Xia, S., Lu, Y., Wang, J., He, C., Hong, S., Serhan, C. N., & Kang, J. X. (2006). Melanoma growth is reduced in fat-1 transgenic mice: impact of omega-6/omega-3 essential fatty acids. *Proceedings of the National Academy of Sciences of the United States of America*, 103(33), 12499-504.

Yao, J. K., van Kammen, D. P., & Welker, J. A. (1994). Red blood cell membrane dynamics in schizophrenia. II. Fatty acid composition. *Schizophrenia Research*, 13(3), 217-226.

In: Omega-3 Fatty Acids
Editor: Md. Wasim Khan

ISBN: 978-1-62948-516-4

Chapter 11

Quality and Shelf Stability of Omega-3-Fat-Supplemented Foods

Ann H. Barrett**, *Danielle J. Anderson, Tshinanne V. Ndou, Katherine Kensil and Karen R. Conca

Performance Optimization Research Team, Combat Feeding Directorate,
US Army Natick Soldier Research, Development and Engineering Center,
Natick, MA, US

Abstract

Omega-3 fats are nutritionally important but lacking in the American diet and, particularly, in the diet consumed by military personnel. These fats provide important physiological benefits including reduction of inflammation and improved psychological health, both of significant importance to deployed Warfighters. However, since omega-3 fatty acids are highly prone to oxidation, incorporating them into foods with a necessarily prolonged shelf-life can be problematic; fortification of military ration components, generally required to have a 2-to-3 year shelf-life, is a particular challenge. Thus, our objectives were to evaluate the sensory quality and loss of omega-3 fatty acids throughout accelerated storage of various supplemented shelf-stable foods representing different chemical and packaging environments. Representative low-moisture products (performance bars, packaged w/o oxygen scavengers), intermediate-moisture products (cakes, packaged with oxygen scavengers), and high-moisture (retorted) products were produced each with two levels of encapsulated omega-3 fat, up to 400 mg/serving, and stored for 4 weeks at 49°C and 3 and 6 months at 37°C. Sensory quality, physical properties, retained omega-3 DHA and EPA, and fluorescent volatile (i.e., Schiff base) production were measured. Sensory quality of the cakes and retorted beef lo mein did not diminish throughout storage regardless of omega-3 content. These items also showed negligible loss of omega-3 fat. Performance bars, however, showed a 30% decrease in overall quality over 6 months and a 17% loss in omega-3 fat due to oxidation.

* Corresponding author: ann.h.barrett.civ@mail.mil.

Inconclusion, extended shelf-life foods supplemented with high, physiologically significant levels of omega-3 fats can be successfully developed, with the provision that oxygen is limited in the carrier products.

Introduction

Omega-3 fatty acids, particularly the long chain fatty acids docosahexaenoic acid (DHA, n-22) and eicosapentaenoic acid (EPA, n-20), are essential nutrients in that these lipids have vital physiological functions; however, their distinguishing double bond at the omega-3 carbon position is not producible *in vivo* and thus must be consumed in the diet. While EPA (and ultimately DHA) can be derived in the body by elongation of the shorter-chain omega-3 lipid α-linolenic acid (ALA, n-18), the rate of conversion — involving the addition of at least one double bonded carbon pair in the interior of the molecule — tends to be limited: conversion is estimated to be 15% or less (Enkin et al., 1994; Pawlosky, et al, 2001) and thus potentially insufficient to maintain healthy physiological levels of EPA + DHA.

The human diet historically provided higher levels of omega-3 fats, particularly when larger amounts of both fish and free-range (i.e., grass-fed) animals were consumed (Eaton and Konner, 1985). However, current American intake of the longer chain omega-3 PUFAs (polyunsaturated fatty acids), estimated at 0.1 to 0.2 g/day (Kris-Etherton et al., 2000), is significantly less than that recommended by several dietary associations. Both the American Heart Association (Kris-Etherton et al., 2002) and the US Department of Agriculture (April 2011) recommend sufficient servings of fatty fish (two meals, or 8 ounces) per week, which provide roughly 250 mg/day of combined EPA and DHA; for individuals with heart disease the recommended intake is 1g/day. International organizations such as the World Health Organization, the North Atlantic Treaty Organization and a number of foreign countries (Canada, Sweden, United Kingdom, Australia, and Japan) recommend daily intake levels ranging from 0.3 to 0.5 g DHA + EPA per day.

The problem of low omega-3 consumption is compounded by the high concentration of omega-6 fats in the American diet, largely due to the abundance of processed foods formulated with omega-6 vegetable oils (Kris-Etherton et al., 2000). While the benefits of omega-3 fatty acids are reportedly due to their anti-inflammatory properties (Song, 2007; Razzak et al., 2008; Chen, 2010; Im, 2012), the ratio of omega-3 to omega-6 fatty acids in the diet is considered as important to health. This ratio influences the proportion of prostaglandins derived respectively from omega-3 and omega-6 substrates, the omega-6-based eicoisanoids being pro-inflammatory and the omega-3-based eicoisanoids serving to blunt this inflammatory effect (Simopoulos, 2008). Estimates of the ratio of omega-6 to omega-3 lipids currently in the American diet, largely due to the high intake of processed food, are as high as 10:1 (Kris-Etherton et al., 2000) or 20:1 (Mahan et al., 2004); this proportion far exceeds the ratio of roughly 1:1 estimated for humans prior to industrialization (Simopoulos, 1999).

Omega-3 fats provide a number of physiological benefits. The most widely known is their contribution to cardiovascular health (Kris-Etherton, 2002; Wang et al., 2009), which is in part attributable to their potent anti-inflammatory properties; this effect is largely due to their function as precursors to species that offset the production of pro-inflammatory cytokines produced from omega-6 fatty acids (Song, 2007; Im, 2012; Mais and Smith, 1998).

In addition to ameliorating heart and inflammation-related disorders, omega-3 fats have also been reported to potentially positively affect mood (Mais and Smith, 1998; Tanskanen, et al., 2001; Arstog et al., 2008; Lucas et al., 2011; Sublette, et al., 2011). Omega-3 fatty acids constitute a significant structural part of the brain, thus affecting the fluidity of cell membranes and consequently the transport of neurotransmitters (Fernstrom, 1999; Song, 2007; Lucas et al., 2011). High doses of omega-3 fats are often advised as an adjuvant treatment for depression, and have also been studied as a treatment for traumatic brain injury (Mills et al., 2011).

The physiological activity of omega-3 fats may, in particular, be useful through multiple mechanisms for deployed Warfighters, who are prone to significant muscle strain from heavy loading, running, and other mission-mandated exertion. Inflammation resulting from strain or injury causes discomfort and thus potential loss in function due to pain. Warfighters can also incur battlefield wounds, including traumatic brain injury (TBI), and are subject to associated psychological trauma. A markedly elevated incidence of suicide among deployed US troops in the last several years has become an issue of great concern (Lewis et al, 2011), precipitating questions about possible dietary factors. Specifically, intake of long chain omega-3 fats was shown by Lewis et al. (2011) to be negatively correlated with rates of suicide in the military. Moreover, consumption of omega-3 fats by military personnel is even lower than that by the general population: a study by Johnston et al. (2013) revealed that the average blood level of omega-3 lipids in the military population is 29% lower than that of civilians.

While there is a consensus that increasing the level of omega-3 lipids in military rations would be beneficial, a significant constraint on such supplementation has been the relative instability of these fats. The high degree of unsaturation of omega-3 fatty acids, particularly in the longer chain DHA and EPA, renders them prone to oxidation and has heretofore precluded their incorporation into ration components, most of which have a mandated 3-year at 27°C shelf life. Fish oil, a rich source of DHA and EPA, when unprotected by antioxidants is particularly susceptible to oxidation; even antioxidant-supplemented fish oil was shown by fluorescence testing to undergo degradation at a rate precluding long-term acceptability (Barrett et al., 2011). Commercially available omega-3 fat-containing products are generally not formulated to withstand such extreme stability requirements. Moreover, a market survey of such fortified items revealed few products having contents in excess of the 30 mg omega-3 fat required for label declaration, a level far below that required to significantly increase Warfighter blood and erythrocyte levels.

Recently, however, advances in encapsulation technology/spray drying have led to the commercial availability of more chemically stable omega-3 fat-containing powders that can be directly added to food formulations. Such powders have led to the development of several food products by the Combat Feeding Directorate (CFD) at the US Army Natick Soldier Research, Development and Engineering Center (NSRDEC, Kansas St., Natick MA) with loading levels up to 400 mg/serving that have retained sensory quality throughout accelerated storage. Standard storage conditions used to predict quality and acceptance after the required 3 years at 27°C are 6 months storage at 38°C, with an additional shorter-term test of 4 weeks storage at 49°C. (While reasonable correlation between sensory results for 3 years at 27°C and 6 months at 38°C has been determined for selected foods, the higher temperature conditions have been found to occasionally over-predict degradation (LeBlanc et al., 2013)).

These studies have included concerted experiments on the stability and omega-3 fat retention in three products that varied in processing history and packaging strategy, representing: an intermediate moisture product packaged with oxygen scavengers (a cake product); a retorted high a_w entrée product (beef lo mein); and a low a_w food packaged without oxygen scavengers (an energy-type bar). These foods are either actual current ration components in use or items under development as ration components. Each product was fortified at two levels (within or close to the recommended dietary ranges cited previously) and was analyzed for omega-3 fat retention and fluorescent volatile production, as well as for sensory quality throughout storage. Results, while varying with specific product and packaging/processing history, indicated the feasibility of including selected high-load omega-3 fortified items into the military ration platform.

Materials and Methods

Product Formulations and Processing Methods for Food Matrices

Gelatin-encapsulated omega-3 lipid, reported to contain 15-20% EPA + DHA, was obtained from Ocean Nutrition Canada Limited (Dartmouth, Nova Scotia, Cannada) as their Meg-3® product. This form of encapsulated omega-3 fat was selected after pre-screening several sources for sensory quality after short-term accelerated storage. Three ration components, orange cake (OC), beef lo mein (BLM), and chocolate First Strike™ bars (FSBs), were produced with three levels each of combined DHA + EPA, assuming the lower level of 15% concentration in the powder. OC contained target levels of 0, 200, and 400 mg per 100 g serving; BLM contained target levels of 0, 200, and 400 mg per 224 g (8 oz.) serving; and FSB contained target levels of 0, 100, and 200 mg per 50 g serving. Respective serving sizes are those actually used by the military for these products. Formulations are shown in Tables 1a-c.

For OC, dry and wet ingredients in 5.75 kg batches were separately mixed and then added together; 100 g portions were baked at 163°C in pans containing 6.5 x 11.5 cm rectangular slots for 18 minutes. Cakes were sealed with 200 cc oxygen scavengers (Multisorb Technologies, Buffalo NY) in 12 x 20 cm trilaminate foil pouches. For BLM, ingredients were combined in 11 kg batches and kettle cooked to a temperature of 88°C; eight ounce portions were packaged in 12 x 20 cm trilaminate foil pouches and retorted to commercial sterility. For FSB, corn syrup and glycerin were heated to 88°C and cooled to 60°C, prior to addition of and mixing in of remaining ingredients to produce 3 kg batches. Fifty g portions were shaped by hand into approximately 4x8x1 cm bars that were individually sealed in 11 x 14 cm trilaminate foil pouch. FSB specimens were packaged without oxygen scavengers, as is standard for this product.

The products were stored under the following conditions and withdrawn at the following times for evaluation: 49°C, pulled at four weeks; and 38°C, pulled at 3 and 6 months.

Table 1a. Orange Cake formulations

Ingredients	Control: (No omega-3 fat) %	Target 200 mg omega-3 fat per serving, %	Target 400 mg omega-3 fat per serving, %
Flour, cake	24.0	24.0	24.0
Sugar, granulated	25.68	24.43	23.18
Starch	1.0	1.0	1.0
Maltodextrin M180	2.5	2.5	2.5
Salt	0.34	0.34	0.34
Baking Powder	0.15	0.15	0.15
Xanthan gum	0.03	0.03	0.03
Encapsulated potassium sorbate	0.07	0.07	0.07
Shortening	16.3	16.3	16.3
Eggs, whole, frozen	17.3	17.3	17.3
Water	6.9	6.9	6.9
Glycerol	3.5	3.5	3.5
Vanilla flavoring 2X	0.11	0.11	0.11
Orange Extract	0.1	0.1	0.1
Orange oil	0.09	0.09	0.09
Citric acid	0.07	0.07	0.07
Meg-3®	0.00	1.25	2.5

Table 1b. Beef Lo Mein formulations

Ingredients	Control: (No omega-3 fat) %	Target 200 mg omega-3 fat per serving, %	Target 400 mg omega-3 fat per serving, %
Beef Broth	9.95	9.35	8.75
Soy Sauce	4.52	4.52	4.52
CHOL-FLO starch	1.16	1.16	1.16
Olive Oil	1.76	1.76	1.76
Water	13.02	13.02	13.02
Green Pepper, Frozen, Diced	13.56	13.56	13.56
Red Pepper, Frozen, Diced	13.56	13.56	13.56
Onion, Frozen, Diced	5.74	5.74	5.74
Green Onion Aquaresin	0.01	0.01	0.01
Garlic Oleoresin	0.01	0.01	0.01
Ginger Oleoresin	0.01	0.01	0.01
Whole Wheat Linguini, dry wt	8.15	8.15	8.15
Beef Strips, Frozen	20.26	20.26	20.26
Mushroom, Canned, Drained	8.29	8.29	8.29
Meg-3®	0	0.6	1.2

Table 1c. First Strike™ Bar formulations

Ingredients	Control: (No omega-3 fat) %	Target 100 mg omega-3 fat per serving, %	Target 200 mg omega-3 fat per serving, %
Date Plum Paste,40% Fructose	14	14	14
Corn Syrup - 42 DE	19.2	19.2	19.2
Glycerin USP or Food Grade	1.9	1.9	1.9
Vanilla Extract, 2X	0.415	0.415	0.415
Fractional Part Hydrogenated, Cotton Seed &/or Soybean	8.3	8.3	8.3
High Roast African Chocolate Liquor	7.2	7.2	7.2
Cocoa Powder, 10-12% Fat	3	3	3
Mixed Tocopherols GT-1	0.003	0.003	0.003
Ascorbyl Palmitate	0.015	0.015	0.015
Lecithin Emulpur-N	0.5	0.5	0.5
Maltodextrin DE 15	9	9	9
Dextrose Monohydrate Powder	5	4.4	3.8
Whey Protein Concentrate	2.3	2.3	2.3
Crystalline Fructose	9	9	9
Crisp Rice, Nutty #13811	10	10	10
Crisp Rice Tiny Round 200	10	10	10
Ascorbic Acid	0.05	0.05	0.05
Vit E acetate	0.0248	0.0248	0.0248
Meg-3®	0	1.25	2.5

Product Evaluation

Analytical Testing

Product water activity was measured using an Aqualab Series 3TE (Decagon Devices, Inc., Pullman WA) a_w meter after grinding the product in a food processor. Cake color was measured on a Hunter LAB Miniscan Spectrocolorimeter, Model MS/s-4500L with Color Analyst Software (Hunter Associates Laboratory, Inc., Reston VA). Analyses were replicated 3 times and results averaged.

Sensory Testing

Technical panels were conducted using trained panelists in order to document detectable differences among products due to level of omega-3 fat. The tests were administered to civilian volunteers trained in sensory analysis from the U.S. Army NSRDEC. Each panel consisted of 12 participants. Technical scores for appearance, odor, flavor, texture, and overall quality were obtained using a nine-point hedonic scale (with 0 corresponding to "dislike extremely" and 9 corresponding to "like extremely"). Varying omega-3 content specimens of the same product were presented together for each pull. Panelists evaluated the products in individualized booths in a dedicated sensory laboratory. Data were collected electronically using SIMS 2000 Sensory Data Collection Software, which automatically calculated statistics for each specimen attribute and determined significance of differences due to treatment.

Omega-3 Fat Quantification

Sample preparation and omega-3 fat extraction and quantification were conducted according to methods described by Curtis et al. (2008). Samples were freeze dried before extraction to prevent water from interfering with the methylation of fatty acids. Nitrogen flushed glove bags were used throughout preparation and extraction of samples to prevent oxidation of omega-3 fatty acids. A methanolic acid hydrolysis in toluene was used to release omega-3 fatty acids from the gelatin micro-encapsulant, methylate and derivatize released fatty acids into fatty acid methyl esters (FAMEs), and extract FAMEs for analysis. FAME extracts were further purified to remove trace water and other non-fatty acid components extracted from food matrices that would have interfered with the analyses. FAME extracts were analyzed using gas chromatography (GC).

Preparation of Samples

Omega-3-fortified ration components were ground using a food processor in a nitrogen-flushed polyethylene glove bag then transferred to freeze drying flasks, capped, weighed and removed from the bag. Samples were frozen and then freeze-dried until constant weight (w/in 0.1g) was achieved (~2-4 days). Percent moisture loss was calculated. Dried specimens were quickly transferred to a nitrogen-flushed polyethylene glove bag, finely ground using a mortar and pestle, and transferred into tri-laminate military food pouches with 100 cc oxygen scavengers. Pouches were vacuum sealed and stored at -20°C until needed. Prior to extraction and analysis, specimen pouches were equilibrated to room temperature, opened in a nitrogen flushed polyethylene glove bag, and between 0.7-1.0 gram of each sample (weight recorded) transferred into a tared 50 mL glass tube with Teflon-lined cap.

Extraction / Chemicals Used

Methyl tricosanoate (Sigma, St. Louis MO., #T9900), used as the internal standard (IS), was dissolved at appropriate concentrations in toluene (Sigma, St. Louis MO., #650579), with solutions freshly prepared prior to extractions. The following extraction procedures were conducted in a nitrogen-flushed custom-made Teflon glove bag (Ingeniven, Hampton, NH): the hydrolysis and methylation solution (10% acetyl chloride (Fluka, St. Louis MO., #00-990) in methanol (Sigma, St. Louis MO., #322415)) was prepared by slowly adding methanol to acetyl chloride in a 100 mL volumetric flask, set in dry ice (to minimize heat and off-gassing); six mL of the acetyl chloride-methanol mixture, 6 mL of toluene, and 1 mL of pre-prepared IS were added to each tube (IS was excluded from blank). Tubes were then capped and removed from the glove bag. Samples were vortexed for 30 seconds, heated at 80 +/- 2°C in a shaking water bath for 2 hours, cooled to room temperature, and 10 mL of a 6% (w/v) sodium carbonate (Fisher Scientific, Pittsburgh, PA, #S293) solution was added to each tube. Samples were again vortexed for 30 seconds, then centrifuged at 5000 rpm for 15 minutes at 4°C. The toluene layer containing FAMEs was decanted and mixed with approximately 1 g of sodium sulfate (Sigma-Aldrich, St. Louis MO., #239313) to remove trace amounts of water. Toluene FAME extracts were further purified using Silica solid phase extraction (SPE) tubes (conditioned with 3 mL toluene, loaded with 1 mL sample, and rinsed with 3 mL toluene). Final extracts (4 mL) were filtered (0.2 um PVDF) into GC vials for analysis.

Analysis of DHA and EPA Methyl Esters / Chemicals Used

Toluene extracts were analyzed by flame ionization detection. EPA and DHA methyl esters were separated and detected according to the GC method described by Na (2011). The specific GC instrumentation and operating parameters are listed in Table 2. DHA and EPA methyl ester standards (in iso-octane (Sigma, St. Louis MO., #59045) w/butylated hydroxytoluene (Sigma, St. Louis MO., #B1378)), at appropriate concentrations, were analyzed at the beginning and end of each set of samples tested.

Each sample set included a blank, IS, and MEG-3® powder controls. Analyses were conducted in triplicate.

Table 2. GC Instrumentation and operating parameters for DHA and EPA analysis

Gas Chromatograph	Agilent 7890A Series
Autosampler	Agilent 7683B Series with sample tray
Column	HP-88, 100m x 0.25mm x 0.2um
Inlet	260°C, 30:1 split ratio
Inlet liner	Split liner, tapered deactivated (p/n 5183-4701)
Injection volume	1 u L
Carrier gas	Nitrogen, constant flow mode, 1mL/min
Oven temperature program	140°C (5min), 4°C/min to 240°C (15min)
Detector	Flame Ionization Detector (FID) @ 280°C

Calculations to Determine EPA and DHA (Mg/Serving) / Chemicals Used

Peak areas (A) and concentrations (C, mg/g) for DHA ($A_{DHA\text{-}R}$ and $C_{DHA\text{-}R}$), EPA ($A_{EPA\text{-}R}$ and $C_{EPA\text{-}R}$), and IS ($A_{IS\text{-}R}$ and $C_{IS\text{-}R}$) methyl ester reference standards (Nu-chek prep, #U-84M-DHA standard; Nu-chek prep, #U-99M-EPA standard) were used to calculate the response factors (RF) for DHA and EPA, according to:

$RF_{DHA} = (A_{IS\text{-}R} \text{ x } C_{DHA\text{-}R})/ (A_{DHA\text{-}R} \text{ x } C_{IS\text{-}R})$; and

$RF_{EPA} = (A_{IS\text{-}R} \text{ x } C_{EPA\text{-}R})/ (A_{EPA\text{-}R} \text{ x } C_{IS\text{-}R})$.

Mass (mg) of DHA and EPA in final solutions (X_{DHA} and X_{EPA}) were calculated using the following formulas, where M is the mass of reference IS (g):

$X_{DHA} = (M_{IS\text{-}R} \text{ x } C_{IS\text{-}R} \text{ x } A_{DHA} \text{ x } RF_{DHA})/A_{IS\text{-}R}$; and

$X_{EPA} = (M_{IS\text{-}R} \text{ x } C_{IS\text{-}R} \text{ x } A_{EPA} \text{ x } RF_{EPA})/A_{IS\text{-}R}$.

Final recovery concentrations of EPA and DHA (mg/serving) were calculated using X_{DHA} and X_{EPA}, the SPE dilution factor, loss of moisture during freeze drying, sample weight for extraction (g), and initial weight of ration component (g).

Fluorescence

Products were analyzed by front-face fluorimetry using a SPEX Industries (Metuchen NJ) fluorescence spectrophotometer according to the procedures of Barrett et al. (2011). Size-appropriate pieces (~0.3 g) of FSB and OC specimens were compressed into the well of the sample holder using a spatula; drops of the sauce of BLM were extracted using a spatula and placed in the sample holder to fill the well. Samples were scanned from 390-520 nm with (xenon lamp) excitation at 380 nm. A ratiometric technique, in which the intensity of the emerging oxidation-related peak (occurring at ~450 nm) was divided by the intensity at the curve's minimum (occurring at ~400 nm), thereby providing an internal standard, was used to evaluate and compare the scans (Porter et al., 1989; Genot et al., 2007; Barrett et al, 2011). The production of fluorescent oxidative compounds in food has been associated with the reaction of volatile unsaturated aldehydes with free amino groups in the food to yield fluorescent Schiff bases (Frankel 1998). Samples were analyzed in triplicate.

Additionally, the Meg-3® powder by itself (sealed in trilaminate pouches) was evaluated after the same storage conditions (3 and 6 months at 38°C) and also after 1 week at 75°C. The high-temperature, short time test was to verify that validity of the emission range used for analysis.

Statistical Analysis of Results

All analyses of effects (analysis of variance and two-sample comparisons) were conducted using Minitab Statistical Software, Release 16.

Results and Discussion

Analytical Testing

Average water activities, across all pulls and omega-3 levels, were 0.757±0.005 and 0.472±0.019 respectively for OC and FSB and did not vary significantly with time or omega-3 content. Water activity of BLM was not measured due to its high moisture content. Average pH across all pulls and all omega-3 levels was 5.43±0.12, 5.30±0.13, and 5.82±0.3 respectively for OC, BLM, and FSB. While pH decreased slightly over time in all products due to equilibration, no difference was seen due to omega-3 level. Color change during storage was only observed in the orange cake. Average L (i.e., lightness) values decreased approximately 12% over time ($p < 0.01$) for all specimens, indicating typical Maillard browning, or non-enzymatic darkening, for items stored at high temperatures for long periods. Omega-3 fat level, however, had no effect on cake color.

Sensory Testing

Results for overall acceptance are shown in Figures 1a-c and results for the other sensory scores are tabulated in Table 3a-c. Even the highest levels of fortification, (nominally 400 mg in the OC and BLM), did not significantly diminish the scores for any sensory attribute. All scores remained above 6.0 even after 6 months storage, indicating retention of quality and high acceptance, with no discernible effect of omega-3 fat level on the sensory quality of these products.

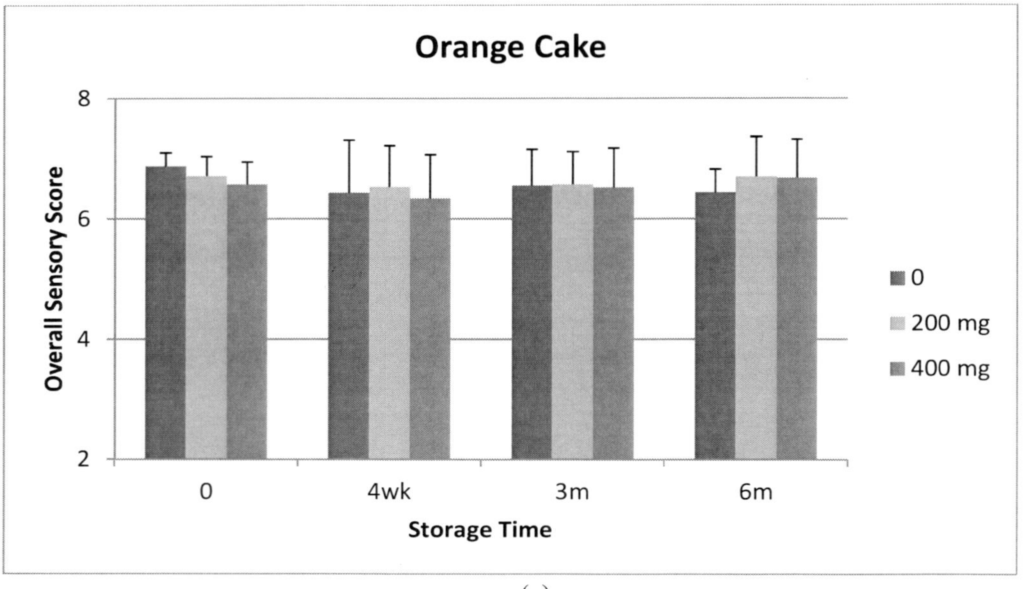

(a)

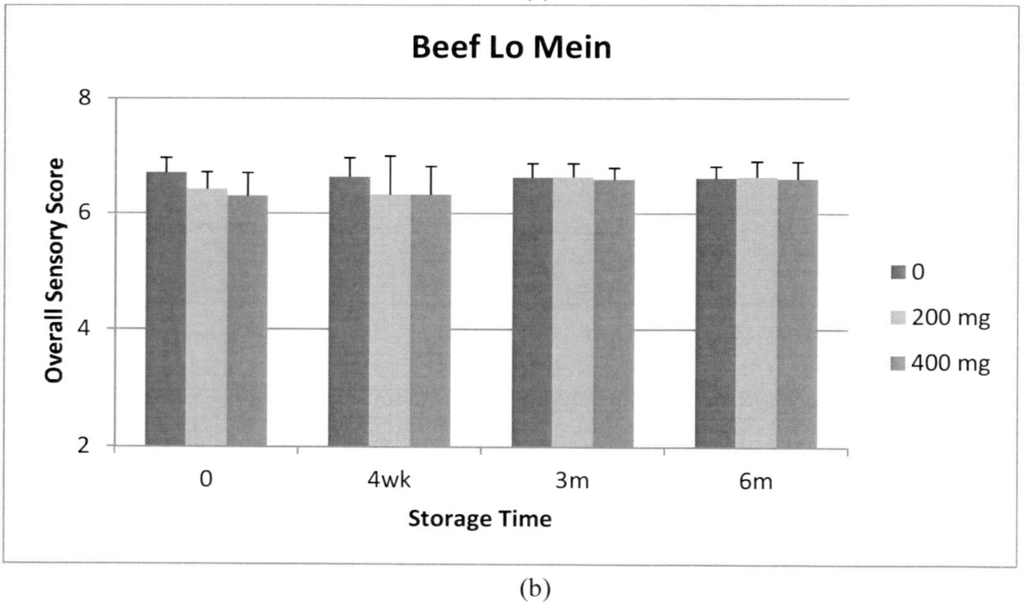

(b)

Figure 1. (Continued).

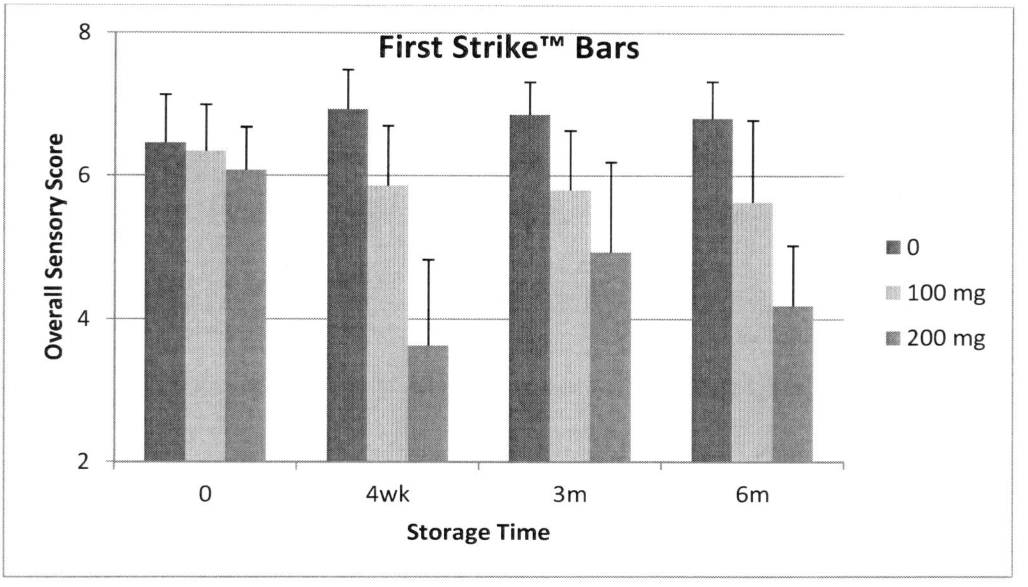

(c)

Figure 1. Overall quality scores prior to storage and after 4 weeks at 49°C and 3 and 6 months at 38°C: (a) Orange Cake; (b) Beef Lo Mein; and (c) Chocolate First Strike™ Bars. Shows reduced quality through storage due to omega-3 fat fortification for the First Strike™ Bar only.

Table 3a. Sensory scores for Orange Cake

OC Control (no omega-3)				
Time	**Appearance**	**Odor**	**Flavor**	**Texture**
0	6.91 (0.23)*	6.91 (0.25)	6.85 (0.23)	6.88 (0.34)
4 weeks**	6.57 (0.85)	6.75 (0.51)	6.55 (0.71)	6.58 (0.76)
3 months	6.64 (0.59)	6.74 (0.26)	6.71 (0.30)	6.63 (0.35)
6 months	6.82 (0.53)	6.81 (0.45)	6.62 (0.49)	6.47 (0.37)
OC 200 mg omega-3				
Time	**Appearance**	**Odor**	**Flavor**	**Texture**
0	6.84 (0.27)	6.76 (0.31)	6.72 (0.32)	6.78 (0.38)
4 weeks	6.65 (0.73)	6.64 (0.60)	6.59 (0.69)	6.61 (0.64)
3 months	6.69 (0.59)	6.70 (0.23)	6.68 (0.28)	6.54 (0.38)
6 months	6.77 (0.61)	6.86 (0.44)	6.72 (0.62)	6.78 (0.69)
OC 400 mg omega-3				
Time	**Appearance**	**Odor**	**Flavor**	**Texture**
0	6.74 (0.23)	6.62 (0.38)	6.58 (0.32)	6.66 (0.27)
4 weeks	6.68 (0.60)	6.45 (0.66)	6.35 (0.79)	6.40 (0.67)
3 months	6.58 (0.70)	6.66 (0.30)	6.58 (0.41)	6.59 (0.36)
6 months	6.68 (0.65)	6.73 (0.53)	6.68 (0.56)	6.66 (0.63)

* Standard Deviation.

** 4 week samples stored at 49°C.

Table 3b. Sensory scores for Beef Lo Mein

BLM control (no omega-3)				
Time	**Appearance**	**Odor**	**Flavor**	**Texture**
0	6.88 (0.26)*	6.70 (0.17)	6.68 (0.29)	6.57 (0.39)
4 weeks**	6.68 (0.33)	6.62 (0.41)	6.67 (0.36)	6.64 (0.29)
3 months	6.67 (0.27)	6.71 (0.21)	6.72 (0.20)	6.55 (0.34)
6 months	6.74 (0.28)	6.77 (0.18)	6.61 (0.20)	6.60 (0.25)
BLM 200 mg omega-3				
Time	**Appearance**	**Odor**	**Flavor**	**Texture**
0	6.49 (0.34)	6.53 (0.33)	6.32 (0.44)	6.52 (0.32)
4 weeks	6.25 (0.83)	6.48 (0.45)	6.44 (0.44)	6.29 (0.66)
3 months	6.73 (0.21)	6.69 (0.22)	6.70 (0.24)	6.55 (0.31)
6 months	6.78 (0.22)	6.75 (0.15)	6.64 (0.30)	6.63 (0.23)
BLM 400 mg omega-3				
Time	**Appearance**	**Odor**	**Flavor**	**Texture**
0	6.46 (0.39)	6.48 (0.33)	6.25 (0.48)	6.43 (0.41)
4 weeks	6.30 (0.54)	6.44 (0.42)	6.41 (0.46)	6.38 (0.50)
3 months	6.69 (0.24)	6.63 (0.23)	6.62 (0.23)	6.58 (0.20)
6 months	6.81 (0.20)	6.72 (0.24)	6.59 (0.31)	6.63 (0.25)

* Standard Deviation.

** 4 week samples stored at 49°C.

Table 3c. Sensory scores for First Strike™ Bars

FSB control (no omega-3)				
Time	**Appearance**	**Odor**	**Flavor**	**Texture**
0	6.86 (0.25)*	6.68 (0.27)	6.41 (0.90)	6.78 (0.29)
4 weeks**	7.04 (0.41)	7.01 (0.54)	6.85 (0.57)	6.99 (0.55)
3 months	6.98 (0.37)	6.93 (0.38)	6.78 (0.50)	6.83 (0.53)
6 months	7.00 (0.37)	6.97 (0.49)	6.76 (0.57)	6.96 (0.39)
FSB 100 mg omega-3				
Time	**Appearance**	**Odor**	**Flavor**	**Texture**
0	6.83 (0.31)	6.58 (0.41)	6.20 (0.92)	6.69 (0.27)
4 weeks	7.01 (0.43)	6.93 (0.60)	5.62 (1.00)	6.72 (0.50)
3 months	6.97 (0.37)	6.52 (0.80)	5.59 (0.93)	6.73 (0.59)
6 months	6.98 (0.40)	6.15 (0.85)	5.36 (1.34)	6.85 (0.38)
FSB 200 mg omega-3				
Time	**Appearance**	**Odor**	**Flavor**	**Texture**
0	6.77 (0.36)	6.48 (0.42)	5.93 (0.68)	6.66 (0.28)
4 weeks	5.61 (1.15)	4.53 (1.40)	3.28 (1.27)	4.74 (1.20)
3 months	6.92 (0.42)	6.03 (1.32)	4.85 (1.31)	6.61 (0.77)
6 months	6.88 (0.48)	5.04 (0.95)	3.83 (0.95)	6.33 (0.93)

* Standard Deviation.

** 4 week samples stored at 49°C.

Sensory scores for the fortified FSB, however, declined markedly during storage with quality degradation proportional to omega-3 content. Flavor and aroma attributes were particularly affected, with scores decreasing 43% and 28% ($p < 0.01$) over 6 months for the highest omega-3 content product. Flavor and odor perceptions were therefore drivers for the 38% decrease in overall quality for this product ($p < 0.01$). The intermediate level of omega-3 fat produced smaller but still significant ($p < 0.01$) decreases in perceived flavor, perceived odor and overall quality (21%, 12%, and 17% respectively).

These differences in retention of sensory quality among the three food products are most likely attributable to the differences in their packaging environments. While none of the products were packaged under vacuum, orange cakes were packaged with oxygen scavengers; scavengers were not included with FS bars (as is typical for these ration components) or with BLM, in which use was precluded by the high moisture content of the product. Presumably, greater headspace oxygen in the FSB packages compared with OC contributed to greater degradation of omega-3 lipid. Furthermore, headspace in the retorted BLM was fairly low, and certainly proportionally lower than that in FSB packages, given the high fill level in BLM samples. While effects of formulation and processing on compatibility with omega-3 incorporation cannot be completely disregarded, it is likely that the packaging environment was an overriding factor in shelf-stability.

Omega-3 Fat Quantification

In-house analysis of the Meg-3® powder determined an average 20.3±1.2% load of EPA + DHA, with more than a 2 to 1 ratio of EPA to DHA. This measured level was slightly higher than the 15-20% content range specified by the supplier, and thus was on average 35% higher than the 15% content assumed when formulating the products. Calculated added levels of EPA + DHA (based upon measurement of Meg-3® loading prior to analysis of each food product) were, respectively: 261 and 507 mg for intermediate and high loading of OC; 272 and 549 mg for intermediate and high loading of BLM; and 128 and 253 mg for intermediate and high loading of the FSB. Measured initial EPA + DHA contents in the food samples, respectively for intermediate and high loads, were: 288±14 and 548±3.2mg for OC; 253±33 and 503±17mg for BLM; and 114 ±3.9 and 187 ±11mg for the FSB. These results indicate: no processing loss (and, actually, higher than calculated loading) in OC; 7 and 8% processing losses in BLM for low and high omega-3 loading, respectively; and 11 and 26% processing losses in FSBs for low and high omega-3 loading, respectively.

Charts showing EPA and DHA levels in each food after the different storage intervals are shown in Figures 2a-c. While OC and BLM showed no significant (NS, $p > 0.05$) loss of DHA and EPA during storage (for any interval), loss of both types of omega-3 fatty acids was marked in FSB. Low and high omega-3-fat content First Strike™ bars respectively showed a reduction of 8.3% EPA (NS) and 20.5% DHA ($p<0.01$) after 3 months storage and 15.4% ($p<0.01$) and 19.1% ($p<0.01$) after 6 months storage. Differences in omega-3 fat retention most likely resulted from the different packaging environments, which were presumably effected by oxygen level. OC samples were packaged with oxygen scavengers and BLM pouches, while not vacuum packed, had little headspace; both factors would serve to limit entrained oxygen. Losses of EPA and DHA combined with low sensory scores for the stored FSBs indicate that the gelatin microencapsulation may not have sufficiently protected the omega-3 fat to prevent degradation. Furthermore, the lower than expected level of EPA + DHA in the fortified FSBs prior to storage (0 time), particularly for the higher concentration sample, suggests that the omega-3-fat capsules may have been disrupted during processing. It is possible that, given the high viscosity of the bar dough, levels of shear developed during mixing were high enough to release a portion of the oil, rendering the product susceptible to oxidation.

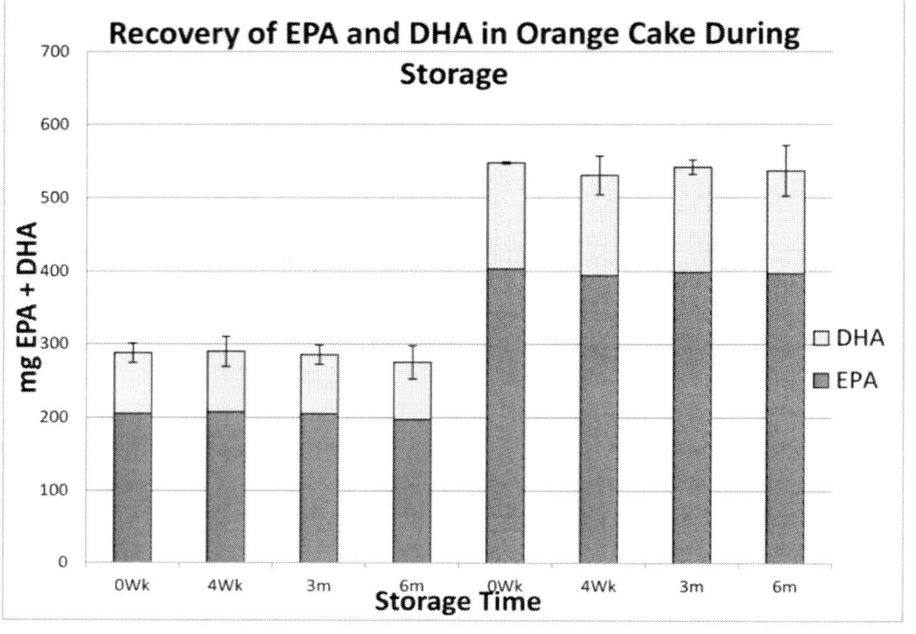

(a)

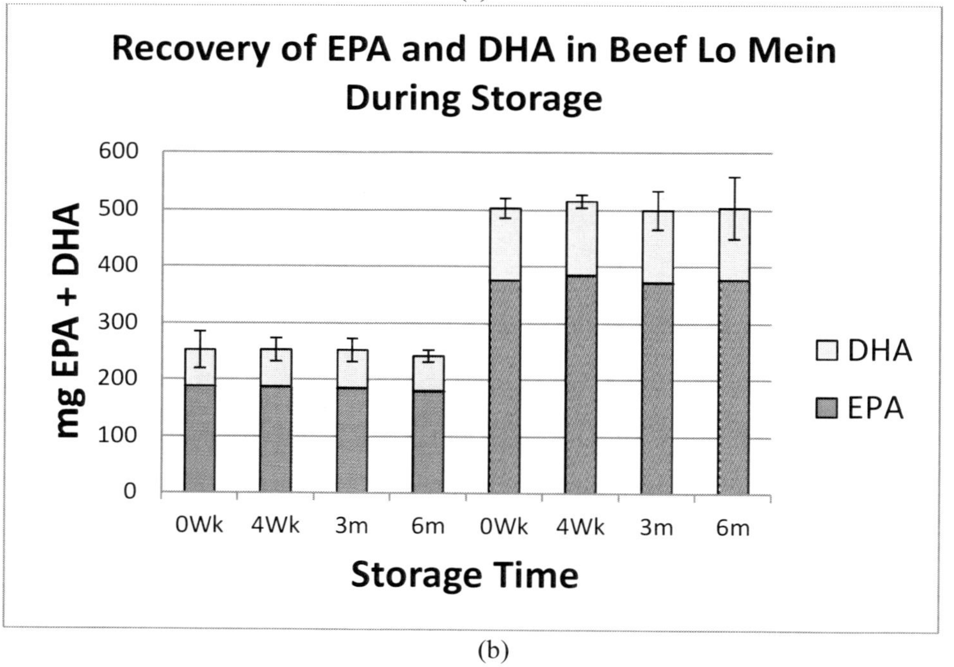

(b)

Figure 2. (Continued)

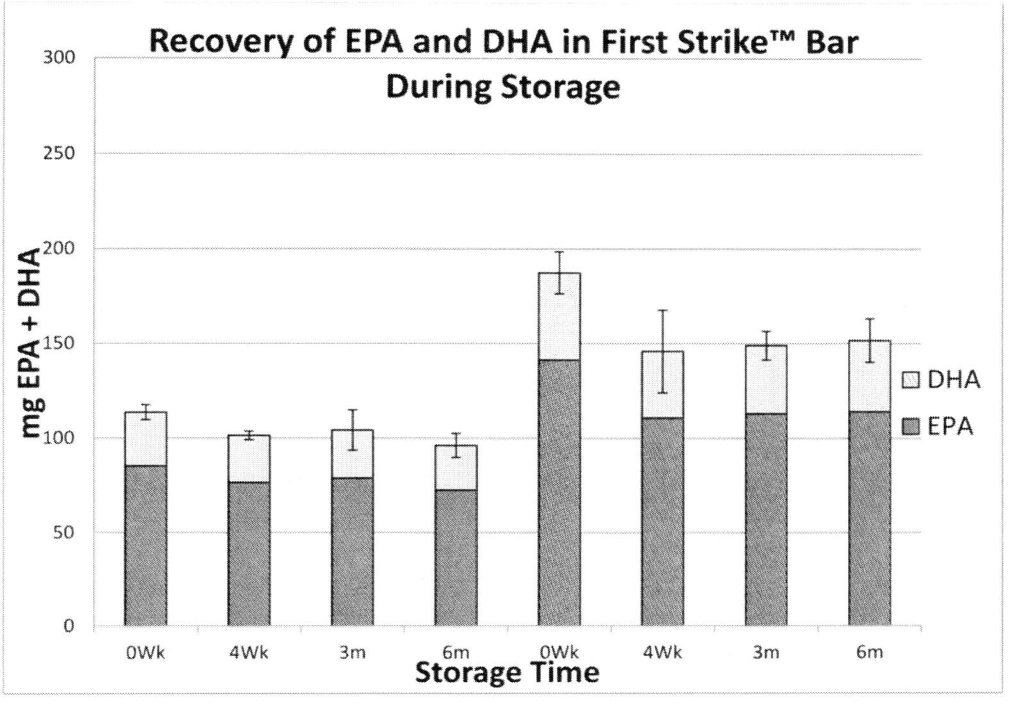

(c)

Figure 2. Stacked bar graphs for EPA and DHA concentrations measured throughout storage for both low and high fortification levels: (a) Orange Cake [target levels 200 and 400 mg total omega 3 fat/serving]; (b) Beef Lo Mein [target levels 200 and 400 mg total omega 3 fat/serving]; and (c) Chocolate First Strike™ Bar [target levels 100 and 200 mg total omega 3 fat/serving]. Left four bars are for the lower concentration; right four bars are for the higher concentration. Shows degradation of omega-3 fat in the First Strike™ Bar only.

Volatile Production

Statistical results for fluorescent ratio determinations are shown in Tables 4a and 4b. Analysis of variance for the individual food products over time and omega-3 fat concentration showed significant effects ($p < 0.01$) of: both omega-3 content and time on volatile production by OC; and time on volatile production by FSB (Table 4a). Paired t-tests for OC showed significant differences ($p < 0.02$) between: 0-time and samples stored for any time interval (across temperature, with average increase in ratio = 29%); and between control samples and samples containing either fortification level (across all pull times, with average increase in ratio = 11%) (Table 4b). Paired t-tests for FSB showed significant differences ($p \leq 0.002$) between: 0-time and samples stored for any time interval (across temperature, with average increase in ratio = 10%); and between control samples and samples containing the highest fortification level (across temperature, with average increase in ratio = 3%). Paired t-tests for BLM showed no significant increases in fluorescence ratio due to either storage time or fortification level (Table 4b).

Table 4a. Significance of ANOVA results for fluorescence ratio: Effects of omega-3-fat level and storage time

Food Matrix	p, omega-3 level	p, storage time
Orange Cake	> 0.001	0.002
Chocolate First Strike™ Bar	> 0.001	(NS)
Beef Lo Mein	(NS)	(NS)

Table 4b. Significance of paired t-test results for fluorescence ratio: Effects of omega-3-fat level and storage time

Comparison	T	p
Cake, 0 time vs. Cake, 4 weeks @49°C	-13.2	0.000
Cake, 0 time vs. Cake 3 months @37°C	-15.2	0.000
Cake, 0 time vs. Cake 6 months @100F	-10.7	0.000
First Strike™ Bar, 0 time vs. First Strike™ Bar, 4 weeks @49°C	-7.64	0.000
First Strike™ Bar, 0 time vs. First Strike™ Bar, 3 months @37°C	-6.75	0.000
First Strike™ Bar, 0 time vs. First Strike™ Bar, 3 months @37°C	-4.59	0.002
Beef Lo Mein, 0 time vs. Beef Lo Mein, 4 weeks @49°C	(NS)	(NS)
Beef Lo Mein, 0 time vs. Beef Lo Mein, 3 months @ 37°C	(NS)	(NS)
Beef Lo Mein, 0 time vs. Beef Lo Mein, 3 months @ 37°C	(NS)	(NS)
Cake, control vs. Cake, intermediate omega-3 level	-5.88	0.000
Cake, control vs. Cake, high omega-3 level	-7.93	0.000
First Strike™ Bar, control vs. First Strike™ Bar, intermediate omega-3 level	(NS)	(NS)
First Strike™ Bar, control vs. First Strike™ Bar, high omega-3 level	-2.71	0.020
Beef Lo Mein, control vs. Beef Lo Mein, intermediate omega-3 level	(NS)	(NS)
Beef Lo Mein control vs. Beef Lo Mein, high omega-3 level	(NS)	(NS)

Analysis of the encapsulated powder by itself, however, showed a much more significant effect of accelerated storage: signals for the omega-3 powder increased 62% and 49% after 38°C for 3 and 6 months, respectively ($p<0.01$). Furthermore, the preliminary 1 week at 75°C test of this material produced more than a 200% increase in fluorescence ratio. The larger effect seen in the encapsulated powder is most likely due to its high concentration of omega-3 oil relative to the food products, in which the fat was diluted by the other ingredients; furthermore, it is possible that other constituents of the foods served as antioxidants.

Increases in fluorescence are due to production of intermediate-species during oxidation, so measurements of fluorescence ratio will expectedly change throughout storage — rising to a peak then falling as these intermediaries transition to more stable molecules. In fact, analysis of all the products, including the encapsulated powder, revealed at least a slightly lower fluorescence ratio after 6 months storage than after 3 months storage. Higher storage temperatures apparently accelerated development of fluorescence, as indicated by 4 weeks storage at 49°C producing the highest fluorescence ratio for both OC and FSB, and 1 week storage at 75°C producing the highest fluorescence ratio for the Meg-3® powder. Interestingly, however, fluorescence ratio was not in keeping with sensory perception for either OC or FSB samples. For OC, it is likely that the level of oxidation, while producing increases in fluorescence, was still reasonably low and masked effectively enough by the orange flavor of the cake to limit reduction in sensory quality, a conclusion supported by the finding of no significant loss in EPA and DHA by the GC analysis. For the FSB, conversely, it is likely that oxidation had progressed far enough so that analyses at 4 weeks and 3 months had missed the peak in fluorescence, a conclusion supported by the GC finding of large losses in the EPA and DHA. Fluorescence analysis, which determines a "snapshot" of oxidation at a specific time, is advisably paired with additional measurements of sample quality and composition.

Conclusion

Recent advances in encapsulation technology have resulted in the feasibility of producing a variety of shelf-stable omega-3-fortified products, provided that appropriate packaging environments, specifically limitation of oxygen, are maintained. The representative long shelf life products discussed comprise high, medium and low water activities in disparate food matrices. While the FSB showed severe loss of both omega-3 fat and sensory quality, it is possible that this item would also be an appropriate vehicle for supplementation if it were packaged with an oxygen scavenger.

Acknowledgment

The authors thank Ocean Nutrition Canada Limited, especially Mr. John Getzinger and Ms. Sherry Siteman, for provision of Meg-3® powder and analysis procedures.

References

Arstog, P., Couthouis, A., Bertrais, S., Arnault, N., Meneton, Pierre, G., Philippe, Alessandri, J.-M., Galan, P., Hercberg, S. (2008). Association of fish and long-chain n-3 polyunsaturated fatty acid intakes with the occurrence of depressive episodes in middle-aged French men and women. *Prostaglandins, Leukotrienes and Essential Fatty Acids*. 7, 171-182.

Barrett, A. H., Porter, W. L., Marando, G., Chinachoti, P. (2011). Effect of various antioxidants, antioxidant levels, and encapsulation on the stability of fish and flaxseed oils: assessment by fluorometric analysis. *J. Food Proc. and Pres*. 35, 349-358.

Chen, C. (2010). COX-2's new role in inflammation. *Nature Chem. Biol.* 6, 401-402.

Curtis, J. M., Berrigan, N., Dauphinee, P. (2008). The Determination of n-3 fatty acid levels in food products containing microencapsulated fish oil using the one-step extraction method. Part 1: Measurement in the raw ingredient and in dry powdered foods". *Journal Am. Oil Chem. Soc.* 84, 297-305.

Eaton, S. B. and Konner, M. (1985). Paleolithic nutrition. A consideration of its nature and current implications. *New Engl. J. Med.* 312, 283-289.

Emkin, F. A., Adlof, R. O., Gulley, R. M. (1994). Dietary linolenic acid influences desaturation and acylation of deuterium-labeled linolenic and linolenic acid in young adult males. *Biochim. Biohphys. Acta.* 1213, 277-288.

Fernstrom, J. D. (1999). Effects of dietary polyunsaturated fatty acids on neuronal function. *Lipids*. 34, 161-169.

Frankel, E. N. (1998). Chapter 8, in *Lipid Oxidation*. The Oily Press Ltd., Bridgewater, U. K.

Im, D.-S. (2012). Omega-3 fatty acids in anti-inflammation (pro-resolution) and GPCRs. *Prog. Lipid Res.* 51, 232-237.

Johnston, D. T., Deuster, P. A., Harris, W. S., MacRae, H., Dretsch, M. N. (2013). Red blood cell omega-3 fatty acid levels and neurocognitive performance in deployed U. S. service members. *Nutritional Neuroscience*. 16, 30-38.

Kris-Etherton, P. M., Taylor, D. S., and Yu-Roth, S. (2000). Polyunsaturated fatty acids in the food chain in the United States. *Am. J. Clin. Nutr.* 71(1 Suppl), 179S-188S.

Kris-Etherton, P. M., Harris, W. S., Appel, L. J. (2002). Fish consumption, fish oil, omega-3 fatty acids and cardiovascular disease. *Circulation*, 106, 2747-2757.

LeBlanc, J., Ross, E. Barrett, A. (2013). *Quality Kinetics for Stability*. US Army Natick Soldier RD&E Center Technical Report. In Press.

Lewis, M. D., Hibbein, J. R., Johnson, J. E., Lin, Y. H., Hyun, D. Y., Loewke, J. D. (2011). Suicide deaths of active duty US military and omega-3 fatty acid status: a case control comparison. *J. Clin. Psychiatry*. 72, 1565-1590.

Lucas, M., Mirzaie, F, O'Reilly, E. J., Willett, W. C., Kawachi, I., Koenen, K. Ascherio, A. (2011). Dietary intake of n-3 and n-6 fatty acids and the risk of clinical depression in women: a 10-y prospective follow-up study. *Am. J. Clin. Nutr.* 93, 1337-1143.

Mahan, L. K. and Escott-Stump, S. (2004). *Krause's Food, Nutrition & Diet Therapy*. Saunders/Elsevier, Philadelphia.

Mais, M. and Smith, R. (1998). Fatty acids, cytokines, and major depression. *Biol. Psychiatry*. 43, 313-314.

Mills, J. D., Hadley, K., Bailes, J. E. (2011). Dietary supplementation with the omega-3 fatty acid docosahexaenoic acid in traumatic brain injury. *Neurosurgery*. 68, 474-481.

Na, S. (2011). Analysis of fatty acids in infant formulas using an Agilent J & W HP-88 capillary GC column. Application Note for Agilent Technologies.

Pawlosky, R. I., Hibbeln, I. R. and Novotnov., I. A. (2001). Physiological compartmental analysis of alpha-linolenic acid metabolism in adult humans. *J. Lipid Res.* 42, 1257-1265.

Razzak, A., Aldrich, C., Babcock, T. A., Saied, A., Espat, J. (2008). Attenuation of iNOS in an LPS-stimulated macrophage model by omega-3 fatty acids is independent of COX-2 derived PGE_2. *J. Surgical Res.* 145, 244-250.

Simopoulos, A. P., (1999). Evolutionary aspects of omega-3 fatty acids in the food supply. *Prostaglandins Leukot. Essent. Fatty Acids*. 60, 421-429.

Simopoulos, A. P. (2008). The omega-6/omega-3 fatty acid ratio, genetic variation, and cardiovascular disease. *Asia Pacific J. Clin. Nutr.* 17 (Suppl), 131-134.

Song, C. (2007). Omega-3 fatty acids in the treatment of neurodegenerative diseases through anti-inflammation and neuroprotection: a review of studies in animal models. In: *Marine Nutraceuticals and Functional Foods*, Fereidoon Shahidi, Ed. CRC Press. Baton Rouge. Pp. 91-113.

Sublette, M. E., Ellis, S. P. Geant, A. L., and Mann, J. J. (2011). Meta-analysis of the effects of eicosapentaenoic acid (EPA) in clinical trials in depression. *J. Clin. Psychiatry*. 72, 1577-1584.

Tanskanen, A., Hibbeln, J. R., Hintikka, J., Haatainen, K., Honkalampi, K., Viinamaka, H. (2001). Fish consumption, depression, and suicidality in a general population. *Arch. Gen. Psychiatry*. 58, 512-513.

Wang, S., Wu, L., Matthan, N. R., Lamon-Fava, S., Lecker, J. L., Lichtenstein, A. H. (2009). Reduction in dietary omega-6 polyunsaturate fatty acids: eicosapentaenoic acid plus docosahexaenoic acid ratio minimizes atherosclerotic lesion formation and inflammatory response in the LDL receptor null mouse. *Atherosclerosis*. 204, 147-155.

In: Omega-3 Fatty Acids
Editor: Md. Wasim Khan

ISBN: 978-1-62948-516-4

Chapter 12

Omega 3 Fatty Acids and GPR120

N. A. Arivarasu[1], Shubha Priyamvada[1] and Medha Priyadarshini[2,*]

[1]Department of Medicine, University of Illinois, Chicago, US
[2]Department of Biochemistry, F/O Life Sciences, AMU, Aligarh, U.P., India

Abstract

Omega-3 fatty acids have various health benefits of which the underlying mechanisms remain unknown. Free fatty acids have been demonstrated to serve as ligands for G protein coupled receptors. GPR120 acts as the receptor for omega-3 fatty acids and mediates most of their physiological effects. Importantly, GPR120 signaling is at the focal point of inflammatory and insulin sensitization pathways, presenting it as a potential therapeutic target. This chapter highlights GPR120 signaling mechanisms with the main emphasis on the physiological effects exerted upon its sensitization with omega-3 fatty acids.

Keywords: ALA, DHA, EPA, arrestin, diabetes, G protein coupled receptors, inflammation, obesity

Introduction

Free fatty acids (FFAs) have traditionally been considered as important energy sources and metabolic substrates (Yaney and Corkey, 2003). Lately experimental evidence has pointed to the role of FFAs in direct activation of siganling pathways affecting various physiological processes (Talukdar et al., 2011; Oh et al., 2010; Mo et al., 2013; Yan et al., 2013). Nuclear receptors (like peroxisome proliferator-activated receptor, PPARs) are FFA

* E-mail: medha.priyadarshini@gmail.com.

sensors. They maintain homeostasis by sustaining the expression of proteins involved in metabolism of lipids (Chawla et al., 2001; Ichimura et al., 2009). Some effects of FFAs, however, have been found to be independent of PPARs (Sauer et al., 2000; Louet et al., 2001). The alternative route taken by FFAs to mediate certain effects is via siganling through cell surface G protein coupled receptors (GPCRs). GPCRs are vital signaling molecules for varied aspects of cellular function. They are members of a large receptor family sharing common structural and functional motifs like seven trans membrane helices and ability to transduce signals by activating heterotrimeric G proteins, such as, Gαs, Gαi and Gαq (Oh et al., 2010). Ligand binding to GPCRs induce various cellular responses through several second messenger pathways, like changes in cAMP levels, phospholipase C, mitogen activated protein kinases (MAPK) (Ulloa-Aguirre et al., 1999; Gether, 2000; Schulte and Fredholm, 2003).

A number of FFA sensing rhodopsin like GPCRs (or FFARs, free fatty acid receptors) have been identified recently. They include GPR40, GPR41, GPR43, GPR84 and GPR120 with different specificities/ affinities for different chain length fatty acids. Their main characterstics are summarized in Table 1.

FFAs have been grouped into different classes depending upon their carbon chain length. The short chain fatty acids (C2-C5), medium chain fatty acids (C6-C12) and the long chain fatty acids (>C12). Omega-3 fatty acids (also called as ω-3 fatty acids or n-3 fatty acids) are essential polyunsaturated fatty acids (PUFA) with a double bond between the third and fourth carbon atoms from the methyl end (also known as the ω end or the n end) of the fatty acid carbon chain (Im, 2012). ω-3 FAs cannot be synthesized in the human body de novo, but are crucial for normal metabolism (Talukdar et al., 2011). ω-3 FAs like eicosapentaenoic acid (EPA, 20:5n-3) and docosahexaenoic acid (DHA, 22:6n-3) are mostly derived from marine sources. EPA can be synthesized in the human body from the plant ω-3 FA, α-linolenic acid (ALA), though not efficiently. Their benfeficial health effects have been known since long, mechanisms, however, the underlying mechanisms of these effects remained unclarified, only till recently. Various groups have independently found that ω-3 FAs signal through GPR120, their physiological receptor (and in some cases through GPR40). This chapter synopsizes the information on GPR120 and its activation via ω-3 FAs and the downstream effects.

1. *Identification and de-orphanisation*

   GPR120 was identified by informatic searches of human NCBI and Celera genome databases as a rhodopsin-family GPCR with limited sequence homology (~10%) to other family members. The gene is located on chromosome 10q23.33 (Fredriksson et al., 2003). It was first isolated from the human genomic DNA by Hirasawa et al. (Hirasawa et al., 2005). The GPR120 gene contains 4 exons and exists in 2 alternatively spliced form, GPR120-S (the short isoform) and GPR120-L (the long isoform) (Mo et al., 2013). Internalization of C-terminally GFP-tagged GPR120 from cell surface and capacity of GPR-120-Gα16 fusion protein to increase intracellular calcium concentration upon stimulation by ALA and DHA revealed that GPR120 is physiological receptor for PUFAs (Hirasawa et al., 2005). In some cases saturated long chain fatty acids could produce an increase in intracellular calcium concentrations via GPR120 but only PUFAs were able to cause the secretion of glucagon like peptide 1 (GLP-1) from enteroendocrine cell line, STC-1 suggesting that ω-3 FAs are specific agonists for this receptor (Jump, 2008; Tanaka et al.,

2008a). Since then this has been reaffirmed by a number of studies (Talukdar et al., 2011; Yan et al., 2013).

2. *Tissue localization*

High levels of GPR120 expression have been reported in the lungs and intestinal tract (Hirasawa et al., 2005). In rat and mouse it is highly expressed in colon (Tanaka et al., 2008b; Ito et al., 2009). GPR120 is expressed in mouse enteroendocrine cell line, STC-1, but not in ither intestinal cell lines like NCI-H716, Caco-2, HT29, IEC-6, etc. and ghrelinoma cell line MGN3-1 (Hirasawa et al., 2005; Janssen et al., 2012; Hwakura et al., 2010). GPR120 expression in enteroendocrine cells, K cells has also been established by the studies suggesting its role in secretion of GLP-1 and cholecystokinin from the small intestine (Matsumura et al., 2007; Parker et al., 2009). GPR120 expression has also been reported in adipose tissue but not in preadipocytes (Gotoh et al., 2007) and adipose/hepatic tissue macrophages (Oh et al., 2010). GPR120 expression has also been detected in the epithelium of circumvallate papillae containing taste buds (Matsumura et al., 2007) suggesting that it might be involved in taste perception.

The short isoform of GPR120 has been detected in pancreatic islets (Morgan and Dhayal, 2009). However, presence of GPR120 in pancreatic tissue is debated. Few studies have shown its complete absence in rat pancreas and beta cell line, MIN6 (Hirasawa et al., 2005; Tanaka et al., 2008) yet others report the presence of the short isoform in isolated mouse and human islets and beta cell lines (BRIN-BD11, NS-1 etc.) (Morgan and Dhayal, 2009; Whalley et al., 2011).

Likewise in cardiac, hepatic tissue and pituitary and breast cancer cells expression of GPR120 has to be confirmed and physiological roles assigned (Cornall et al., 2011; Soto-Guzman et al., 2008).

3. *Ligands, signal transduction and regulation*

Based on the recommendations of the International Union of Basic and Clinical Pharmacology Committee on Receptor Nomenclature and Drug Classification (NC-IUPHAR) the cognate ligands for GPR120 are ω-3 FAs, ALA, DHA and saturated long chain FA, palmitoleic acid (PA) (IUPHAR-DB ID no. 1049; IUPHAR-DB ID no. 1051; Davenport et al., 2013) linked to the Gq/11 pathway (Hirasawa et al., 2005). Oh and colleagues (Oh et al., 2010) have shown that GPR120 is a physiological receptor for ω-3 FAs in differentiated adipocytes and macrophages mediating the anti-inflammatory and insulin sensitizing effects. Some natural compunds like grifolic acid and grifolic acid methyl ether have been demonstrated as partial agonists for GPR120 activating the calcium response and ERK (extracellular signal regulated kinase) signaling and competing with ALA (Hara et al., 2009).

The main signal transduction pathways of GPR120 are depicted in Figure 1. GPR120 couples to heterotrimeric Gαq/11 protein in specific cell types activating phospholipase C (PLC). This results in the hydrolysis of plasma membrane phospholipid phosphatidylinositol bisphosphate into diacylglycerol (DAG) and inositol triphosphate (IP3). DAG activates protein kinase C and IP3 liberates calcium from intracellular stores (endoplasmic reticulum). This coupling has been shown to result in release of GLP-1 (Hirasawa et al., 2005). Utilizing STC-1 cells, Hirasawa et al., showed that stimulation by DHA and ALA caused GLP-1 secretion. This was reduced in presence of calcium free medium but remained unaffected in presence of inhibitors of L-type calcium channels/ phospholipase C or ERK MAPK cascade. Transfection of these cells with GPR120 cDNA increased GLP-1 secretion but was unchanged on addition of protein kinase C activator. Small interfering RNA (Si RNA) knockdown of GPR120 abrogated ALA induced increase in GLP-1 secretion. It has also been shown that si-RNA knockdown of Gαq/11 inhibits the DHA and GW9508 (synthetic GPR120 agonist) induced glucose uptake in adipocytes (Oh et al., 2010). In this case binding of the DHA/agonist to GPR120 results in activation of phosphoinositide 3 kinase (PI3K) which causes downstream activation/ phosphorylation of protein kinase B (Akt) required for the glucose transporter translocation to the cell membrane and increased glucose uptake (insulin sensitization). FFA mediated activation of GPR120 prevents apoptosis in STC-1 cells by a similar transduction mechanism (Katsuma et al., 2005).

ω-3 FAs signal through GPR120 by an additional mechanism also. Ligand sensitization of GPR120 may also activate mitogen activated kinases (Figure 1). Upon stimulation by ω-3 FA, PLC mediates activation of PKC a key enzyme in ERK signaling (Werry et al., 2005). Increased phosphorylation of ERK1/2 upon GPR120 stimulation in human embryonic kidney cell line (HEK293) has been demonstrated by Hirasawa et al. (2005). ERK1/2 activation has also been reported in other cell lines upon stimulation by GPR120 ligands (Soto-Guzman et al., 2008; Tanaka et al., 2008a; Iakoubov et al., 2007; Katsuma et al., 2005).

ω-3 FAs induce anti-inflammatory effects by siganling through GPR120 via β-arrestin 2 dependent mechanisms (Oh et al., 2010) (Figure 1). After ligand stimulation β-arrestin 2 (an adapter protein) can translocate to a number of GPCRs where it mediates receptor internalization and signaling (Barak et al., 1997). The mechanism of GPR120 mediated anti-inflammation involves inhibition of TAK1 (transforming growth factor β activated kinase 1). β-arrestin 2 binds to the C-terminal binding sites in GPR120 after sensitization with ω-3 FAs. This is followed by receptor and β-arrestin 2 internalization and association of β-arrestin 2 with TAB1 (TAK1 binding protein). This prevents TAB1 mediated activation of TAK1 and downstream signaling to IKKβ/NFκB (inhibitor of nuclear factor kappa-B kinase subunit beta/ nuclear factor kappa-B) and JNK/AP1 (c-Jun N-terminal kinase/ activator protein1) system. This in turn prevents the inflammatory responses to tumor necrosis factor α (TNF α), lipopolysaccharide (LPS) binding to toll like receptor 4 (TLR4) etc. (Oh et al., 2010). These effects of GPR120 are completely independent of Gαq/11. This points to the fact that multiple signaling pathways can be undertaken by a single GPCR. Activated/phosphorylated GPCRs recruit arrestins/adapter proteins and get internalized. They may be recycled back to cell surface or may undergo lysosomal degradation (Watson et al., 2012).

4. *ω-3 FAs physiological effects through GPR120 signaling*

(a) Anti-inflammatory effects

Chronic low grade metabolic inflammation (also called as metainflammation) (Gregor and Hotamisligil, 2011) has been found to be strongly associated with obesity, diabetes, insulin resistance, Alzheimer's disease, Parkinson's disease (Talukdar et al., 2011; Priyadarshini et al., 2012). ω-3 FAs, mainly EPA and DHA, possess inflammation suppressive properties and their consumption mitigates the severity in various inflammatory diseases (Yan et al., 2013; Zhang and Spite, 2012).

The mechanisms for anti-inflammatory effects of ω-3 FAs, however, are not fully elucidated. Recently, Yan et al., (2013) demonstrated that ω-3 FAs (EPA and DHA) prevented inflammation and metabolic disease by inhibiting the activation of NLRP3 inflammasome in GPR120 (and GPR40) β arrestin 2 dependent fashion. The inflammasome is a cytosolic protein complex composed of nucleotide-binding domain and leucine-rich repeat containing proteins (NLRs), adaptor protein ASC and inflammatory protease caspase 1 (Davis et al., 2011). Inflammasome is assembled in response to infection/danger signals and promotes the maturation and release of several proinflammatory cytokines. In their breakthrough study, Yan and coworkers found that ω-3 FAs suppressed LPS induced priming and inflammasome (NLRP3) activation in macrophages (Yan et al., 2013). Further extending their work in vivo, they found that ω-3 FA supplementation suppress high fat diet (HFD) induced NLRP3 inflammasome activation and prevent NLRP3 inflammasome dependent insulin resistance in WT (wild type) mice but not in $Nlrp3^{-/-}$ mice. Since NLRP3 inflammasome activation and interleukin1β production is implicated in diabetes, gout, autoinflammatory and inflammasome driven diseases, ω-3 FAs, may present potential therapeutic options against them.

Another study demonstrated that DHA abrogates LPS induced inflammatory responses (detailed in signal transduction section) via GPR120-β arrestin 2 pathway but not in GPR120 or β arrestin 2 knockdown cells. Furthermore GPR120 activation in primary adipose tissue or mature adipocytes (3T3-L1) led to an increase in glucose uptake and insulin sensitization via GLUT4 and Gαq/11 dependent mechanisms (Oh et al., 2010).

In the same study, GPR120 knockout (KO) and WT mice when fed on standard chow, HFD and HFD supplemented with ω-3 FAs showed different phenotypes. The GPR120 KO mice were more insulin resistant on standard chow. While both WT and KO were equally insulin resistant on HFD. ω-3 FA supplementation had no effect on the phenotype of KO mice. When macrophage specific knockout mice of GPR120 were studied, ω-3 FA supplementation was effective only in WT mice (Oh et al., 2010). This study also points at the possible clinical use of ω-3 FAs or synthetic GPR120 agonists in treatment of disorders associated with inflammatoy states.

A recent study has revealed that same signaling pathway operates in the hypothalamus. Cintra et al. (Cintra et al., 2012) have shown that i.c.v. administration of ω-3 FAs in obese rats produces local anti-inflammatory effects. Interaction of ω-3 FA with GPR120 expressed in the hypothalamic region results in reduction in weight and improved insulin sensitivity.

Recently, it was shown that metabolic products of ω-3 FAs like protectins, resolvins, etc. may play a role in resolution of inflammation and attenuation of insulin resistance and obesity (Gonzalez-Periz et al., 2009).

Table 1. Characterstics of FFA sensing GPCRs

	GPR40 (FFAR1)	GPR41 (FFAR3)	GPR43 (FFAR2)	GPR120 (FFAR4)	GPR84	References
Agonist (FFA)	Medium-long (C12-C16)	Short chain (C3>C4>>C2)	Short chain (C3~C4~C2)	Polyunsaturated fatty acids, ω-3 fatty acids (main endogenous ligands)	Medium	Stoddart et al., 2008 Davenport et al., 2013 Oh and Lagakos, 2011
Antagonist	-	-	-	-	-	-
G-protein coupling	Gq/11	Gi/o	Gq/11 Gi/o	Gq/11	Gi/o	Stoddart et al., 2008 Davenport et al., 2013 Oh et al., 2010
Chromosomal location	19q13.1	19q13.1	19q13.1	10q23.33	12q13.13	Stoddart et al., 2008 Davenport et al., 2013
Tissue distribution	Pancreatic beta cell	Adipocytes, spleen, colon, pancreas, nervous system, muscle, heart	Adipocytes, enteroendocrine cells, immune cells, pancreas, muscle, heart	Enteroendocrine cells, immune cells, adipocytes	Spleen, monocytes, macrophages	Stoddart et al., 2008 Davenport et al., 2013 Regard et al., 2007
Physiological role	Insulin secretion	Leptin production	Immune responses, glucose homeostasis	GLP-1 secretion, anti-inflammation, insulin sensitization	Regulation of microglia, neuroinflammatory processe, immune response	Stoddart et al., 2008 Davenport et al., 2013 Tolhurst et al., 2012 Kimura et al., 2011 Ichimura et al., 2009

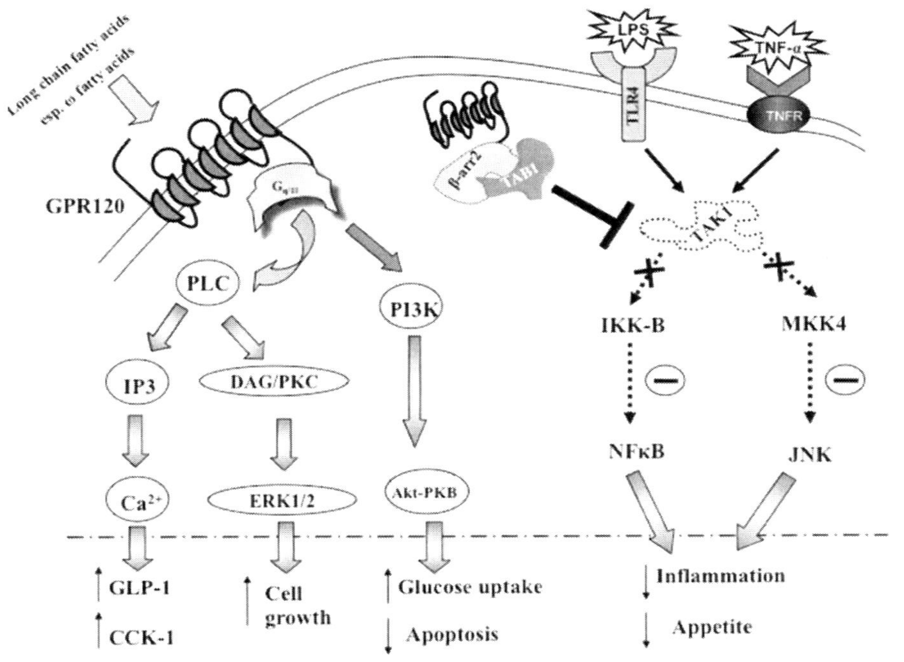

Redrawn on basis of Mo et al., 2013.

Figure 1. Signal transduction through GPR120 and consequent physiological effects.

(b) Effects on incretin secretion

GPR120 is highly expressed in intestinal tract, enteroendocrine cells and K-cells. Hirasawa et al. (2005) showed that ALA stimulates the secretion of GLP-1 in STC-1 cells, an effect abrogated by siRNA knockdown of GPR120. The secretion of cholecystokinin (CCK) is also regulated by GPR120 (Tanaka et al., 2008a; Sidhu et al., 2000). GPR120 also participates in regulation of ghrelin secretion. Specific natural agonist of GPR120, grifolic acid (Hara et al., 2009) increases plasma ghrelin levels in mice (Janssen et al., 2012). The role of GPR120 and ω-3 FAs on ghrelin secretion needs to be further clarified.

(c) Effects on insulin secretion and sensitivity

Signaling through GPR120 increases GLP-1 secretion. GLP-1 is a natural modulator of glucose dependent insulin secretion and beta cell proliferation. Thus, GPR120 indirectly affects insulin secretion and beta cell proliferation. Dietary ALA has been shown to increase the levels of plasma GLP-1 and insulin in mice fed on normal chow or HFD (Hirasawa *et al., 2005; Tanaka et al., 2008b).*

Ichimura et al., (2012) showed that *Gpr120* KO mice on HFD exhibit insulin resistant, glucose intolerant, hyperinsulinemic, insulin insensitive phenotype with decreased insulin signaling in white adipose tissue and liver, increased inflammatory and decreased insulin-signaling gene expression. The study highlights the importance of GPR120 signaling. Another recent study revealed a strong positive association between GPR120 expression and insulin secretion in human islets (Taneera et al., 2012). The same group showed differential expression of *GPR120* in islets from diabetic vs. nondiabetic subjects, and normoglycemic (HbA_{1c}<6%) vs. hyperglycemic (HbA_{1c}>6%) subjects. They further went on to show that treatment of human islets with EPA abrogates the proapoptotic affects palmitate and linoleic acid and increases cell viability, the effect reduced to 50% in GPR120 knockdown islet suggesting that signaling through GPR120 protects islets from lipotoxicity.

(d) *Effects on adipogenesis and energy homeostasis*

GPR120 is involved in adipocyte maturation and differentiation. A higher expression of GPR120 has been detected in mature adipocytes then the preadipocytes. Silencing GPR120 reduces the gene expression of adipogenic factors like PPARγ2 (Gotoh et al., 2007).

Ichimura et al. (Ichimura et al., 2012) showed that activation of GPR120 by ω-3 FAs exerts anti-obesity effects. As mentioned previously, in GPR120 KO mice on HFD, severely deregulated adipogenesis was observed with excessive weight gain, inflammation and insulin resistance in comparison to the WT mice. This suggests that proper functioning of GPR120 is required to prevent HFD induced obesity and obesity induced metabolic syndrome in mice. The authors further conducted genetic analysis of 312 obese children and adults revealed two nonsynonyous GPR120 variants (R270H and R67C). Further, larger groups of subjects were genotyped for these variants. It was demonstrated that R270H polymorphism is strongly associated with obesity and insulin resistance. On expressing these mutant variants in cell lines a functionally defective GPR120 phenotype was demonstrated. Thus, the key feature of this study if the identification of a loss-of-function GPR120 human variant associated with obesity and insulin resistance.

(e) *Effects in other tissues*

ω-3 FAs may also govern the taste and food preference. In a recent study, Cartoni et al. (Cartoni et al., 2010) demonstrated that GPR120 KO mice exhibit reduced prefernce for FFA intake and weaker glossopharyngeal response to long chain FFAs, suggesting the role for GPR120 signaling in taste perception. It has also been revelaed that GPR120 signaling may play role in osteoclastogeneis (Cornish et al., 2008).

Conclusion

GPR120 functions as an ω-3 FA receptor/sensor and mediates strong anti-inflammatory, anti-apoptotic, anti-diabetic, anti-obesity and insulin sensitizing effects. A robust association of GPR120 mutation and metabolic defects in humans has been pointed out recently. GPR120 thus has emerged as an important drug target. ω-3 FAs and their synthetic derivatives might be potential new therapeutic approaches for the treatment of obesity and metainflammation.

References

Barak, L. S., Ferguson, S. S., Zhang, J., Caron, M. G. (1997), A beta-arrestin/green fluorescent protein biosensor for detecting G protein-coupled receptor activation. *J. Biol. Chem.*, 272(44), 27497-500.

Cartoni, C., Yasumatsu, K., Ohkuri, T., Shigemura, N., Yoshida, R., Godinot N., et al. (2010), Taste preference for fatty acids is mediated by GPR40 and GPR120. *J. Neurosci.,* 30, 8376–8382.

Cintra, D. E., Ropelle, E. R., Moraes, J. C., Pauli, J. R., Morari, J., de Souza, C. T., et al. (2012), Unsaturated fatty acids revert diet-induced hypothalamic inflammation in obesity. *PLoS One,* 7, e30571.

Chawla, A., Repa, J. J., Evans, R. M., Mangelsdorf, D. J. (2001). Nuclear receptors and lipid physiology: opening the X-files. *Science,* 294(5548), 1866-70. DOI:10.1126/science. 294.5548.1866.

Cornall, L. M., Mathai, M. L., Hryciw, D. H., McAinch, A. J. (2011), Diet-induced obesity up-regulates the abundance of GPR43 and GPR120 in a tissue specific manner. *Cell Physiol. Biochem.,* 28, 949–958.

Cornish, J., MacGibbon, A., Lin, J. M., Watson, M., Callon, K. E., Tong, P. C., et al. (2008), Modulation of osteoclastogenesis by fatty acids. *Endocrinology,* 149, 5688–5695.

Davenport A. P., Alexander S. P., Sharman J. L., Pawson A. J., Benson H. E., Monaghan A. E., Liew W. C., Mpamhanga C. P., Bonner T. I., Neubig R. R., Pin J. P., Spedding M., Harmar A. J. (2013), International Union of Basic and Clinical Pharmacology. LXXXVIII. G protein-coupled receptor list: recommendations *Pharmacol. Rev.,* 65(3), 967-86. doi: 10.1124/pr.112.007179.

Davis, B. K., Wen, H., Ting, J. P. (2011), The inflammasome NLRs in immunity, inflammation, and associated diseases. *Annu. Rev. Immunol.,* 29, 707-35. doi: 10.1146/ annurev-immunol-031210-101405.

Fredriksson, R., Hoglund, P. J., Gloriam, D. E. I., Lagerstrom, M. C., Schioth, H. B. (2003), Seven evolutionarily conserved human rhodopsin G protein-coupled receptors lacking close relatives. *FEBS Lett.,* 554, 381–388.

Gether, U. (2000) Uncovering molecular mechanisms involved in activation of G protein-coupled receptors *Endocr. Rev.,* 21(1), 90-113.

González-Périz, A., Horrillo, R., Ferré ,N., Gronert, K., Dong, B., Morán-Salvador, E., Titos, E., Martínez-Clemente, M., López-Parra, M., Arroyo, V., Clària ,J. (2009), Obesity-induced insulin resistance and hepatic steatosis are alleviated by omega-3 fatty acids: a role forresolvins and protectins. *FASEB J.,* 23(6), 1946-57. doi: 10.1096/fj.08-125674.

Gotoh, C., Hong, Y. H., Iga, T., Hishikawa, D, Suzuki, Y., Song, S. H., et al. (2007), The regulation of adipogenesis through GPR120. *Biochem. Biophys. Res. Commun.*, 354, 591–597.

Gregor, M. F. and Hotamisligil, G. S. (2011), Inflammatory mechanisms in obesity *Annu. Rev. Immunol.*, 29, 415-45. doi: 10.1146/annurev-immunol-031210-101322.

Hara, T., Hirasawa, A., Sun, Q., Sadakane, K., Itsubo, C., Iga, T. et al. (2009), Novel selective ligands for free fatty acid receptors GPR120 and GPR40. *Naunyn. Schmiedebergs Arch. Pharmacol.*, 380, 247–255.

Hirasawa, A., Tsumaya, K., Awaji, T., Katsuma, S., Adachi, T., Yamada, M. et al. (2005), Free fatty acids regulate gut incretin glucagon-like peptide-1 secretion through GPR120. *Nat. Med.,* 11, 90–94.

Hwakura, H., Li, Y., Ariyasu, H., Hosoda, H., Kanamoto, N., Bando, M., et al. (2010), Establishment of a novel ghrelin-producing cell line. *Endocrinology,* 151, 2940–2945.

Iakoubov, R., Izzo, A., Yeung, A., Whiteside, C. I., Brubaker, P. L. (2007), Protein kinase C zeta is required for oleic acid-induced secretion of glucagon-like peptide-1 by intestinal endocrine L cells. *Endocrinology*, 148, 1089–1098.

Ichimura, A., Hirasawa, A., Poulain-Godefroy, O., Bonnefond, A., Hara, T., Yeng, L., et al. (2012), Dysfunction of lipid sensor GPR120 leads to obesity in both mouse and human. *Nature*, 483, 350–354.

Ichimura, A., Hirasawa, A., Hara, T., Tsujimoto, G. (2009), Free fatty acid receptors act as nutrient sensors to regulate energy homeostasis. *Prostaglandins Other Lipid Mediat.,* 89(3-4), 82-8. doi: 10.1016/j.prostaglandins.2009.05.003.

Im, D. S. (2012), Omega-3 fatty acids in anti-inflammation (pro-resolution) and GPCRs. *Prog. Lipid. Res.,* 51(3), 232-7. doi: 10.1016/j.plipres.2012.02.003.

Ito, J., Ito, M., Nambu, H., Fujikawa, T., Tanaka, K., Iwaasa H., et al. (2009), Anatomical and histological profiling of orphan G-protein-coupled receptor expression in gastrointestinal tract of C57BL/6J mice. *Cell Tissue Res.,* 338, 257–269.

Janssen, S., Laermans, J., Iwakura, H., Tack, J., Depoortere, I. (2012), Sensing of fatty acids for octanoylation of ghrelin involves a gustatory g-protein. *PLoS One*, 7, e40168.

Jump, D. B. (2008), N-3 polyunsaturated fat *Curr. Opin. Lipidol.,* 19(3), 242-7. doi: 10.1097/MOL.0b013e3282ffaf6a.

Katsuma, S., Hatae, N., Yano, T., Ruike, Y., Kimura, M., Hirasawa, A., Tsujimoto, G. (2005), Free fatty acids inhibit serum deprivation-induced apoptosis through GPR120 in a murine enteroendocrine cell line STC-1. *J. Biol. Chem*., 280(20), 19507-15.

Kimura, I., Inoue, D., Maeda, T., Hara, T., Ichimura, A., Miyauchi, S., Kobayashi, M., Hirasawa, A., Tsujimoto, G. (2011), Short-chain fatty acids *Proc. Natl. Acad. Sci. U S A,* 108(19), 8030-5. doi: 10.1073/pnas.1016088108.

Louet, J. F., Chatelain, F., Decaux, J. F., et al. (2001), Long-chain fatty acids regulate liver carnitine palmitoyltransferase I gene (L-CPT I) expression through a peroxisome-proliferator-activated receptor alpha (PPARalpha)-independent pathway. *Biochem. J.*, 354, 189–97.

Matsumura, S., Mizushige, T., Yoneda, T., Iwanaga, T., Tsuzuki, S., Inoue, K., et al. (2007), GPR expression in the rat taste bud relating to fatty acid sensing. *Biomed. Res*., 28, 49–5.

Mo, X. L., Wei, H. K., Peng, J., Tao, Y. X. (2013) Free fatty acid *Prog. Mol. Biol. Transl. Sci.,* 114:251-76. doi: 10.1016/B978-0-12-386933-3.00007-8.

Morgan, N. G., Dhayal. S. (2009), G-protein coupled receptors mediating long chain fatty acid signalling in the pancreatic beta-cell. *Biochem. Pharmacol*., 78, 1419–1427.

Oh, D. Y., Talukdar, S., Bae, E. J., Imamura, T., Morinaga, H., Fan, W., Li, P., Lu, W. J., Watkins, S. M., Olefsky, J. M. (2010). GPR120 is an omega-3 fatty acid receptor mediating potent anti-inflammatory and insulin-sensitizing effects. *Cell*, 142, 687–698. doi: 10.1016/j.cell.2010.07.041.

Oh, D. Y. and Lagakos, W. S. (2011), The role of G-protein-coupled receptors in mediating the effect of fatty acids on inflammation and insulin sensitivity. *Curr. Opin. Clin. Nutr. Metab. Care,* 14(4), 322-7. doi: 10.1097/MCO.0b013e3283479230.

Parker, H. E., Habib, A. M., Rogers, G. J., Gribble, F. M., Reimann, F. (2009), Nutrient-dependent secretion of glucose-dependent insulinotropic polypeptide from primary murine K cells. *Diabetologia*, 52, 289–298.

Priyadarshini, M., Kamal, M. A., Greig, N. H., Realef, M., Abuzenadah, A. M., Chaudhary, A. G., Damanhouri, G. A. (2012), Alzheimer's disease and type 2 diabetes *CNS Neurol. Disord. Drug Targets,* 11(4):482-9.

Regard, J. B., Kataoka, H., Cano, D. A., Camerer, E., Yin, L., Zheng, Y. W., Scanlan, T. S., Hebrok, M., Coughlin, S. R. (2007) Probing cell type-specific functions of Gi in vivo identifies GPCR regulators of insulin secretion. *J. Clin. Invest.*, 117(12), 4034-43.

Sauer, L. A., Dauchy, R. T., Blask, D. E. (2000), Mechanism for the antitumor and anticachectic effects of n−3 fatty acids. *Cancer Res.*, 60, 5289–95.

Schulte, G., Fredholm, B. B. (2003) Signalling from adenosine *Cell Signal*, 15(9), 813-27.

Sidhu, S. S., Thompson, D. G., Warhurst, G., Case, R. M., Benson, R. S. (2000), Fatty Acid-Induced Cholecystokinin Secretion And Changes In Intracellular Ca2+ In Two Enteroendocrine Cell Lines, STC-1 And Glutag. *J. Physiol.*, 528 Pt 1, 165-76.

Soto-Guzman, A., Robledo, T., Lopez-Perez, M., Salazar, E. P. (2008), Oleic acid induces ERK1/2 activation and AP-1 DNA binding activity through a mechanism involving Src kinase and EGFR transactivation in breast cancer cells. *Mol. Cell Endocrinol.*, 294, 81–91.

Stoddart, L. A., Smith, N. J., Milligan, G. (2008), International Union of Pharmacology. LXXI. Free fatty acid receptors FFA1, -2, and -3: pharmacology and pathophysiological functions. *Pharmacol. Rev.,* 60(4), 405-17. doi: 10.1124/pr.108.00802.

Talukdar, S., Olefsky, J. M., Osborn, O. (2011). Targeting GPR120 and other fatty acid-sensing GPCRs ameliorates insulin resistance and inflammatory diseases. *Trends Pharmacol. Sci.,* 32, 543–550. doi: 10.1016/j.tips.2011.04.004.

Tanaka, T., Katsuma, S., Adachi, T., Koshimizu, T. A., Hirasawa, A., Tsujimoto, G. (2008a), Free fatty acids induce cholecystokinin secretion through GPR120. *Naunyn. Schmiedebergs Arch. Pharmacol.*, 377(4-6), 523-7.

Tanaka, T., Yano, T., Adachi, T., Koshimizu, T. A., Hirasawa, A., Tsujimoto. G. (2008b), Cloning and characterization of the rat free fatty acid receptor GPR120: in vivo effect of the natural ligand on GLP-1 secretion and proliferation of pancreatic beta cells. *Naunyn. Schmiedebergs Arch. Pharmacol.*, 377, 515–522.

Taneera, J., Lang, S., Sharma, A., Fadista, J., Zhou, Y., Ahlqvist, E., et al. (2012), A systems genetics approach identifies genes and pathways for type 2 diabetes in human islets. *Cell Metab.*, 16, 122–134.

Tolhurst, G., Heffron, H., Lam, Y. S., Parker, H. E., Habib, A. M., Diakogiannaki, E., Cameron, J., Grosse, J., Reimann, F., Gribble, F. M. (2012), Short-chain fatty acids stimulate glucagon-like peptide-1 secretion via the G-protein-coupled receptor FFAR2. *Diabetes,* 61(2), 364-71. doi: 10.2337/db11-1019.

Ulloa-Aguirre, A., Stanislaus, D., Janovick, J. A., Conn, P. M. (1999) Structure-activity relationships of G protein-coupled receptors. *Arch. Med. Res.*, 30(6), 420-35.

Whalley, N. M., Pritchard, L. E., Smith, D. M., White, A. (2011), Processing of proglucagon to GLP-1 in pancreatic alpha-cells: is this a paracrine mechanism enabling GLP-1 to act on beta-cells? *J. Endocrinol.,* 211, 99–106.

Werry, T. D., Sexton, P. M., Christopoulos, A. (2005), "Ins and outs" of seven-transmembrane receptor signalling *Trends Endocrinol. Metab*., 16(1), 26-33.

Watson, S. J., Brown, A. J. H., Holliday, N. D. (2012), Differential signaling by splice variants of the human free fatty acid receptor GPR120. *Mol. Pharmacol*., 81, 631–642.

Yan, Y., Jiang, W., Spinetti, T., Tardivel, A., Castillo, R., Bourquin, C., Guarda, G., Tian, Z., Tschopp, J., Zhou, R. (2013). Omega-3 Fatty Acids Prevent Inflammation and Metabolic *Immunity*, 38(6):1154-63. doi: 10.1016/j.immuni.2013.05.015.

Yaney, G. C., and Corkey, B. E. (2003). Fatty acid metabolism and insulin secretion in pancreatic beta cells. *Diabetologia,* 46, 1297–1312. doi: 10.1007/s00125-003-1207-4.

Zhang, M. J. and Spite, M. (2012) Resolvins: anti-inflammatory and proresolving mediators derived from omega-3 polyunsaturated fatty acids. *Annu. Rev. Nutr*., 32, 203-27. doi: 10.1146/annurev-nutr-071811-150726.

Index

#

A

B

C

D

E

F

G

H

I

J

K

L

M

N

O

P

Q

R

S

V

W

X

Y

Z